RESIDENTIAL ELECTRICAL INSTALLATIONS

2020 Edition

TIM McCLINTOCK
JEFF SARGENT

NATIONAL FIRE PROTECTION ASSOCIATION
The leading information and knowledge resource on fire, electrical and related hazards

Product Manager: Debra Rose
Project Editor: Tracy Gaudet
Composition: Shepherd, Inc.
Printer: Command Digital

Copyright © 2019 All rights reserved.
National Fire Protection Association
One Batterymarch Park
Quincy, Massachusetts 02169-7471

The following are registered trademarks of the National Fire Protection Association:

National Fire Protection Association®
NFPA®
National Electrical Code®
NEC®

NFPA No.: PGNECR20
ISBN: 978-1-4559-2309-0
ISSN: 2152-5986

Printed in the United States of America

19 20 21 22 23   5 4 3 2 1

IMPORTANT NOTICES AND DISCLAIMERS OF LIABILITY

This *Pocket Guide* is a compilation of extracts from the 2020 edition of the *National Electrical Code®* (*NEC®*). These extracts have been selected and arranged by the Authors of the *Pocket Guide*, who have also supplied introductory material located at the beginning of each chapter.

The *NEC*, like all NFPA® codes and standards, is a document that is developed through a consensus standards development process approved by the American National Standards Institute. This process brings together volunteers representing varied viewpoints and interests to achieve consensus on fire, electrical, and other safety issues. While the NFPA administers the process and establishes rules to promote fairness in the development of consensus, it does not independently test, evaluate, or verify the accuracy of any information or the soundness of any judgments contained in its codes and standards.

The NFPA and the Authors disclaim liability for any personal injury, property or other damages of any nature whatsoever, whether special, indirect, consequential, or compensatory, directly or indirectly resulting from the publication, use of, or reliance on the *NEC* or this *Pocket Guide*. The NFPA and the Authors also make no guaranty or warranty as to the accuracy or completeness of any information published herein.

In issuing and making the *NEC* and this *Pocket Guide* available, the NFPA and the Author are not undertaking to render professional or other services for or on behalf of any person or entity. Nor is the NFPA or the Authors undertaking to perform any duty owed by any person or entity to someone else. Anyone using these materials should rely on his or her own independent judgment or, as appropriate, seek the advice of a competent professional in determining the exercise of reasonable care in any given circumstances.

This *Pocket Guide* is not intended to be a substitute for the *NEC* itself, which should always be consulted to ensure full *Code* compliance.

Additional Important Notices and Legal Disclaimers concerning the use of the *NEC* can be viewed at www.nfpa.org/disclaimers.

NOTICE CONCERNING CODE INTERPRETATIONS

The introductory materials and the selection and arrangement of the *NEC* extracts contained in this *Pocket Guide* reflect the personal opinions and judgments of the Authors and do not necessarily represent the official position of the NFPA (which can only be obtained through Formal Interpretations processed in accordance with NFPA rules).

REMINDER: UPDATING OF NFPA DOCUMENTS

NFPA 70, National Electrical Code, like all NFPA codes, standards, recommended practices, and guides ("NFPA Standards"), may be amended from time to time through the issuance of Tentative Interim Amendments or corrected by Errata. An official NFPA Standard at any point in time consists of the current edition of the document together with any Tentative Interim Amendment and any Errata then in effect.

In order to determine whether an NFPA Standard has been amended through the issuance of Tentative Interim Amendments or corrected by Errata, visit the "Codes & Standards" section on NFPA's website. There, the document information pages located at the "List of NFPA Codes & Standards" provide up-to-date, document-specific information including any issued Tentative Interim Amendments and Errata.

To view the document information page for a specific NFPA Standard, go to http://www.nfpa.org/docinfo to choose from the list of NFPA Standards or use the search feature to select the NFPA Standard number (e.g., *NFPA 70*). The document information page includes postings of all existing Tentative Interim Amendments and Errata. It also includes the option to register for an "Alert" feature to receive an automatic email notification when new updates and other information are posted regarding the document.

CONTENTS/TABS

CHAPTER 1	INTRODUCTION TO THE NATIONAL ELECTRICAL CODE	1
CHAPTER 2	DEFINITIONS APPLICABLE TO DWELLING UNITS	5
CHAPTER 3	GENERAL INSTALLATION REQUIREMENTS	21
CHAPTER 4	TEMPORARY WIRING	33
CHAPTER 5	SERVICE AND FEEDER LOAD CALCULATIONS	39
CHAPTER 6	SERVICES	55
CHAPTER 7	FEEDERS	77
CHAPTER 8	PANELBOARDS AND DISCONNECTS	81
CHAPTER 9	OVERCURRENT PROTECTION	87
CHAPTER 10	GROUNDED (NEUTRAL) CONDUCTORS	97
CHAPTER 11	GROUNDING AND BONDING	103
CHAPTER 12	BOXES AND ENCLOSURES	143
CHAPTER 13	CABLES	163
CHAPTER 14	RACEWAYS	177

CONTENTS

CHAPTER 1	INTRODUCTION TO THE NATIONAL ELECTRICAL CODE	1
CHAPTER 2	DEFINITIONS AND PRACTICE TO DWELLING UNITS	9
CHAPTER 3	GENERAL INSTALLATION REQUIREMENTS	21
CHAPTER 4	TEMPORARY WIRING	33
CHAPTER 5	SERVICE AND FEEDER LOAD CALCULATIONS	39
CHAPTER 6		49
CHAPTER 7		77
CHAPTER 8		85
CHAPTER 9		91
CHAPTER 10	GROUNDED (NEUTRAL) CONDUCTORS	97
CHAPTER 11	RACEWAYS AND BOXES	101
CHAPTER 12	BOXES AND ENCLOSURES	149
CHAPTER 13	CABLES	163
CHAPTER 14	RACEWAYS	171

CONTENTS/TABS

CHAPTER 15	WIRING METHODS AND CONDUCTORS	199
CHAPTER 16	BRANCH CIRCUITS	223
CHAPTER 17	RECEPTACLE PROVISIONS	235
CHAPTER 18	SWITCHES	249
CHAPTER 19	LIGHTING REQUIREMENTS	255
CHAPTER 20	APPLIANCES	269
CHAPTER 21	ELECTRIC SPACE-HEATING EQUIPMENT	279
CHAPTER 22	AIR-CONDITIONING AND REFRIGERATING EQUIPMENT	289
CHAPTER 23	COMMUNICATIONS CIRCUITS	299
CHAPTER 24	SEPARATE BUILDINGS OR STRUCTURES	325
CHAPTER 25	SWIMMING POOLS, FOUNTAINS, AND SIMILAR INSTALLATIONS	331
CHAPTER 26	INTERACTIVE AND STAND-ALONE PHOTOVOLTAIC (PV) SYSTEMS	367
CHAPTER 27	OPTIONAL STANDBY SYSTEMS	405
CHAPTER 28	CONDUIT AND TUBING FILL TABLES	411

CONTENTS

CHAPTER 16	WIRING METHODS AND CONDUCTORS	155
CHAPTER 17	BRANCH CIRCUITS	201
CHAPTER 18	RECEPTACLE PROVISIONS	219
CHAPTER 19	LIGHTING	235
CHAPTER 20	HEATING PROVISIONS	253
CHAPTER 21	APPLIANCES	269
CHAPTER 22	ELECTRICAL SERVICE EQUIPMENT	279
CHAPTER 23	AIR-CONDITIONING AND HEAT PUMPS	293
CHAPTER 24	SWIMMING POOLS AND SIMILAR INSTALLATIONS	305
CHAPTER 25	STANDARD FUEL GAS APPLIANCES AND SIMILAR INSTALLATIONS	321
CHAPTER 26	ALTERNATIVE AND STAND-ALONE PHOTOVOLTAIC (PV) SYSTEMS	387
CHAPTER 27	OPTIONAL STANDBY SYSTEMS	409
CHAPTER 28	CONDUIT AND TUBING FILL TABLES	411

CHAPTER/ARTICLE CROSS-REFERENCE TABLE

Note: *NEC* article number refers to selected requirements from that article.

Chapter No.	*NEC* Article No.
1	90
2	100
3	110
4	590
5	220
6	230
7	215
8	408, 404
9	240
10	200
11	250
12	314, 312
13	334, 330, 336, 338, 340
14	358, 344, 352, 362, 348, 350, 356
15	300, 310
16	210
17	210, 406
18	404
19	210, 410, 411
20	422
21	424
22	440
23	800
24	225, 250
25	680
26	690, 705
27	702
28	Chapter 9 Tables, Annex C

ARTICLE/CHAPTER CROSS-REFERENCE TABLE

Note: *NEC* article number refers to selected requirements from that article.

NEC Article No.	Chapter No.
90	1
100	2
110	3
200	10
210	16, 17, 19
215	7
220	5
225	24
230	6
240	9
250	11, 24
300	15
310	15
312	12
314	12
330	13
334	13
336	13
338	13
340	13
344	14
348	14
350	14
352	14
356	14
358	14
362	14
404	8, 18
406	17
408	8
410	19
411	19
422	20
424	21
440	22
590	4
680	25
690	26
702	27
705	26
800	23
Annex C	28
Chapter 9 Tables	28

INTRODUCTION

The *Pocket Guide to Residential Electrical Installations* has been developed to provide a portable and practical comprehensive field reference source for electricians, contractors, inspectors, and others responsible for applying the rules of the *National Electrical Code*. Together with its companion volume, *Pocket Guide to Commercial and Industrial Electrical Installations,* the most widely applied general requirements of the *NEC* are compiled in these two extremely useful on-the-job, on-the-go documents.

The compiled requirements follow an order similar to the progression of an electrical system installation on the construction site, rather than the normal sequence of requirements in the *NEC*. Introductory chapters are followed by the requirements pertaining to temporary electrical installations. The *Pocket Guide* then continues with branch circuit, feeder, and service load calculation requirements from Article 220 and then proceeds into other general requirements extracted from Chapters 2, 3, and 4 of the *NEC*.

Due to its compact size, the *Pocket Guide* does not contain all of the *Code* requirements — it is a complement to, not a replacement of, the 2020 *NEC*. When paired with the entire volume of electrical safety requirements contained in the 2020 *Code*, the *Pocket Guide* is an effective tool that everyone concerned with safe electrical installations will want to have. In-depth explanations of *NEC* requirements are available in NFPA's *National Electrical Code Handbook*, the only *NEC* handbook with the entire text of the *NEC*, plus useful explanation and commentary developed by the NFPA staff.

Users of the *Pocket Guide to Residential Electrical Installations* must also understand that some adopting jurisdictions enact requirements to address unique local environmental, geographical, or other considerations. Always consult with the local authority having jurisdiction for information on locally enacted requirements.

The highly specialized requirements in Chapters 5 through 8 of the *Code* are not included in this compilation because this material is outside of the intended purpose of this document. The requirements included in the *Pocket Guide* apply to virtually all residential electrical installations and are based on a practical working knowledge of how a residential electrical construction project evolves from beginning to completion. As indicated above, however, the *Pocket Guide* is not intended to be a substitute for the *NEC* itself, which should always be consulted to ensure full *Code* compliance.

It is hoped that this 2020 edition of the *Pocket Guide to Residential Electrical Installations* proves to be one of the most valuable tools that you use on a daily basis in performing safe, *NEC*-compliant electrical installations.

CHAPTER 1
INTRODUCTION TO THE NATIONAL ELECTRICAL CODE

INTRODUCTION

This chapter contains requirements from Article 90 — Introduction, which contains foundational elements necessary for proper application and use of the *National Electrical Code*®. Besides providing the purpose of the *Code*, Article 90 specifies the document scope, describes how the *Code* is arranged, provides information on enforcement, and contains other global requirements such as equipment evaluation and how metric conversions are applied. Section 90.1(A) states that "the purpose of this *Code* is the practical safeguarding of persons and property from hazards arising from the use of electricity." Safety — for persons *and* property — has always been the primary objective of the *Code*. Section 90.1(B) specifies that compliance with the *Code* and proper maintenance thereafter results in an installation essentially free from electrical hazards, but not necessarily efficient, convenient, or adequate for good service or future expansion of electrical use. The *Code* is a compilation of installation requirements focused on electrical safety, and its primary focus is not effective or efficient electrical system design. Although state and local jurisdictions typically adopt the *Code* exactly as written, there are some that amend the *Code* or incorporate requirements to address specific local conditions. Therefore, obtain a copy of any amendments or additional rules and regulations for the area(s) where you work.

1

ARTICLE 90
Introduction

90.4 Enforcement. This *Code* is intended to be suitable for mandatory application by governmental bodies that exercise legal jurisdiction over electrical installations, including signaling and communications systems, and for use by insurance inspectors. The authority having jurisdiction for enforcement of the *Code* has the responsibility for making interpretations of the rules, for deciding on the approval of equipment and materials, and for granting the special permission contemplated in a number of the rules.

By special permission, the authority having jurisdiction may waive specific requirements in this *Code* or permit alternative methods where it is assured that equivalent objectives can be achieved by establishing and maintaining effective safety.

This *Code* may require new products, constructions, or materials that may not yet be available at the time the *Code* is adopted. In such event, the authority having jurisdiction may permit the use of the products, constructions, or materials that comply with the most recent previous edition of this *Code* adopted by the jurisdiction.

90.5 Mandatory Rules, Permissive Rules, and Explanatory Material.

(A) Mandatory Rules. Mandatory rules of this *Code* are those that identify actions that are specifically required or prohibited and are characterized by the use of the terms *shall* or *shall not*.

(B) Permissive Rules. Permissive rules of this *Code* are those that identify actions that are allowed but not required, are normally used to describe options or alternative methods, and are characterized by the use of the terms *shall be permitted* or *shall not be required*.

(C) Explanatory Material. Explanatory material, such as references to other standards, references to related sections of this *Code*, or information related to a *Code* rule, is included in this *Code* in the form of informational notes. Such notes are informational only and are not enforceable as requirements of this *Code*.

Brackets containing section references to another NFPA document are for informational purposes only and are provided as a guide to indicate the source of the extracted text. These bracketed references immediately follow the extracted text.

Informational Note: The format and language used in this *Code* follows guidelines established by NFPA and published in the *NEC Style Manual*. Copies of this manual can be obtained from NFPA.

(D) Informative Annexes. Nonmandatory information relative to the use of the *NEC* is provided in informative annexes. Informative annexes are not part of the enforceable requirements of the *NEC*, but are included for information purposes only.

90.8 Wiring Planning.

(A) Future Expansion and Convenience. Plans and specifications that provide ample space in raceways, spare raceways, and additional spaces allow for future increases in electric power and communications circuits. Distribution centers located in readily accessible locations provide convenience and safety of operation.

(B) Number of Circuits in Enclosures. It is elsewhere provided in this *Code* that the number of circuits confined in a single enclosure be varyingly restricted. Limiting the number of circuits in a single enclosure minimizes the effects from a short circuit or ground fault.

CHAPTER 2
DEFINITIONS APPLICABLE TO DWELLING UNITS

INTRODUCTION

The definition of a word or phrase used in the *NEC* may differ from that found in a dictionary. In fact, some of these terms are not defined in a standard dictionary. Understanding the *Code* without a thorough understanding of the terms is difficult. Therefore, having an understanding of Article 100 is an important step toward proper application of the *Code*.

As stated in the scope of Article 100, only definitions essential to the proper application of the *Code* are included. Since commonly defined general terms can be found in most dictionaries, they are not included. Also not included are commonly defined technical terms. These terms can be found in a dictionary of electrical and electronic terms.

Since a number of terms in Article 100 do not apply to dwelling units, they are not included in this book. Likewise, because of this guide's compact design, other terms and informational notes are also omitted. Reference the 2017 *National Electrical Code* for a complete listing.

Article 100 contains terms used in two or more articles. Other terms that pertain to one article may be defined in that particular article. For example, Article 680 — Swimming Pools, Fountains, and Similar Installations, includes terms that are not used anywhere else in the *Code*. Therefore, those terms are defined in Article 680 rather than in Article 100.

ARTICLE 100
Definitions

PART I. GENERAL

Accessible (as applied to equipment). Capable of being reached for operation, renewal, and inspection. (CMP-1)

Accessible (as applied to wiring methods). Capable of being removed or exposed without damaging the building structure or finish or not permanently closed in by the structure or finish of the building. (CMP-1)

Accessible, Readily (Readily Accessible). Capable of being reached quickly for operation, renewal, or inspections without requiring those to whom ready access is requisite to take actions such as to use tools (other than keys), to climb over or under, to remove obstacles, or to resort to portable ladders, and so forth. (CMP-1)

> Informational Note: Use of keys is a common practice under controlled or supervised conditions and a common alternative to the ready access requirements under such supervised conditions as provided elsewhere in the NEC.

Ampacity. The maximum current, in amperes, that a conductor can carry continuously under the conditions of use without exceeding its temperature rating. (CMP-6)

Appliance. Utilization equipment, generally other than industrial, that is normally built in standardized sizes or types and is installed or connected as a unit to perform one or more functions such as clothes washing, air-conditioning, food mixing, deep frying, and so forth. (CMP-17)

Approved. Acceptable to the authority having jurisdiction. (CMP-1)

Arc-Fault Circuit Interrupter (AFCI). A device intended to provide protection from the effects of arc faults by recognizing characteristics unique to arcing and by functioning to de-energize the circuit when an arc fault is detected. (CMP-2)

Attachment Fitting. A device that, by insertion into a locking support and mounting receptacle, establishes a connection between the conductors of the attached utilization equipment and the branch-circuit conductors connected to the locking support and mounting receptacle. (CMP-18)

Definitions Applicable to Dwelling Units — Article 100

> Informational Note: An attachment fitting is different from an attachment plug because no cord is associated with the fitting. An attachment fitting in combination with a locking support and mounting receptacle secures the associated utilization equipment in place and supports its weight.

Attachment Plug (Plug Cap) (Plug). A device that, by insertion in a receptacle, establishes a connection between the conductors of the attached flexible cord and the conductors connected permanently to the receptacle. (CMP-18)

Authority Having Jurisdiction (AHJ). An organization, office, or individual responsible for enforcing the requirements of a code or standard, or for approving equipment, materials, an installation, or a procedure. (CMP-1)

> Informational Note: The phrase "authority having jurisdiction," or its acronym AHJ, is used in NFPA documents in a broad manner, since jurisdictions and approval agencies vary, as do their responsibilities. Where public safety is primary, the authority having jurisdiction may be a federal, state, local, or other regional department or individual such as a fire chief; fire marshal; chief of a fire prevention bureau, labor department, or health department; building official; electrical inspector; or others having statutory authority. For insurance purposes, an insurance inspection department, rating bureau, or other insurance company representative may be the authority having jurisdiction. In many circumstances, the property owner or his or her designated agent assumes the role of the authority having jurisdiction; at government installations, the commanding officer or departmental official may be the authority having jurisdiction.

Automatic. Performing a function without the necessity of human intervention. (CMP-1)

Bathroom. An area including a sink (basin) with one or more of the following fixtures: a toilet, a urinal, a tub, a shower, a bidet, or similar plumbing fixtures. (CMP-2)

Bonded (Bonding). Connected to establish electrical continuity and conductivity. (CMP-5)

Bonding Conductor or Jumper. A reliable conductor to ensure the required electrical conductivity between metal parts required to be electrically connected. (CMP-5)

Article 100 — Chapter 2

Bonding Jumper, Equipment. The connection between two or more portions of the equipment grounding conductor. (CMP-5)

Bonding Jumper, Main. The connection between the grounded circuit conductor and the equipment grounding conductor, or the supply-side bonding jumper, or both, at the service. (CMP-5)

Branch Circuit. The circuit conductors between the final overcurrent device protecting the circuit and the outlet(s). (CMP-2)

Branch Circuit, Appliance. A branch circuit that supplies energy to one or more outlets to which appliances are to be connected and that has no permanently connected luminaires that are not a part of an appliance. (CMP-2)

Branch Circuit, General-Purpose. A branch circuit that supplies two or more receptacles or outlets for lighting and appliances. (CMP-2)

Branch Circuit, Individual. A branch circuit that supplies only one utilization equipment. (CMP-2)

Branch Circuit, Multiwire. A branch circuit that consists of two or more ungrounded conductors that have a voltage between them, and a grounded conductor that has equal voltage between it and each ungrounded conductor of the circuit and that is connected to the neutral or grounded conductor of the system. (CMP-2)

Building. A structure that stands alone or that is separated from adjoining structures by fire walls. (CMP-1)

Cabinet. An enclosure that is designed for either surface mounting or flush mounting and is provided with a frame, mat, or trim in which a swinging door or doors are or can be hung. (CMP-9)

Cable Routing Assembly. A single channel or connected multiple channels, as well as associated fittings, forming a structural system that is used to support and route communications wires and cables, optical fiber cables, data cables associated with information technology and communications equipment, Class 2, Class 3, and Type PLTC cables, and power-limited fire alarm cables in plenum, riser, and general-purpose applications. (CMP-16)

Circuit Breaker. A device designed to open and close a circuit by nonautomatic means and to open the circuit automatically on a predetermined overcurrent without damage to itself when properly applied within its rating. (CMP-10)

Definitions Applicable to Dwelling Units — Article 100

Informational Note: The automatic opening means can be integral, direct acting with the circuit breaker, or remote from the circuit breaker.

Adjustable (as applied to circuit breakers). A qualifying term indicating that the circuit breaker can be set to trip at various values of current, time, or both, within a predetermined range.

Instantaneous Trip (as applied to circuit breakers). A qualifying term indicating that no delay is purposely introduced in the tripping action of the circuit breaker.

Inverse Time (as applied to circuit breakers). A qualifying term indicating that there is purposely introduced a delay in the tripping action of the circuit breaker, which delay decreases as the magnitude of the current increases.

Nonadjustable (as applied to circuit breakers). A qualifying term indicating that the circuit breaker does not have any adjustment to alter the value of the current at which it will trip or the time required for its operation.

Setting (of circuit breakers). The value of current, time, or both, at which an adjustable circuit breaker is set to trip.

Clothes Closet. A nonhabitable room or space intended primarily for storage of garments and apparel. (CMP-1)

Communications Equipment. The electronic equipment that performs the telecommunications operations for the transmission of audio, video, and data, and includes power equipment (e.g., dc converters, inverters, and batteries), technical support equipment (e.g., computers), and conductors dedicated solely to the operation of the equipment. (CMP-16)

Informational Note: As the telecommunications network transitions to a more data-centric network, computers, routers, servers, and their powering equipment, are becoming essential to the transmission of audio, video, and data and are finding increasing application in communications equipment installations.

Concealed. Rendered inaccessible by the structure or finish of the building. (CMP-1)

Informational Note: Wires in concealed raceways are considered concealed, even though they may become accessible by withdrawing them.

Conductor, Bare. A conductor having no covering or electrical insulation whatsoever. (CMP-6)

Conductor, Covered. A conductor encased within material of composition or thickness that is not recognized by this *Code* as electrical insulation. (CMP-6)

Conductor, Insulated. A conductor encased within material of composition and thickness that is recognized by this *Code* as electrical insulation. (CMP-6)

Conduit Body. A separate portion of a conduit or tubing system that provides access through a removable cover(s) to the interior of the system at a junction of two or more sections of the system or at a terminal point of the system.

  Boxes such as FS and FD or larger cast or sheet metal boxes are not classified as conduit bodies. (CMP-9)

Connector, Pressure (Solderless). A device that establishes a connection between two or more conductors or between one or more conductors and a terminal by means of mechanical pressure and without the use of solder. (CMP-1)

Continuous Load. A load where the maximum current is expected to continue for 3 hours or more. (CMP-2)

Cooking Unit, Counter-Mounted. A cooking appliance designed for mounting in or on a counter and consisting of one or more heating elements, internal wiring, and built-in or mountable controls. (CMP-2)

Copper-Clad Aluminum Conductors. Conductors drawn from a copper-clad aluminum rod, with the copper metallurgically bonded to an aluminum core, where the copper forms a minimum of 10 percent of the cross-sectional area of a solid conductor or each strand of a stranded conductor. (CMP-6)

Cutout Box. An enclosure designed for surface mounting that has swinging doors or covers secured directly to and telescoping with the walls of the enclosure. (CMP-9)

Dead Front. Without live parts exposed to a person on the operating side of the equipment. (CMP-9)

Demand Factor. The ratio of the maximum demand of a system, or part of a system, to the total connected load of a system or the part of the system under consideration. (CMP-2)

Definitions Applicable to Dwelling Units — Article 100

Device. A unit of an electrical system, other than a conductor, that carries or controls electric energy as its principal function. (CMP-1)

Disconnecting Means. A device, or group of devices, or other means by which the conductors of a circuit can be disconnected from their source of supply. (CMP-1)

Dwelling, One-Family. A building that consists solely of one dwelling unit. (CMP-1)

Dwelling, Two-Family. A building that consists solely of two dwelling units. (CMP-1)

Dwelling, Multifamily. A building that contains three or more dwelling units. (CMP-1)

Dwelling Unit. A single unit, providing complete and independent living facilities for one or more persons, including permanent provisions for living, sleeping, cooking, and sanitation. (CMP-2)

Effective Ground-Fault Current Path. An intentionally constructed, low-impedance electrically conductive path designed and intended to carry current under ground-fault conditions from the point of a ground fault on a wiring system to the electrical supply source and that facilitates the operation of the overcurrent protective device or ground-fault detectors. (CMP-5)

Electric Power Production and Distribution Network. Power production, distribution, and utilization equipment and facilities, such as electric utility systems that are connected to premises wiring and are external to and not controlled by an interactive system. (CMP-13)

Electric Sign. A fixed, stationary, or portable self-contained, electrically operated and/or electrically illuminated utilization equipment with words or symbols designed to convey information or attract attention. (CMP-18)

Electric-Discharge Lighting. Systems of illumination utilizing fluorescent lamps, high-intensity discharge (HID) lamps, or neon tubing. (CMP-18)

Enclosed. Surrounded by a case, housing, fence, or wall(s) that prevents persons from accidentally contacting energized parts. (CMP-1)

Enclosure. The case or housing of apparatus, or the fence or walls surrounding an installation to prevent personnel from accidentally contacting energized parts or to protect the equipment from physical damage. (CMP-1)

Informational Note: See Table 110.28 for examples of enclosure types.

Energized. Electrically connected to, or is, a source of voltage. (CMP-1)

Equipment. A general term, including fittings, devices, appliances, luminaires, apparatus, machinery, and the like used as a part of, or in connection with, an electrical installation. (CMP-1)

Exposed (as applied to live parts). Capable of being inadvertently touched or approached nearer than a safe distance by a person. (CMP-1)

Informational Note: This term applies to parts that are not suitably guarded, isolated, or insulated.

Exposed (as applied to wiring methods). On or attached to the surface or behind panels designed to allow access. (CMP-1)

Feeder. All circuit conductors between the service equipment, the source of a separately derived system, or other power supply source and the final branch-circuit overcurrent device. (CMP-10)

Festoon Lighting. A string of outdoor lights that is suspended between two points. (CMP-18)

Fitting. An accessory such as a locknut, bushing, or other part of a wiring system that is intended primarily to perform a mechanical rather than an electrical. (CMP-1)

Ground. The earth. (CMP-5)

Grounded (Grounding). Connected (connecting) to ground or to a conductive body that extends the ground connection. (CMP-5)

Grounded, Solidly. Connected to ground without inserting any resistor or impedance device. (CMP-5)

Grounded Conductor. A system or circuit conductor that is intentionally grounded. (CMP-5)

Informational Note: Although an equipment grounding conductor is grounded, it is not considered a grounded conductor.

Ground-Fault Circuit Interrupter (GFCI). A device intended for the protection of personnel that functions to de-energize a circuit or portion thereof within an established period of time when a ground-fault current exceeds the values established for a Class A device. (CMP-2)

Definitions Applicable to Dwelling Units — Article 100

Informational Note: Class A ground-fault circuit interrupters trip when the ground-fault current is 6 mA or higher and do not trip when the ground-fault current is less than 4 mA. For further information, see UL 943, *Standard for Ground-Fault Circuit Interrupters*.

Grounding Conductor, Equipment (EGC). A conductive path(s) that is part of an effective ground-fault current path and connects normally non–current-carrying metal parts of equipment together and to the system grounded conductor or to the grounding electrode conductor, or both. (CMP-5)

Informational Note No. 1: It is recognized that the equipment grounding conductor also performs bonding.

Informational Note No. 2: See 250.118 for a list of acceptable equipment grounding conductors.

Grounding Electrode. A conducting object through which a direct connection to earth is established. (CMP-5)

Grounding Electrode Conductor. A conductor used to connect the system grounded conductor or the equipment to a grounding electrode or to a point on the grounding electrode system. (CMP-5)

Guarded. Covered, shielded, fenced, enclosed, or otherwise protected by means of suitable covers, casings, barriers, rails, screens, mats, or platforms to remove the likelihood of approach or contact by persons or objects to a point of danger. (CMP-1)

Guest Room. An accommodation combining living, sleeping, sanitary, and storage facilities within a compartment. (CMP-2)

Guest Suite. An accommodation with two or more contiguous rooms comprising a compartment, with or without doors between such rooms, that provides living, sleeping, sanitary, and storage facilities. (CMP-2)

Habitable Room. A room in a building for living, sleeping, eating, or cooking, but excluding bathrooms, toilet rooms, closets, hallways, storage or utility spaces, and similar areas. (CMP-2)

Handhole Enclosure. An enclosure for use in underground systems, provided with an open or closed bottom, and sized to allow personnel to reach into, but not enter, for the purpose of installing, operating, or maintaining equipment or wiring or both. (CMP-9)

Hermetic Refrigerant Motor-Compressor. A combination consisting of a compressor and motor, both of which are enclosed in the same housing, with no external shaft or shaft seals, with the motor operating in the refrigerant. (CMP-11)

Identified (as applied to equipment). Recognizable as suitable for the specific purpose, function, use, environment, application, and so forth, where described in a particular *Code* requirement. (CMP-1)

> Informational Note: Some examples of ways to determine suitability of equipment for a specific purpose, environment, or application include investigations by a qualified testing laboratory (listing and labeling), an inspection agency, or other organizations concerned with product evaluation.

Intersystem Bonding Termination. A device that provides a means for connecting intersystem bonding conductors for communications systems to the grounding electrode system. (CMP-16)

Kitchen. An area with a sink and permanent provisions for food preparation and cooking. (CMP-2)

Labeled. Equipment or materials to which has been attached a label, symbol, or other identifying mark of an organization that is acceptable to the authority having jurisdiction and concerned with product evaluation, that maintains periodic inspection of production of labeled equipment or materials, and by whose labeling the manufacturer indicates compliance with appropriate standards or performance in a specified manner. (CMP-1)

> Informational Note: If a listed product is of such a size, shape, material, or surface texture that it is not possible to apply legibly the complete label to the product, the complete label may appear on the smallest unit container in which the product is packaged.

Laundry Area. An area containing or designed to contain a laundry tray, clothes washer, or clothes dryer. (CMP-2)

Lighting Outlet. An outlet intended for the direct connection of a lampholder or luminaire. (CMP-18)

Lighting Track (Track Lighting). A manufactured assembly designed to support and energize luminaires that are capable of being readily repositioned on the track. Its length can be altered by the addition or subtraction of sections of track. (CMP-18)

Definitions Applicable to Dwelling Units — Article 100

Listed. Equipment, materials, or services included in a list published by an organization that is acceptable to the authority having jurisdiction and concerned with evaluation of products or services, that maintains periodic inspection of production of listed equipment or materials or periodic evaluation of services, and whose listing states that either the equipment, material, or service meets appropriate designated standards or has been tested and found suitable for a specified purpose. (CMP-1)

Informational Note: The means for identifying listed equipment may vary for each organization concerned with product evaluation, some of which do not recognize equipment as listed unless it is also labeled. Use of the system employed by the listing organization allows the authority having jurisdiction to identify a listed product.

Location, Damp. Locations protected from weather and not subject to saturation with water or other liquids but subject to moderate degrees of moisture. (CMP-1)

Informational Note: Examples of such locations include partially protected locations under canopies, marquees, roofed open porches, and like locations, and interior locations subject to moderate degrees of moisture, such as some basements, some barns, and some cold-storage warehouses.

Location, Dry. A location not normally subject to dampness or wetness. A location classified as dry may be temporarily subject to dampness or wetness, as in the case of a building under construction. (CMP-1)

Location, Wet. Installations underground or in concrete slabs or masonry in direct contact with the earth; in locations subject to saturation with water or other liquids, such as vehicle washing areas; and in unprotected locations exposed to weather. (CMP-1)

Luminaire. A complete lighting unit consisting of a light source such as a lamp or lamps, together with the parts designed to position the light source and connect it to the power supply. It may also include parts to protect the light source or the ballast or to distribute the light. A lampholder itself is not a luminaire. (CMP-18)

Multioutlet Assembly. A type of surface, flush, or freestanding raceway designed to hold conductors and receptacles, assembled in the field or at the factory. (CMP-18)

Neutral Conductor. The conductor connected to the neutral point of a system that is intended to carry current under normal conditions. (CMP-5)

Neutral Point. The common point on a wye-connection in a polyphase system or midpoint on a single-phase, 3-wire system, or midpoint of a single-phase portion of a 3-phase delta system, or a midpoint of a 3-wire, direct-current system. (CMP-5)

> Informational Note: At the neutral point of the system, the vectorial sum of the nominal voltages from all other phases within the system that utilize the neutral, with respect to the neutral point, is zero potential.

Outlet. A point on the wiring system at which current is taken to supply utilization equipment. (CMP-1)

Outline Lighting. An arrangement of incandescent lamps, electric-discharge lighting, or other electrically powered light sources to outline or call attention to certain features such as the shape of a building or the decoration of a window. (CMP-18)

Overcurrent. Any current in excess of the rated current of equipment or the ampacity of a conductor. It may result from overload, short circuit, or ground fault. (CMP-10)

> Informational Note: A current in excess of rating may be accommodated by certain equipment and conductors for a given set of conditions. Therefore, the rules for overcurrent protection are specific for particular situations.

Overcurrent Protective Device, Branch-Circuit. A device capable of providing protection for service, feeder, and branch circuits and equipment over the full range of overcurrents between its rated current and its interrupting rating. Such devices are provided with interrupting ratings appropriate for the intended use but no less than 5000 amperes. (CMP-10)

Overcurrent Protective Device, Supplementary. A device intended to provide limited overcurrent protection for specific applications and utilization equipment such as luminaires and appliances. This limited protection is in addition to the protection provided in the required branch circuit by the branch-circuit overcurrent protective device. (CMP-10)

Overload. Operation of equipment in excess of normal, full-load rating, or of a conductor in excess of its ampacity that, when it persists for a sufficient length of time, would cause damage or dangerous overheating. A fault, such as a short circuit or ground fault, is not an overload. (CMP-10)

Definitions Applicable to Dwelling Units — Article 100

Panelboard. A single panel or group of panel units designed for assembly in the form of a single panel, including buses and automatic overcurrent devices, and equipped with or without switches for the control of light, heat, or power circuits; designed to be placed in a cabinet or cutout box placed in or against a wall, partition, or other support; and accessible only from the front. (CMP-9)

Photovoltaic (PV) System. The total components, circuits, and equipment up to and including the PV system disconnecting means that, in combination, convert solar energy into electric energy. (CMP-4)

Premises Wiring (System). Interior and exterior wiring, including power, lighting, control, and signal circuit wiring together with all their associated hardware, fittings, and wiring devices, both permanently and temporarily installed. This includes (a) wiring from the service point or power source to the outlets or (b) wiring from and including the power source to the outlets where there is no service point.

Such wiring does not include wiring internal to appliances, luminaires, motors, controllers, motor control centers, and similar equipment. (CMP-1)

> Informational Note: Power sources include, but are not limited to, interconnected or stand-alone batteries, solar photovoltaic systems, other distributed generation systems, or generators.

Raceway. An enclosed channel designed expressly for holding wires, cables, or busbars, with additional functions as permitted in this *Code*. (CMP-8)

> Informational Note: A raceway is identified within specific article definitions.

Rainproof. Constructed, protected, or treated so as to prevent rain from interfering with the successful operation of the apparatus under specified test conditions. (CMP-1)

Raintight. Constructed or protected so that exposure to a beating rain will not result in the entrance of water under specified test conditions. (CMP-1)

Receptacle. A contact device installed at the outlet for the connection of an attachment plug, or for the direct connection of electrical utilization equipment designed to mate with the corresponding contact device. A single receptacle is a single contact device with no other contact device on the same yoke or strap. A multiple receptacle is two or more contact devices on the same yoke or strap. (CMP-18)

Informational Note: A duplex receptacle is an example of a multiple receptacle that has two receptacles on the same yoke or strap.

Receptacle Outlet. An outlet where one or more receptacles are installed. (CMP-18)

Remote-Control Circuit. Any electrical circuit that controls any other circuit through a relay or an equivalent device. (CMP-3)

Retrofit Kit. A general term for a complete subassembly of parts and devices for field conversion of utilization equipment. (CMP-18)

Separately Derived System. An electrical source, other than a service, having no direct connection(s) to circuit conductors of any other electrical source other than those established by grounding and bonding connections. (CMP-5)

Service. The conductors and equipment connecting the serving utility to the wiring system of the premises served. (CMP-10)

Service Cable. Service conductors made up in the form of a cable. (CMP-10)

Service Conductors. The conductors from the service point to the service disconnecting means. (CMP-10)

Service Conductors, Overhead. The overhead conductors between the service point and the first point of connection to the service-entrance conductors at the building or other structure. (CMP-10)

Service Conductors, Underground. The underground conductors between the service point and the first point of connection to the service-entrance conductors in a terminal box, meter, or other enclosure, inside or outside the building wall. (CMP-10)

Informational Note: Where there is no terminal box, meter, or other enclosure, the point of connection is considered to be the point of entrance of the service conductors into the building.

Service Drop. The overhead conductors between the serving utility and the service point. (CMP-10)

Service-Entrance Conductors, Overhead System. The service conductors between the terminals of the service equipment and a point usually outside the building, clear of building walls, where joined by tap or splice to the service drop or overhead service conductors. (CMP-10)

Definitions Applicable to Dwelling Units — Article 100

Service-Entrance Conductors, Underground System. The service conductors between the terminals of the service equipment and the point of connection to the service lateral or underground service conductors. (CMP-10)

> Informational Note: Where service equipment is located outside the building walls, there may be no service-entrance conductors or they may be entirely outside the building.

Service Equipment. The necessary equipment, consisting of a circuit breaker(s) or switch(es) and fuse(s) and their accessories, connected to the serving utility and intended to constitute the main control and disconnect of the serving utility. (CMP-10)

Service Lateral. The underground conductors between the utility electric supply system and the service point. (CMP-10)

Service Point. The point of connection between the facilities of the serving utility and the premises wiring. (CMP-10)

> Informational Note: The service point can be described as the point of demarcation between where the serving utility ends and the premises wiring begins. The serving utility generally specifies the location of the service point based on the conditions of service.

Structure. That which is built or constructed, other than equipment. (CMP-1)

Surge Arrester. A protective device for limiting surge voltages by discharging or bypassing surge current; it also prevents continued flow of follow current while remaining capable of repeating these functions. (CMP-10)

Switch, General-Use. A switch intended for use in general distribution and branch circuits. It is rated in amperes, and it is capable of interrupting its rated current at its rated voltage. (CMP-9)

Switch, General-Use Snap. A form of general-use switch constructed so that it can be installed in device boxes or on box covers, or otherwise used in conjunction with wiring systems recognized by this *Code*. (CMP-9)

Switch, Transfer. An automatic or nonautomatic device for transferring one or more load conductor connections from one power source to another. (CMP-13)

Ungrounded. Not connected to ground or to a conductive body that extends the ground connection. (CMP-5)

Utilization Equipment. Equipment that utilizes electric energy for electronic, electromechanical, chemical, heating, lighting, or similar purposes. (CMP-1)

Voltage (of a circuit). The greatest root-mean-square (rms) (effective) difference of potential between any two conductors of the circuit concerned. (CMP-1)

> Informational Note: Some systems, such as 3-phase 4-wire, single-phase 3-wire, and 3-wire direct current, may have various circuits of various voltages.

Voltage, Nominal. A nominal value assigned to a circuit or system for the purpose of conveniently designating its voltage class (e.g., 120/240 volts, 480Y/277 volts, 600 volts). (CMP-1)

> Informational Note No. 1: The actual voltage at which a circuit operates can vary from the nominal within a range that permits satisfactory operation of equipment.
>
> Informational Note No. 2: See ANSI C84.1-2011, *Voltage Ratings for Electric Power Systems and Equipment (60 Hz)*.
>
> Informational Note No. 3: Certain battery units may be considered to be rated at nominal 48 volts dc, but may have a charging float voltage up to 58 volts. In dc applications, 60 volts is used to cover the entire range of float voltages.

Voltage to Ground. For grounded circuits, the voltage between the given conductor and that point or conductor of the circuit that is grounded; for ungrounded circuits, the greatest voltage between the given conductor and any other conductor of the circuit. (CMP-1)

Watertight. Constructed so that moisture will not enter the enclosure under specified test conditions. (CMP-1)

Weatherproof. Constructed or protected so that exposure to the weather will not interfere with successful operation. (CMP-1)

> Informational Note: Rainproof, raintight, or watertight equipment can fulfill the requirements for weatherproof where varying weather conditions other than wetness, such as snow, ice, dust, or temperature extremes, are not a factor.

CHAPTER 3
GENERAL INSTALLATION REQUIREMENTS

INTRODUCTION

This chapter contains requirements from Article 110 — Requirements for Electrical Installations. These important general requirements address fundamental safety concepts that are essential to the installation of conductors and equipment. The requirements of Article 110 apply to all electrical installations covered within the scope of the *Code*. Included in Article 110 are requirements for the approval, examination, identification, installation, use, and access to and spaces about electrical equipment (and conductors). Compliance with the requirements covering the working space around electric equipment is essential to the safety of workers. Sufficient access and working space must be provided and maintained around all electrical equipment to permit ready and safe operation and maintenance of such equipment. Section 110.26 provides specific dimensions for the height, width, and depth of the working space for equipment rated up to 1000 volts. The work space requirements of 110.26(A) apply to all equipment that is likely to require examination, adjustment, servicing, or maintenance while energized; the dedicated space requirements of 110.26(E) are applicable only to switchboards, panelboards, distribution boards, and motor control centers.

ARTICLE 110

Requirements for Electrical Installations

PART I. GENERAL

110.1 Scope. This article covers general requirements for the examination and approval, installation and use, access to and spaces about electrical conductors and equipment; enclosures intended for personnel entry; and tunnel installations.

Informational Note: See Informative Annex J for information regarding ADA accessibility design.

110.2 Approval. The conductors and equipment required or permitted by this *Code* shall be acceptable only if approved.

Informational Note: See 90.7, Examination of Equipment for Safety, and 110.3, Examination, Identification, Installation, and Use of Equipment. See definitions of *Approved, Identified, Labeled,* and *Listed*.

110.3 Examination, Identification, Installation, Use, and Listing (Product Certification) of Equipment.

(A) Examination. In judging equipment, considerations such as the following shall be evaluated:

(1) Suitability for installation and use in conformity with this *Code*

Informational Note No. 1: Equipment may be new, reconditioned, refurbished, or remanufactured.

Informational Note No. 2: Suitability of equipment use may be identified by a description marked on or provided with a product to identify the suitability of the product for a specific purpose, environment, or application. Special conditions of use or other limitations and other pertinent information may be marked on the equipment, included in the product instructions, or included in the appropriate listing and labeling information. Suitability of equipment may be evidenced by listing or labeling.

(2) Mechanical strength and durability, including, for parts designed to enclose and protect other equipment, the adequacy of the protection thus provided
(3) Wire-bending and connection space

General Installation Requirements 110.11

(4) Electrical insulation
(5) Heating effects under normal conditions of use and also under abnormal conditions likely to arise in service
(6) Arcing effects
(7) Classification by type, size, voltage, current capacity, and specific use
(8) Other factors that contribute to the practical safeguarding of persons using or likely to come in contact with the equipment

(B) Installation and Use. Equipment that is listed, labeled, or both shall be installed and used in accordance with any instructions included in the listing or labeling.

(C) Listing. Product testing, evaluation, and listing (product certification) shall be performed by recognized qualified electrical testing laboratories and shall be in accordance with applicable product standards recognized as achieving equivalent and effective safety for equipment installed to comply with this *Code*.

> Informational Note: The Occupational Safety and Health Administration (OSHA) recognizes qualified electrical testing laboratories that perform evaluations, testing, and certification of certain products to ensure that they meet the requirements of both the construction and general industry OSHA electrical standards. If the listing (product certification) is done under a qualified electrical testing laboratory program, this listing mark signifies that the tested and certified product complies with the requirements of one or more appropriate product safety test standards.

110.5 Conductors. Conductors used to carry current shall be of copper, aluminum, or copper-clad aluminum unless otherwise provided in this *Code*. Where the conductor material is not specified, the sizes given in this *Code* shall apply to copper conductors. Where other materials are used, the size shall be changed accordingly.

110.6 Conductor Sizes. Conductor sizes are expressed in American Wire Gage (AWG) or in circular mils.

110.7 Wiring Integrity. Completed wiring installations shall be free from short circuits, ground faults, or any connections to ground other than as required or permitted elsewhere in this *Code*.

110.11 Deteriorating Agents. Unless identified for use in the operating environment, no conductors or equipment shall be located in damp or wet locations; where exposed to gases, fumes, vapors, liquids, or other

110.12 Chapter 3

agents that have a deteriorating effect on the conductors or equipment; or where exposed to excessive temperatures.

Equipment not identified for outdoor use and equipment identified only for indoor use, such as "dry locations," "indoor use only," "damp locations," or enclosure Types 1, 2, 5, 12, 12K, and/or 13, shall be protected against damage from the weather during construction.

110.12 Mechanical Execution of Work. Electrical equipment shall be installed in a neat and workmanlike manner.

> Informational Note: Accepted industry practices are described in ANSI/NECA 1-2015, *Standard for Good Workmanship in Electrical Construction,* and other ANSI-approved installation standards.

(A) Unused Openings. Unused openings, other than those intended for the operation of equipment, those intended for mounting purposes, or those permitted as part of the design for listed equipment, shall be closed to afford protection substantially equivalent to the wall of the equipment. Where metallic plugs or plates are used with nonmetallic enclosures, they shall be recessed at least 6 mm (¼ in.) from the outer surface of the enclosure.

(B) Integrity of Electrical Equipment and Connections. Internal parts of electrical equipment, including busbars, wiring terminals, insulators, and other surfaces, shall not be damaged or contaminated by foreign materials such as paint, plaster, cleaners, abrasives, or corrosive residues. There shall be no damaged parts that may adversely affect safe operation or mechanical strength of the equipment such as parts that are broken; bent; cut; or deteriorated by corrosion, chemical action, or overheating.

(C) Cables and Conductors. Cables and conductors installed exposed on the surfaces of ceilings and sidewalls shall be supported by the building structure in such a manner that the cables and conductors will not be damaged by normal building use. Such cables and conductors shall be secured by hardware including straps, staples, cable ties, hangers, or similar fittings designed and installed so as not to damage the cable. The installation shall also conform with 300.4 and 300.11. Nonmetallic cable ties and other nonmetallic cable accessories used to secure and support cables in other spaces used for environmental air (plenums) shall be listed as having low smoke and heat release properties.

General Installation Requirements 110.14

110.13 Mounting and Cooling of Equipment.

(A) Mounting. Electrical equipment shall be firmly secured to the surface on which it is mounted. Wooden plugs driven into holes in masonry, concrete, plaster, or similar materials shall not be used.

110.14 Electrical Connections.
Because of different characteristics of dissimilar metals, devices such as pressure terminal or pressure splicing connectors and soldering lugs shall be identified for the material of the conductor and shall be properly installed and used. Conductors of dissimilar metals shall not be intermixed in a terminal or splicing connector where physical contact occurs between dissimilar conductors (such as copper and aluminum or aluminum and copper-clad aluminum), unless the device is identified for the purpose and conditions of use. Materials such as solder, fluxes, inhibitors, and compounds, where employed, shall be suitable for the use and shall be of a type that will not adversely affect the conductors, installation, or equipment.

Connectors and terminals for conductors more finely stranded than Class B and Class C stranding as shown in Chapter 9, Table 10, shall be identified for the specific conductor class or classes.

(A) Terminals. Connection of conductors to terminal parts shall ensure a thoroughly good connection without damaging the conductors and shall be made by means of pressure connectors (including set-screw type), solder lugs, or splices to flexible leads. Connection by means of wire-binding screws or studs and nuts that have upturned lugs or the equivalent shall be permitted for 10 AWG or smaller conductors.

Terminals for more than one conductor and terminals used to connect aluminum shall be so identified.

(B) Splices. Conductors shall be spliced or joined with splicing devices identified for the use or by brazing, welding, or soldering with a fusible metal or alloy. Soldered splices shall first be spliced or joined so as to be mechanically and electrically secure without solder and then be soldered. All splices and joints and the free ends of conductors shall be covered with an insulation equivalent to that of the conductors or with an identified insulating device.

Wire connectors or splicing means installed on conductors for direct burial shall be listed for such use.

(C) Temperature Limitations. The temperature rating associated with the ampacity of a conductor shall be selected and coordinated so as not to exceed the lowest temperature rating of any connected termination,

25

conductor, or device. Conductors with temperature ratings higher than specified for terminations shall be permitted to be used for ampacity adjustment, correction, or both.

(1) Equipment Provisions. The determination of termination provisions of equipment shall be based on 110.14(C)(1)(a) or (C)(1)(b). Unless the equipment is listed and marked otherwise, conductor ampacities used in determining equipment termination provisions shall be based on Table 310.16 as appropriately modified by 310.12.

(a) Termination provisions of equipment for circuits rated 100 amperes or less, or marked for 14 AWG through 1 AWG conductors, shall be used only for one of the following:

(1) Conductors rated 60°C (140°F).
(2) Conductors with higher temperature ratings, provided the ampacity of such conductors is determined based on the 60°C (140°F) ampacity of the conductor size used.
(3) Conductors with higher temperature ratings if the equipment is listed and identified for use with such conductors.
(4) For motors marked with design letters B, C, or D, conductors having an insulation rating of 75°C (167°F) or higher shall be permitted to be used, provided the ampacity of such conductors does not exceed the 75°C (167°F) ampacity.

(b) Termination provisions of equipment for circuits rated over 100 amperes, or marked for conductors larger than 1 AWG, shall be used only for one of the following:

(1) Conductors rated 75°C (167°F)
(2) Conductors with higher temperature ratings, provided the ampacity of such conductors does not exceed the 75°C (167°F) ampacity of the conductor size used, or up to their ampacity if the equipment is listed and identified for use with such conductors

(2) Separate Connector Provisions. Separately installed pressure connectors shall be used with conductors at the ampacities not exceeding the ampacity at the listed and identified temperature rating of the connector.

Informational Note: With respect to 110.14(C)(1) and (C)(2), equipment markings or listing information may additionally restrict the sizing and temperature ratings of connected conductors.

General Installation Requirements 110.21

(D) Terminal Connection Torque. Tightening torque values for terminal connections shall be as indicated on equipment or in installation instructions provided by the manufacturer. An approved means shall be used to achieve the indicated torque value.

Informational Note No. 1: Examples of approved means of achieving the indicated torque values include torque tools or devices such as shear bolts or breakaway-style devices with visual indicators that demonstrate that the proper torque has been applied.

Informational Note No. 2: The equipment manufacturer can be contacted if numeric torque values are not indicated on the equipment or if the installation instructions are not available. Informative Annex I of UL Standard 486A-486B, *Standard for Safety-Wire Connectors*, provides torque values in the absence of manufacturer's recommendations.

Informational Note No. 3: Additional information for torquing threaded connections and terminations can be found in Section 8.11 of NFPA 70B-2019, *Recommended Practice for Electrical Equipment Maintenance*.

110.21 Marking.

(B) Field-Applied Hazard Markings. Where caution, warning, or danger signs or labels are required by this *Code*, the labels shall meet the following requirements:

(1) The marking shall warn of the hazards using effective words, colors, symbols, or any combination thereof.

Informational Note: ANSI Z535.4-2011, *Product Safety Signs and Labels*, provides guidelines for suitable font sizes, words, colors, symbols, and location requirements for labels.

(2) The label shall be permanently affixed to the equipment or wiring method and shall not be handwritten.

Exception to (2): Portions of labels or markings that are variable, or that could be subject to changes, shall be permitted to be handwritten and shall be legible.

(3) The label shall be of sufficient durability to withstand the environment involved.

110.22 Identification of Disconnecting Means.

(A) General. Each disconnecting means shall be legibly marked to indicate its purpose unless located and arranged so the purpose is evident. In other than one- or two-family dwellings, the marking shall include the identification of the circuit source that supplies the disconnecting means. The marking shall be of sufficient durability to withstand the environment involved.

110.25 Lockable Disconnecting Means. If a disconnecting means is required to be lockable open elsewhere in this *Code*, it shall be capable of being locked in the open position. The provisions for locking shall remain in place with or without the lock installed.

Exception: Locking provisions for a cord-and-plug connection shall not be required to remain in place without the lock installed.

PART II. 1000 VOLTS, NOMINAL, OR LESS

110.26 Spaces About Electrical Equipment. Access and working space shall be provided and maintained about all electrical equipment to permit ready and safe operation and maintenance of such equipment.

(A) Working Space. Working space for equipment operating at 1000 volts, nominal, or less to ground and likely to require examination, adjustment, servicing, or maintenance while energized shall comply with the dimensions of 110.26(A)(1), (A)(2), (A)(3), and (A)(4) or as required or permitted elsewhere in this *Code*.

(1) Depth of Working Space. The depth of the working space in the direction of live parts shall not be less than that specified in Table 110.26(A)(1) unless the requirements of 110.26(A)(1)(a), (A)(1)(b), or (A)(1)(c) are met. Distances shall be measured from the exposed live parts or from the enclosure or opening if the live parts are enclosed.

(a) *Dead-Front Assemblies.* Working space shall not be required in the back or sides of assemblies, such as dead-front switchboards, switchgear, or motor control centers, where all connections and all renewable or adjustable parts, such as fuses or switches, are accessible from locations other than the back or sides. Where rear access is required to work on nonelectrical parts on the back of enclosed equipment, a minimum horizontal working space of 762 mm (30 in.) shall be provided.

General Installation Requirements — 110.26

Table 110.26(A)(1) Working Spaces

Nominal Voltage to Ground	Minimum Clear Distance		
	Condition 1	Condition 2	Condition 3
0–150	900 mm (3 ft)	900 mm (3 ft)	900 mm (3 ft)
151–600	900 mm (3 ft)	1.0 m (3 ft 6 in.)	1.2 m (4 ft)
601–1000	900 mm (3 ft)	1.2 m (4 ft)	1.5 m (5 ft)

Note: Where the conditions are as follows:

Condition 1 — Exposed live parts on one side of the working space and no live or grounded parts on the other side of the working space, or exposed live parts on both sides of the working space that are effectively guarded by insulating materials.

Condition 2 — Exposed live parts on one side of the working space and grounded parts on the other side of the working space. Concrete, brick, or tile walls shall be considered as grounded.

Condition 3 — Exposed live parts on both sides of the working space.

(b) *Low Voltage.* By special permission, smaller working spaces shall be permitted where all exposed live parts operate at not greater than 30 volts rms, 42 volts peak, or 60 volts dc.

(c) *Existing Buildings.* In existing buildings where electrical equipment is being replaced, Condition 2 working clearance shall be permitted between dead-front switchboards, switchgear, panelboards, or motor control centers located across the aisle from each other where conditions of maintenance and supervision ensure that written procedures have been adopted to prohibit equipment on both sides of the aisle from being open at the same time and qualified persons who are authorized will service the installation.

(2) Width of Working Space. The width of the working space in front of the electrical equipment shall be the width of the equipment or 762 mm (30 in.), whichever is greater. In all cases, the work space shall permit at least a 90 degree opening of equipment doors or hinged panels.

(3) Height of Working Space. The work space shall be clear and extend from the grade, floor, or platform to a height of 2.0 m (6½ ft) or the height of the equipment, whichever is greater. Within the height requirements of this section, other equipment or support structures, such as concrete pads, associated with the electrical installation and located above or below the electrical equipment shall be permitted to extend not more than 150 mm (6 in.) beyond the front of the electrical equipment.

Exception No. 1: On battery systems mounted on open racks, the top clearance shall comply with 480.10(D).

Exception No. 2: In existing dwelling units, service equipment or panelboards that do not exceed 200 amperes shall be permitted in spaces where the height of the working space is less than 2.0 m (6½ ft).

Exception No. 3: Meters that are installed in meter sockets shall be permitted to extend beyond the other equipment. The meter socket shall be required to follow the rules of this section.

(4) Limited Access. Where equipment operating at 1000 volts, nominal, or less to ground and likely to require examination, adjustment, servicing, or maintenance while energized is required by installation instructions or function to be located in a space with limited access, all of the following shall apply:

(1) Where equipment is installed above a lay-in ceiling, there shall be an opening not smaller than 559 mm × 559 mm (22 in. × 22 in.), or in a crawl space, there shall be an accessible opening not smaller than 559 mm × 762 mm (22 in. × 30 in.).
(2) The width of the working space shall be the width of the equipment enclosure or a minimum of 762 mm (30 in.), whichever is greater.
(3) All enclosure doors or hinged panels shall be capable of opening a minimum of 90 degrees.
(4) The space in front of the enclosure shall comply with the depth requirements of Table 110.26(A)(1). The maximum height of the working space shall be the height necessary to install the equipment in the limited space. A horizontal ceiling structural member or access panel shall be permitted in this space.

(B) Clear Spaces. Working space required by this section shall not be used for storage. When normally enclosed live parts are exposed for inspection or servicing, the working space, if in a passageway or general open space, shall be suitably guarded.

(D) Illumination. Illumination shall be provided for all working spaces about service equipment, switchboards, switchgear, panelboards, or motor control centers installed indoors. Control by automatic means shall not be permitted to control all illumination within the working space. Additional lighting outlets shall not be required where the work space is illuminated by an adjacent light source or as permitted by 210.70(A)(1), Exception No. 1, for switched receptacles.

General Installation Requirements 110.26

(E) Dedicated Equipment Space. All switchboards, switchgear, panelboards, and motor control centers shall be located in dedicated spaces and protected from damage.

Exception: Control equipment that by its very nature or because of other rules of the Code must be adjacent to or within sight of its operating machinery shall be permitted in those locations.

(1) Indoor. Indoor installations shall comply with 110.26(E)(1)(a) through (E)(1)(d).

(a) *Dedicated Electrical Space.* The space equal to the width and depth of the equipment and extending from the floor to a height of 1.8 m (6 ft) above the equipment or to the structural ceiling, whichever is lower, shall be dedicated to the electrical installation. No piping, ducts, leak protection apparatus, or other equipment foreign to the electrical installation shall be located in this zone.

Exception: Suspended ceilings with removable panels shall be permitted within the 1.8-m (6-ft) zone.

(b) *Foreign Systems.* The area above the dedicated space required by 110.26(E)(1)(a) shall be permitted to contain foreign systems, provided protection is installed to avoid damage to the electrical equipment from condensation, leaks, or breaks in such foreign systems.

(c) *Sprinkler Protection.* Sprinkler protection shall be permitted for the dedicated space where the piping complies with this section.

(d) *Suspended Ceilings.* A dropped, suspended, or similar ceiling that does not add strength to the building structure shall not be considered a structural ceiling.

(2) Outdoor. Outdoor installations shall comply with 110.26(E)(2)(a) through (E)(2)(c).

(a) *Installation Requirements.* Outdoor electrical equipment shall be the following:

(1) Installed in identified enclosures
(2) Protected from accidental contact by unauthorized personnel or by vehicular traffic
(3) Protected from accidental spillage or leakage from piping systems

(b) *Work Space.* The working clearance space shall include the zone described in 110.26(A). No architectural appurtenance or other equipment shall be located in this zone.

Chapter 3

(c) *Dedicated Equipment Space.* The space equal to the width and depth of the equipment, and extending from grade to a height of 1.8 m (6 ft) above the equipment, shall be dedicated to the electrical installation. No piping or other equipment foreign to the electrical installation shall be located in this zone.

110.27 Guarding of Live Parts.

(B) Prevent Physical Damage. In locations where electrical equipment is likely to be exposed to physical damage, enclosures or guards shall be so arranged and of such strength as to prevent such damage.

CHAPTER 4
TEMPORARY WIRING

INTRODUCTION

This chapter contains requirements from Article 590 — Temporary Installations. Article 590 applies to temporary electrical power and lighting installations. New construction and many renovation or addition projects generally require the installation of temporary electrical equipment, which, particularly in the case of new construction, often involves setting a pole and building a temporary service. Although the feeders, branch circuits, and receptacles have specific requirements in Article 590, the service must be installed in accordance with Article 230 (see Chapter 6 of this *Pocket Guide*). For example, the minimum clearance of overhead service-drop conductors must comply with 230.24. Of course, not all electrical projects will require the installation of a temporary service. It is important to note that all of the general requirements in Chapters 1 through 4 of the *Code* apply to temporary installations unless there is a specific modification to a general rule specified in Article 590.

Because of the inherent hazards associated with construction sites, ground-fault protection of receptacles with certain ratings that are in use by personnel is required for all temporary wiring installations, not just for new construction sites with temporary services. If a receptacle(s) is installed or exists as part of the permanent wiring or the building or structure and is used for temporary electric power, ground fault circuit-interrupter (GFCI) protection for personnel must be provided. This may require replacing the existing receptacle with a GFCI receptacle or the use of a portable GFCI device. All other outlets (besides 125-volt, single-phase, 15-, 20-, and 30-ampere receptacle outlets) must have GFCI protection for personnel, special purpose ground-fault protection, or an assured equipment grounding conductor program in accordance with 590.6(B)(2).

ARTICLE 590
Temporary Installations

590.2 All Wiring Installations.

(A) Other Articles. Except as specifically modified in this article, all other requirements of this *Code* for permanent wiring shall apply to temporary wiring installations.

(B) Approval. Temporary wiring methods shall be acceptable only if approved based on the conditions of use and any special requirements of the temporary installation.

590.3 Time Constraints.

(A) During the Period of Construction. Temporary electric power and lighting installations shall be permitted during the period of construction, remodeling, maintenance, repair, or demolition of buildings, structures, equipment, or similar activities.

(D) Removal. Temporary wiring shall be removed immediately upon completion of construction or purpose for which the wiring was installed.

590.4 General.

(A) Services. Services shall be installed in conformance with Parts I through VIII of Article 230, as applicable.

(B) Feeders. Overcurrent protection shall be provided in accordance with 240.4, 240.5, 240.100, and 240.101. Conductors shall be permitted within cable assemblies or within multiconductor cords or cables of a type identified in Table 400.4 for hard usage or extra-hard usage. For the purpose of this section, the following wiring methods shall be permitted:

(1) Type NM, Type NMC, and Type SE cables shall be permitted to be used in any dwelling, building, or structure without any height limitation or limitation by building construction type and without concealment within walls, floors, or ceilings.
(2) Type SE cable shall be permitted to be installed in a raceway in an underground installation.

(C) Branch Circuits. All branch circuits shall originate in an approved power outlet, switchgear, switchboard or panelboard, motor control center, or fused switch enclosure. Conductors shall be permitted within cable

Temporary Wiring 590.4

assemblies or within multiconductor cord or cable of a type identified in Table 400.4 for hard usage or extra-hard usage. Conductors shall be protected from overcurrent as provided in 240.4, 240.5, and 240.100. For the purposes of this section, the following wiring methods shall be permitted:

(1) Type NM, Type NMC, and Type SE cables shall be permitted to be used in any dwelling, building, or structure without any height limitation or limitation by building construction type and without concealment within walls, floors, or ceilings.
(2) Type SE cable shall be permitted to be installed in a raceway in an underground installation.

(D) Receptacles.

(1) All Receptacles. All receptacles shall be of the grounding type. Unless installed in a continuous metal raceway that qualifies as an equipment grounding conductor in accordance with 250.118 or a continuous metal-covered cable that qualifies as an equipment grounding conductor in accordance with 250.118, all branch circuits shall include a separate equipment grounding conductor, and all receptacles shall be electrically connected to the equipment grounding conductor(s). Receptacles on construction sites shall not be installed on any branch circuit that supplies temporary lighting.

(2) Receptacles in Wet Locations. All 15- and 20-ampere, 125- and 250-volt receptacles installed in a wet location shall comply with 406.9(B)(1).

(E) Disconnecting Means. Suitable disconnecting switches or plug connectors shall be installed to permit the disconnection of all ungrounded conductors of each temporary circuit. Multiwire branch circuits shall be provided with a means to disconnect simultaneously all ungrounded conductors at the power outlet or panelboard where the branch circuit originated. Identified handle ties shall be permitted.

(F) Lamp Protection. All lamps for general illumination shall be protected from accidental contact or breakage by a suitable luminaire or lampholder with a guard.

Brass shell, paper-lined sockets, or other metal-cased sockets shall not be used unless the shell is connected to the circuit equipment grounding conductor.

(G) Splices. A box, conduit body, or other enclosure, with a cover installed, shall be required for all splices.

Exception No. 1: On construction sites, a box, conduit body, or other enclosure shall not be required for either of the following conditions:

(1) The circuit conductors being spliced are all from nonmetallic multiconductor cord or cable assemblies, provided that the equipment grounding continuity is maintained with or without the box.
(2) The circuit conductors being spliced are all from metal-sheathed cable assemblies terminated in listed fittings that mechanically secure the cable sheath to maintain effective electrical continuity.

Exception No. 2: On construction sites, branch-circuits that are permanently installed in framed walls and ceilings and are used to supply temporary power or lighting, and that are GFCI protected, the following shall be permitted:

(1) A box cover shall not be required for splices installed completely inside of junction boxes with plaster rings.
(2) Listed pigtail-type lampholders shall be permitted to be installed in ceiling-mounted junction boxes with plaster rings.
(3) Finger safe devices shall be permitted for supplying and connection of devices.

(H) Protection from Accidental Damage. Flexible cords and cables shall be protected from accidental damage. Sharp corners and projections shall be avoided. Where passing through doorways or other pinch points, protection shall be provided to avoid damage.

(I) Termination(s) at Devices. Flexible cords and cables entering enclosures containing devices requiring termination shall be secured to the box with fittings listed for connecting flexible cords and cables to boxes designed for the purpose.

(J) Support. Cable assemblies and flexible cords and cables shall be supported in place at intervals that ensure that they will be protected from physical damage. Support shall be in the form of staples, cable ties, straps, or similar type fittings installed so as not to cause damage. Cable assemblies and flexible cords and cables installed as branch circuits or feeders shall not be installed on the floor or on the ground. Extension cords shall not be required to comply with 590.4(J). Vegetation shall not be used for support of overhead spans of branch circuits or feeders.

Exception: For holiday lighting in accordance with 590.3(B), where the conductors or cables are arranged with strain relief devices, tension

Temporary Wiring 590.6

take-up devices, or other approved means to avoid damage from the movement of the live vegetation, trees shall be permitted to be used for support of overhead spans of branch-circuit conductors or cables.

590.6 Ground-Fault Protection for Personnel. Ground-fault protection for personnel for all temporary wiring installations shall be provided to comply with 590.6(A) and (B). This section shall apply only to temporary wiring installations used to supply temporary power to equipment used by personnel during construction, remodeling, maintenance, repair, or demolition of buildings, structures, equipment, or similar activities. This section shall apply to power derived from an electric utility company or from an on-site-generated power source.

(A) Receptacle Outlets. Temporary receptacle installations used to supply temporary power to equipment used by personnel during construction, remodeling, maintenance, repair, or demolition of buildings, structures, equipment, or similar activities shall comply with the requirements of 590.6(A)(1) through (A)(3), as applicable.

(1) Receptacle Outlets Not Part of Permanent Wiring. All 125-volt, single-phase, 15-, 20-, and 30-ampere receptacle outlets that are not a part of the permanent wiring of the building or structure and that are in use by personnel shall have ground-fault circuit-interrupter protection for personnel. In addition to this required ground-fault circuit-interrupter protection for personnel, listed cord sets or devices incorporating listed ground-fault circuit-interrupter protection for personnel identified for portable use shall be permitted.

(2) Receptacle Outlets Existing or Installed as Permanent Wiring. Ground-fault circuit-interrupter protection for personnel shall be provided for all 125-volt, single-phase, 15-, 20-, and 30-ampere receptacle outlets installed or existing as part of the permanent wiring of the building or structure and used for temporary electric power. Listed cord sets or devices incorporating listed ground-fault circuit-interrupter protection for personnel identified for portable use shall be permitted.

(3) Receptacles on 15-kW or less Portable Generators. All 125-volt and 125/250-volt, single-phase, 15-, 20-, and 30-ampere receptacle outlets that are a part of a 15-kW or smaller portable generator shall have listed ground-fault circuit-interrupter protection for personnel. All 15- and 20-ampere, 125- and 250-volt receptacles, including those that are part of a portable generator, used in a damp or wet location shall comply with 406.9(A) and (B). Listed cord sets or devices incorporating listed ground-fault circuit-interrupter protection for personnel identified for portable use shall be permitted for use with 15-kW or less portable generators manufactured or remanufactured prior to January 1, 2011.

(B) Other Receptacle Outlets. For temporary wiring installations, receptacles, other than those covered by 590.6(A)(1) through (A)(3) used to supply temporary power to equipment used by personnel during construction, remodeling, maintenance, repair, or demolition of buildings, structures, or equipment, or similar activities, shall have protection in accordance with 590.6(B)(1) or the assured equipment grounding conductor program in accordance with 590.6(B)(2).

(1) GFCI Protection. Ground-fault circuit-interrupter protection for personnel.

(2) Assured Equipment Grounding Conductor Program. A written assured equipment grounding conductor program continuously enforced at the site by one or more designated persons to ensure that equipment grounding conductors for all cord sets, receptacles that are not a part of the permanent wiring of the building or structure, and equipment connected by cord and plug are installed and maintained in accordance with the applicable requirements of 250.114, 250.138, 406.4(C), and 590.4(D).

590.8 Overcurrent Protective Devices.

(A) Where Reused. Where overcurrent protective devices that have been previously used are installed in a temporary installation, these overcurrent protective devices shall be examined to ensure these devices have been properly installed, properly maintained, and there is no evidence of impending failure.

CHAPTER 5

SERVICE AND FEEDER LOAD CALCULATIONS

INTRODUCTION

Selecting the size service and panelboard(s) is one of the first steps in residential installations. Load calculations are required to size the service and/or feeder overcurrent protection, as well as the ungrounded, grounded, and equipment grounding conductors. Article 220 contains detailed requirements for calculating branch-circuit, feeder, and service loads for one-family, two-family, and multifamily dwellings. Two load calculation methods are provided, Standard (Part III) and Optional (Part IV).

While certain installations will have a service with a single panelboard, other installations may have multiple panelboards. For dwellings with panelboards that are not part of the service, additional load calculations are necessary — one for the service and one for each feeder.

Collecting specific information and applying the appropriate demand factors will provide answers to questions such as: What is the minimum size service or feeder? What is the minimum size ungrounded (hot) conductor? What is the minimum size neutral conductor? What is the minimum size grounding electrode conductor (for services) or equipment grounding conductor (for feeders)?

ARTICLE 220
Branch-Circuit, Feeder, and Service Load Calculations

PART I. GENERAL

220.5 Calculations.

(A) Voltages. Unless other voltages are specified, for purposes of calculating branch-circuit and feeder loads, nominal system voltages of 120, 120/240, 208Y/120, 240, 347, 480Y/277, 480, 600Y/347, and 600 volts shall be used.

(B) Fractions of an Ampere. Calculations shall be permitted to be rounded to the nearest whole ampere, with decimal fractions smaller than 0.5 dropped.

PART II. BRANCH-CIRCUIT LOAD CALCULATIONS

220.10 General. Branch-circuit loads shall be calculated as shown in 220.12, 220.14, and 220.16.

220.11 Floor Area. The floor area for each floor shall be calculated from the outside dimensions of the building, dwelling unit, or other area involved. For dwelling units, the calculated floor area shall not include open porches, garages, or unused or unfinished spaces not adaptable for future use.

220.14 Other Loads — All Occupancies. In all occupancies, the minimum load for each outlet for general-use receptacles and outlets not used for general illumination shall not be less than that calculated in 220.14(A) through (M), the loads shown being based on nominal branch-circuit voltages.

(A) Specific Appliances or Loads. An outlet for a specific appliance or other load not covered in 220.14(B) through (M) shall be calculated based on the ampere rating of the appliance or load served.

(B) Electric Dryers and Electric Cooking Appliances in Dwellings and Household Cooking Appliances Used in Instructional

Service and Feeder Load Calculations 220.16

Programs. Load calculations shall be permitted as specified in 220.54 for electric dryers and in 220.55 for electric ranges and other cooking appliances.

(C) Motor Outlets. Loads for motor outlets shall be calculated in accordance with the requirements in 430.22, 430.24, and 440.6.

(D) Luminaires. An outlet supplying luminaire(s) shall be calculated based on the maximum volt-ampere rating of the equipment and lamps for which the luminaire(s) is rated.

(J) Dwelling Units. In one-family, two-family, and multifamily dwellings, the minimum unit load shall be not less than 33 volt-amperes/m² (3 volt-amperes/ft²). The lighting and receptacle outlets specified in 220.14(J)(1), (J)(2), and (J)(3) are included in the minimum unit load. No additional load calculations shall be required for such outlets. The minimum lighting load shall be determined using the minimum unit load and the floor area as determined in 220.11 for dwelling occupancies. Motors rated less than ⅛ hp and connected to a lighting circuit shall be considered part of the minimum lighting load.

(1) All general-use receptacle outlets of 20-ampere rating or less, including receptacles connected to the circuits in 210.11(C)(3) and 210.11(C)(4)
(2) The receptacle outlets specified in 210.52(E) and (G)
(3) The lighting outlets specified in 210.70

220.16 Loads for Additions to Existing Installations.

(A) Dwelling Units. Loads added to an existing dwelling unit(s) shall comply with the following as applicable:

(1) Loads for structural additions to an existing dwelling unit or for a previously unwired portion of an existing dwelling unit shall be calculated in accordance with 220.14.
(2) Loads for new circuits or extended circuits in previously wired dwelling units shall be calculated in accordance with 220.14.

PART III. FEEDER AND SERVICE LOAD CALCULATIONS

220.40 General. The calculated load of a feeder or service shall not be less than the sum of the loads on the branch circuits supplied, as determined by Part II of this article, after any applicable demand factors permitted by Part III or IV or required by Part V have been applied.

220.42 General Lighting. The demand factors specified in Table 220.42 shall apply to that portion of the total branch-circuit load calculated for general illumination. They shall not be applied in determining the number of branch circuits for general illumination.

Table 220.42 Lighting Load Demand Factors

Type of Occupancy	Portion of Lighting Load to Which Demand Factor Applies (Volt-Amperes)	Demand Factor (%)
Dwelling units	First 3000 at	100
	From 3001 to 120,000 at	35
	Remainder over 120,000 at	25
Hotels and motels, including apartment houses without provision for cooking by tenants*	First 20,000 or less at	60
	From 20,001 to 100,000 at	50
	Remainder over 100,000 at	35
Warehouses (storage)	First 12,500 or less at	100
	Remainder over 12,500 at	50
All others	Total volt-amperes	100

*The demand factors of this table shall not apply to the calculated load of feeders or services supplying areas in hotels and motels where the entire lighting is likely to be used at one time, as in ballrooms or dining rooms.

220.50 Motors. Motor loads shall be calculated in accordance with 430.24, 430.25, and 430.26 and with 440.6 for hermetic refrigerant motor-compressors.

Service and Feeder Load Calculations 220.54

220.51 Fixed Electric Space Heating. Fixed electric space-heating loads shall be calculated at 100 percent of the total connected load. However, in no case shall a feeder or service load current rating be less than the rating of the largest branch circuit supplied.

220.52 Small-Appliance and Laundry Loads — Dwelling Unit.

(A) Small-Appliance Circuit Load. In each dwelling unit, the load shall be calculated at 1500 volt-amperes for each 2-wire small-appliance branch circuit as covered by 210.11(C)(1). Where the load is subdivided through two or more feeders, the calculated load for each shall include not less than 1500 volt-amperes for each 2-wire small-appliance branch circuit. These loads shall be permitted to be included with the general lighting load and subjected to the demand factors provided in Table 220.42.

Exception: The individual branch circuit permitted by 210.52(B)(1), Exception No. 2, shall be permitted to be excluded from the calculation required by 220.52.

(B) Laundry Circuit Load. A load of not less than 1500 volt-amperes shall be included for each 2-wire laundry branch circuit installed as covered by 210.11(C)(2). This load shall be permitted to be included with the general lighting load and shall be subjected to the demand factors provided in Table 220.42.

220.53 Appliance Load — Dwelling Unit(s). It shall be permissible to apply a demand factor of 75 percent to the nameplate rating load of four or more appliances rated ¼ hp or greater, or 500 watts or greater, that are fastened in place, and that are served by the same feeder or service in a one-family, two-family, or multifamily dwelling. This demand factor shall not apply to:

(1) Household electric cooking equipment that is fastened in place
(2) Clothes dryers
(3) Space heating equipment
(4) Air-conditioning equipment

220.54 Electric Clothes Dryers — Dwelling Unit(s). The load for household electric clothes dryers in a dwelling unit(s) shall be either 5000 watts (volt-amperes) or the nameplate rating, whichever is larger, for each dryer served. The use of the demand factors in Table 220.54

shall be permitted. Where two or more single-phase dryers are supplied by a 3-phase, 4-wire feeder or service, the total load shall be calculated on the basis of twice the maximum number connected between any two phases. Kilovolt-amperes (kVA) shall be considered equivalent to kilowatts (kW) for loads calculated in this section.

Table 220.54 Demand Factors for Household Electric Clothes Dryers

Number of Dryers	Demand Factor (%)
1–4	100
5	85
6	75
7	65
8	60
9	55
10	50
11	47
12–23	47% minus 1% for each dryer exceeding 11
24–42	35% minus 0.5% for each dryer exceeding 23
43 and over	25%

220.55 Electric Cooking Appliances in Dwelling Units and Household Cooking Appliances Used in Instructional Programs. The load for household electric ranges, wall-mounted ovens, counter-mounted cooking units, and other household cooking appliances individually rated in excess of 1¾ kW shall be permitted to be calculated in accordance with Table 220.55. Kilovolt-amperes (kVA) shall be considered equivalent to kilowatts (kW) for loads calculated under this section.

Where two or more single-phase ranges are supplied by a 3-phase, 4-wire feeder or service, the total load shall be calculated on the basis of twice the maximum number connected between any two phases.

220.60 Noncoincident Loads. Where it is unlikely that two or more noncoincident loads will be in use simultaneously, it shall be permissible to use only the largest load(s) that will be used at one time for calculating the total load of a feeder or service. Where a motor is part of the

Service and Feeder Load Calculations

Table 220.55 Demand Factors and Loads for Household Electric Ranges, Wall-Mounted Ovens, Counter-Mounted Cooking Units, and Other Household Cooking Appliances over 1¾ kW Rating (Column C to be used in all cases except as otherwise permitted in Note 3.)

	Demand Factor (%) (See Notes)		Column C
	Column A	Column B	Maximum Demand (kW) (See Notes)
Number of Appliances	(Less than 3½ kW Rating)	(3½ kW through 8¾ kW Rating)	(Not over 12 kW Rating)
1	80	80	8
2	75	65	11
3	70	55	14
4	66	50	17
5	62	45	20
6	59	43	21
7	56	40	22
8	53	36	23
9	51	35	24
10	49	34	25
11	47	32	26
12	45	32	27
13	43	32	28

(continued)

Table 220.55 Demand Factors and Loads for Household Electric Ranges, Wall-Mounted Ovens, Counter-Mounted Cooking Units, and Other Household Cooking Appliances over 1¾ kW Rating (Column C to be used in all cases except as otherwise permitted in Note 3.) — *Continued*

Number of Appliances	Demand Factor (%) (See Notes)		Column C Maximum Demand (kW) (See Notes) (Not over 12 kW Rating)
	Column A (Less than 3½ kW Rating)	Column B (3½ kW through 8¾ kW Rating)	
14	41	32	29
15	40	32	30
16	39	28	31
17	38	28	32
18	37	28	33
19	36	28	34
20	35	28	35
21	34	26	36
22	33	26	37
23	32	26	38
24	31	26	39
25	30	26	40
26–30	30	24	15 kW + 1 kW for each range
31–40	30	22	

41–50	30	
51–60	30	25 kW + ¾ kW
61 and over	30	for each range

	20
	18
	16

Notes:

1. Over 12 kW through 27 kW ranges all of same rating. For ranges individually rated more than 12 kW but not more than 27 kW, the maximum demand in Column C shall be increased 5 percent for each additional kilowatt of rating or major fraction thereof by which the rating of individual ranges exceeds 12 kW.

2. Over 8¾ kW through 27 kW ranges of unequal ratings. For ranges individually rated more than 8¾ kW and of different ratings, but none exceeding 27 kW, an average value of rating shall be calculated by adding together the ratings of all ranges to obtain the total connected load (using 12 kW for any range rated less than 12 kW) and dividing by the total number of ranges. Then the maximum demand in Column C shall be increased 5 percent for each kilowatt or major fraction thereof by which this average value exceeds 12 kW.

3. Over 1¾ kW through 8¾ kW. In lieu of the method provided in Column C, it shall be permissible to add the nameplate ratings of all household cooking appliances rated more than 1¾ kW but not more than 8¾ kW and multiply the sum by the demand factors specified in Column A or Column B for the given number of appliances. Where the rating of cooking appliances falls under both Column A and Column B, the demand factors for each column shall be applied to the appliances for that column, and the results added together.

4. Branch-Circuit Load. It shall be permissible to calculate the branch-circuit load for one range in accordance with Table 220.55. The branch-circuit load for one wall-mounted oven or one counter-mounted cooking unit shall be the nameplate rating of the appliance. The branch-circuit load for a counter-mounted cooking unit and not more than two wall-mounted ovens, all supplied from a single branch circuit and located in the same room, shall be calculated by adding the nameplate rating of the individual appliances and treating this total as equivalent to one range.

5. This table shall also apply to household cooking appliances rated over 1¾ kW and used in instructional programs.

Informational Note No. 1: See the examples in Informative Annex D.

Informational Note No. 2: See Table 220.56 for commercial cooking equipment.

noncoincident load and is not the largest of the noncoincident loads, 125 percent of the motor load shall be used in the calculation if it is the largest motor.

220.61 Feeder or Service Neutral Load.

(A) Basic Calculation. The feeder or service neutral load shall be the maximum unbalance of the load determined by this article. The maximum unbalanced load shall be the maximum net calculated load between the neutral conductor and any one ungrounded conductor.

Exception: For 3-wire, 2-phase or 5-wire, 2-phase systems, the maximum unbalanced load shall be the maximum net calculated load between the neutral conductor and any one ungrounded conductor multiplied by 140 percent.

(B) Permitted Reductions. A service or feeder supplying the following loads shall be permitted to have an additional demand factor of 70 percent applied to the amount in 220.61(B)(1) or portion of the amount in 220.61(B)(2) determined by the following basic calculations:

(1) A feeder or service supplying household electric ranges, wall-mounted ovens, counter-mounted cooking units, and electric dryers, where the maximum unbalanced load has been determined in accordance with Table 220.55 for ranges and Table 220.54 for dryers
(2) That portion of the unbalanced load in excess of 200 amperes where the feeder or service is supplied from a 3-wire dc or single-phase ac system; or a 4-wire, 3-phase system; or a 3-wire, 2-phase system; or a 5-wire, 2-phase system

Informational Note: See Examples D1(a), D1(b), D2(b), D4(a), and D5(a) in Informative Annex D.

PART IV. OPTIONAL FEEDER AND SERVICE LOAD CALCULATIONS

220.82 Dwelling Unit.

(A) Feeder and Service Load. This section applies to a dwelling unit having the total connected load served by a single 120/240-volt or 208Y/120-volt set of 3-wire service or feeder conductors with an ampacity of 100 or greater. It shall be permissible to calculate the feeder and service loads in accordance with this section instead of the method specified in Part III of this article. The calculated load shall be the result of adding the loads from 220.82(B) and (C). Feeder and service-entrance

Service and Feeder Load Calculations 220.82

conductors whose calculated load is determined by this optional calculation shall be permitted to have the neutral load determined by 220.61.

(B) General Loads. The general calculated load shall be not less than 100 percent of the first 10 kVA plus 40 percent of the remainder of the following loads:

(1) 33 volt-amperes/m^2 or 3 volt-amperes/ft^2 for general lighting and general-use receptacles. The floor area for each floor shall be calculated from the outside dimensions of the dwelling unit. The calculated floor area shall not include open porches, garages, or unused or unfinished spaces not adaptable for future use.
(2) 1500 volt-amperes for each 2-wire, 20-ampere small-appliance branch circuit and each laundry branch circuit covered in 210.11(C)(1) and (C)(2).
(3) The nameplate rating of the following:

   a. All appliances that are fastened in place, permanently connected, or located to be on a specific circuit
   b. Ranges, wall-mounted ovens, counter-mounted cooking units
   c. Clothes dryers that are not connected to the laundry branch circuit specified in item (2)
   d. Water heaters

(4) The nameplate ampere or kVA rating of all permanently connected motors not included in item (3).

(C) Heating and Air-Conditioning Load. The largest of the following six selections (load in kVA) shall be included:

(1) 100 percent of the nameplate rating(s) of the air conditioning and cooling.
(2) 100 percent of the nameplate rating(s) of the heat pump when the heat pump is used without any supplemental electric heating.
(3) 100 percent of the nameplate rating(s) of the heat pump compressor and 65 percent of the supplemental electric heating for central electric space-heating systems. If the heat pump compressor is prevented from operating at the same time as the supplementary heat, it does not need to be added to the supplementary heat for the total central space heating load.
(4) 65 percent of the nameplate rating(s) of electric space heating if less than four separately controlled units.
(5) 40 percent of the nameplate rating(s) of electric space heating if four or more separately controlled units.

(6) 100 percent of the nameplate ratings of electric thermal storage and other heating systems where the usual load is expected to be continuous at the full nameplate value. Systems qualifying under this selection shall not be calculated under any other selection in 220.82(C).

220.83 Existing Dwelling Unit. This section shall be permitted to be used to determine if the existing service or feeder is of sufficient capacity to serve additional loads. Where the dwelling unit is served by a 120/240-volt or 208Y/120-volt, 3-wire service, it shall be permissible to calculate the total load in accordance with 220.83(A) or (B).

(A) Where Additional Air-Conditioning Equipment or Electric Space-Heating Equipment Is Not to Be Installed. The following percentages shall be used for existing and additional new loads.

Load (kVA)	Percent of Load
First 8 kVA of load at	100
Remainder of load at	40

Load calculations shall include the following:

(1) General lighting and general-use receptacles at 33 volt-amperes/m² or 3 volt-amperes/ft² as determined by 220.12
(2) 1500 volt-amperes for each 2-wire, 20-ampere small-appliance branch circuit and each laundry branch circuit covered in 210.11(C)(1) and (C)(2)
(3) The nameplate rating of the following:

  a. All appliances that are fastened in place, permanently connected, or located to be on a specific circuit
  b. Ranges, wall-mounted ovens, counter-mounted cooking units
  c. Clothes dryers that are not connected to the laundry branch circuit specified in item (2)
  d. Water heaters

(B) Where Additional Air-Conditioning Equipment or Electric Space-Heating Equipment Is to Be Installed. The following percentages shall be used for existing and additional new loads. The larger connected load of air conditioning or space heating, but not both, shall be used.

Service and Feeder Load Calculations 220.84

Load	Percent of Load
Air-conditioning equipment	100
Central electric space heating	100
Less than four separately controlled space-heating units	100
First 8 kVA of all other loads	100
Remainder of all other loads	40

Other loads shall include the following:

(1) General lighting and general-use receptacles at 33 volt-amperes/m² or 3 volt-amperes/ft² as determined by 220.12
(2) 1500 volt-amperes for each 2-wire, 20-ampere small-appliance branch circuit and each laundry branch circuit covered in 210.11(C)(1) and (C)(2)
(3) The nameplate rating of the following:

   a. All appliances that are fastened in place, permanently connected, or located to be on a specific circuit
   b. Ranges, wall-mounted ovens, counter-mounted cooking units
   c. Clothes dryers that are not connected to the laundry branch circuit specified in item (2)
   d. Water heaters

220.84 Multifamily Dwelling.

(A) Feeder or Service Load. It shall be permissible to calculate the load of a feeder or service that supplies three or more dwelling units of a multifamily dwelling in accordance with Table 220.84 instead of Part III of this article if all the following conditions are met:

(1) No dwelling unit is supplied by more than one feeder.
(2) Each dwelling unit is equipped with electric cooking equipment.

Exception: When the calculated load for multifamily dwellings without electric cooking in Part III of this article exceeds that calculated under Part IV for the identical load plus electric cooking (based on 8 kW per unit), the lesser of the two loads shall be permitted to be used.

(3) Each dwelling unit is equipped with either electric space heating or air conditioning, or both. Feeders and service conductors whose calculated load is determined by this optional calculation shall be permitted to have the neutral load determined by 220.61.

(B) House Loads. House loads shall be calculated in accordance with Part III of this article and shall be in addition to the dwelling unit loads calculated in accordance with Table 220.84.

Table 220.84 Optional Calculations — Demand Factors for Three or More Multifamily Dwelling Units

Number of Dwelling Units	Demand Factor (%)
3–5	45
6–7	44
8–10	43
11	42
12–13	41
14–15	40
16–17	39
18–20	38
21	37
22–23	36
24–25	35
26–27	34
28–30	33
31	32
32–33	31
34–36	30
37–38	29
39–42	28
43–45	27
46–50	26
51–55	25
56–61	24
62 and over	23

(C) Calculated Loads. The calculated load to which the demand factors of Table 220.84 apply shall include the following:

(1) 33 volt-amperes/m^2 or 3 volt-amperes/ft^2 for general lighting and general-use receptacles

Service and Feeder Load Calculations 220.85

(2) 1500 volt-amperes for each 2-wire, 20-ampere small-appliance branch circuit and each laundry branch circuit covered in 210.11(C)(1) and (C)(2)

(3) The nameplate rating of the following:
   a. All appliances that are fastened in place, permanently connected, or located to be on a specific circuit
   b. Ranges, wall-mounted ovens, counter-mounted cooking units
   c. Clothes dryers that are not connected to the laundry branch circuit specified in item (2)
   d. Water heaters

(4) The nameplate ampere or kVA rating of all permanently connected motors not included in item (3)

(5) The larger of the air-conditioning load or the fixed electric space-heating load

220.85 Two Dwelling Units. Where two dwelling units are supplied by a single feeder and the calculated load under Part III of this article exceeds that for three identical units calculated under 220.84, the lesser of the two loads shall be permitted to be used.

CHAPTER 6
SERVICES

INTRODUCTION

This chapter contains requirements from Article 230 — Services. Article 230 provides installation requirements for conductors and equipment that supply, control, and protect services. Contained in this *Pocket Guide* are seven of the eight parts that make up Article 230: Part I General, Part II Overhead Service Conductors, Part III Underground Service Conductors, Part IV Service-Entrance Conductors, Part V Service Equipment — General, Part VI Service Equipment — Disconnecting Means, and Part VII Service Equipment — Overcurrent Protection.

While all service installations must comply with Parts I, IV, V, VI, and VII, compliance with Parts II and III depends on the type of service. If the service is brought to the premises using overhead conductors, Part II applies; if the service is brought to the premises using underground conductors, Part III applies.

ARTICLE 230
Services

PART I. GENERAL

230.2 Number of Services. A building or other structure served shall be supplied by only one service unless permitted in 230.2(A) through (D). For the purpose of 230.40, Exception No. 2 only, underground sets of conductors, 1/0 AWG and larger, running to the same location and connected together at their supply end but not connected together at their load end shall be considered to be supplying one service.

(A) Special Conditions. Additional services shall be permitted to supply the following:

(1) Fire pumps
(2) Emergency systems
(3) Legally required standby systems
(4) Optional standby systems
(5) Parallel power production systems
(6) Systems designed for connection to multiple sources of supply for the purpose of enhanced reliability

(B) Special Occupancies. By special permission, additional services shall be permitted for either of the following:

(1) Multiple-occupancy buildings where there is no available space for service equipment accessible to all occupants
(2) A single building or other structure sufficiently large to make two or more services necessary

(C) Capacity Requirements. Additional services shall be permitted under any of the following:

(1) Where the capacity requirements are in excess of 2000 amperes at a supply voltage of 1000 volts or less
(2) Where the load requirements of a single-phase installation are greater than the serving agency normally supplies through one service
(3) By special permission

(D) Different Characteristics. Additional services shall be permitted for different voltages, frequencies, or phases, or for different uses, such as for different rate schedules.

Services 230.8

(E) Identification. Where a building or structure is supplied by more than one service, or any combination of branch circuits, feeders, and services, a permanent plaque or directory shall be installed at each service disconnect location denoting all other services, feeders, and branch circuits supplying that building or structure and the area served by each. See 225.37.

230.3 One Building or Other Structure Not to Be Supplied Through Another. Service conductors supplying a building or other structure shall not pass through the interior of another building or other structure.

230.6 Conductors Considered Outside the Building. Conductors shall be considered outside of a building or other structure under any of the following conditions:

(1) Where installed under not less than 50 mm (2 in.) of concrete beneath a building or other structure
(2) Where installed within a building or other structure in a raceway that is encased in concrete or brick not less than 50 mm (2 in.) thick
(3) Where installed in any vault that meets the construction requirements of Article 450, Part III
(4) Where installed in conduit and under not less than 450 mm (18 in.) of earth beneath a building or other structure
(5) Where installed within rigid metal conduit (Type RMC) or intermediate metal conduit (Type IMC) used to accommodate the clearance requirements in 230.24 and routed directly through an eave but not a wall of a building

230.7 Other Conductors in Raceway or Cable. Conductors other than service conductors shall not be installed in the same service raceway or service cable in which the service conductors are installed.

Exception No. 1: Grounding electrode conductors or supply side bonding jumpers or conductors shall be permitted within service raceways.

Exception No. 2: Load management control conductors having overcurrent protection shall be permitted within service raceways.

230.8 Raceway Seal. Where a service raceway enters a building or structure from an underground distribution system, it shall be sealed in accordance with 300.5(G). Spare or unused raceways shall also be sealed. Sealants shall be identified for use with the cable insulation, shield, or other components.

230.9 Clearances on Buildings. Service conductors and final spans shall comply with 230.9(A), (B), and (C).

(A) Clearances. Service conductors installed as open conductors or multiconductor cable without an overall outer jacket shall have a clearance of not less than 900 mm (3 ft) from windows that are designed to be opened, doors, porches, balconies, ladders, stairs, fire escapes, or similar locations.

Exception: Conductors run above the top level of a window shall be permitted to be less than the 900 mm (3 ft) requirement.

(B) Vertical Clearance. The vertical clearance of final spans above, or within 900 mm (3 ft) measured horizontally of platforms, projections, or surfaces that will permit personal contact shall be maintained in accordance with 230.24(B).

(C) Building Openings. Overhead service conductors shall not be installed beneath openings through which materials may be moved, such as openings in farm and commercial buildings, and shall not be installed where they obstruct entrance to these building openings.

230.10 Vegetation as Support. Vegetation such as trees shall not be used for support of overhead service conductors or service equipment.

PART II. OVERHEAD SERVICE CONDUCTORS

230.22 Insulation or Covering. Individual conductors shall be insulated or covered.

Exception: The grounded conductor of a multiconductor cable shall be permitted to be bare.

230.23 Size and Ampacity.

(A) General. Conductors shall have sufficient ampacity to carry the current for the load as calculated in accordance with Article 220 and shall have adequate mechanical strength.

(B) Minimum Size. The conductors shall not be smaller than 8 AWG copper or 6 AWG aluminum or copper-clad aluminum.

Exception: Conductors supplying only limited loads of a single branch circuit — such as small polyphase power, controlled water heaters, and similar loads — shall not be smaller than 12 AWG hard-drawn copper or equivalent.

Services 230.24

(C) Grounded Conductors. The grounded conductor shall not be less than the minimum size as required by 250.24(C).

230.24 Clearances. Overhead service conductors shall not be readily accessible and shall comply with 230.24(A) through (E) for services not over 1000 volts, nominal.

(A) Above Roofs. Conductors shall have a vertical clearance of not less than 2.5 m (8 ft) above the roof surface. The vertical clearance above the roof level shall be maintained for a distance of not less than 900 mm (3 ft) in all directions from the edge of the roof.

Exception No. 1: The area above a roof surface subject to pedestrian or vehicular traffic shall have a vertical clearance from the roof surface in accordance with the clearance requirements of 230.24(B).

Exception No. 2: Where the voltage between conductors does not exceed 300 and the roof has a slope of 100 mm in 300 mm (4 in. in 12 in.) or greater, a reduction in clearance to 900 mm (3 ft) shall be permitted.

Exception No. 3: Where the voltage between conductors does not exceed 300, a reduction in clearance above only the overhanging portion of the roof to not less than 450 mm (18 in.) shall be permitted if (1) not more than 1.8 m (6 ft) of overhead service conductors, 1.2 m (4 ft) horizontally, pass above the roof overhang, and (2) they are terminated at a through-the-roof raceway or approved support.

Informational Note: See 230.28 for mast supports.

Exception No. 4: The requirement for maintaining the vertical clearance 900 mm (3 ft) from the edge of the roof shall not apply to the final conductor span where the service drop or overhead service conductors are attached to the side of a building.

Exception No. 5: Where the voltage between conductors does not exceed 300 and the roof area is guarded or isolated, a reduction in clearance to 900 mm (3 ft) shall be permitted.

(B) Vertical Clearance for Overhead Service Conductors. Overhead service conductors, where not in excess of 1000 volts, nominal, shall have the following minimum clearance from final grade:

(1) 3.0 m (10 ft) — at the electrical service entrance to buildings, also at the lowest point of the drip loop of the building electrical entrance, and above areas or sidewalks accessible only to pedestrians, measured from final grade or other accessible surface only for

overhead service conductors supported on and cabled together with a grounded bare messenger where the voltage does not exceed 150 volts to ground

(2) 3.7 m (12 ft) — over residential property and driveways, and those commercial areas not subject to truck traffic where the voltage does not exceed 300 volts to ground

(3) 4.5 m (15 ft) — for those areas listed in the 3.7 m (12 ft) classification where the voltage exceeds 300 volts to ground

(4) 5.5 m (18 ft) — over public streets, alleys, roads, parking areas subject to truck traffic, driveways on other than residential property, and other land such as cultivated, grazing, forest, and orchard

(D) Clearance from Swimming Pools. See 680.9.

230.26 Point of Attachment. The point of attachment of the overhead service conductors to a building or other structure shall provide the minimum clearances as specified in 230.9 and 230.24. In no case shall this point of attachment be less than 3.0 m (10 ft) above finished grade.

230.27 Means of Attachment. Multiconductor cables used for overhead service conductors shall be attached to buildings or other structures by fittings identified for use with service conductors. Open conductors shall be attached to fittings identified for use with service conductors or to noncombustible, nonabsorbent insulators securely attached to the building or other structure.

230.28 Service Masts as Supports. Only power service-drop or overhead service conductors shall be permitted to be attached to a service mast. Service masts used for the support of service-drop or overhead service conductors shall be installed in accordance with 230.28(A) and (B).

(A) Strength. The service mast shall be of adequate strength or be supported by braces or guy wires to withstand safely the strain imposed by the service-drop or overhead service conductors. Hubs intended for use with a conduit that serves as a service mast shall be identified for use with service-entrance equipment.

(B) Attachment. Service-drop or overhead service conductors shall not be attached to a service mast between a weatherhead or the end of the conduit and a coupling, where the coupling is located above the last point of securement to the building or other structure or is located above the building or other structure.

Services

230.29 Supports over Buildings. Service conductors passing over a roof shall be securely supported by substantial structures. For a grounded system, where the substantial structure is metal, it shall be bonded by means of a bonding jumper and listed connector to the grounded overhead service conductor. Where practicable, such supports shall be independent of the building.

PART III. UNDERGROUND SERVICE CONDUCTORS

230.30 Installation.

(A) Insulation. Underground service conductors shall be insulated for the applied voltage.

Exception: A grounded conductor shall be permitted to be uninsulated as follows:

(1) Bare copper used in a raceway
(2) Bare copper for direct burial where bare copper is approved for the soil conditions
(3) Bare copper for direct burial without regard to soil conditions where part of a cable assembly identified for underground use
(4) Aluminum or copper-clad aluminum without individual insulation or covering where part of a cable assembly identified for underground use in a raceway or for direct burial

(B) Wiring Methods. Underground service conductors shall be installed in accordance with the applicable requirements of this *Code* covering the type of wiring method used and shall be limited to the following methods:

(1) Type RMC conduit
(2) Type IMC conduit
(3) Type NUCC conduit
(4) Type HDPE conduit
(5) Type PVC conduit
(6) Type RTRC conduit
(7) Type IGS cable
(8) Type USE conductors or cables
(9) Type MV or Type MC cable identified for direct burial applications
(10) Type MI cable, where suitably protected against physical damage and corrosive conditions

230.31 Size and Ampacity.

(A) General. Underground service conductors shall have sufficient ampacity to carry the current for the load as calculated in accordance with Article 220 and shall have adequate mechanical strength.

(B) Minimum Size. The conductors shall not be smaller than 8 AWG copper or 6 AWG aluminum or copper-clad aluminum.

Exception: Conductors supplying only limited loads of a single branch circuit — such as small polyphase power, controlled water heaters, and similar loads — shall not be smaller than 12 AWG copper or 10 AWG aluminum or copper-clad aluminum.

(C) Grounded Conductors. The grounded conductor shall not be less than the minimum size required by 250.24(C).

230.32 Protection Against Damage.
Underground service conductors shall be protected against damage in accordance with 300.5. Service conductors entering a building or other structure shall be installed in accordance with 230.6 or protected by a raceway wiring method identified in 230.43.

230.33 Spliced Conductors.
Service conductors shall be permitted to be spliced or tapped in accordance with 110.14, 300.5(E), 300.13, and 300.15.

PART IV. SERVICE-ENTRANCE CONDUCTORS

230.40 Number of Service-Entrance Conductor Sets.
Each service drop, set of overhead service conductors, set of underground service conductors, or service lateral shall supply only one set of service-entrance conductors.

Exception No. 1: A building with more than one occupancy shall be permitted to have one set of service-entrance conductors for each service, as permitted in 230.2, run to each occupancy or group of occupancies. If the number of service disconnect locations for any given classification of service does not exceed six, the requirements of 230.2(E) shall apply at each location. If the number of service disconnect locations exceeds six for any given supply classification, all service disconnect locations for all supply characteristics, together with any branch circuit or feeder supply sources, if applicable, shall be clearly described using graphics or text, or both, on one or more plaques located in an approved, readily accessible location(s) on the building or structure served and as near as practicable to the point(s) of attachment or entry(ies) for each service drop or service lateral and for each set of overhead or underground service conductors.

Exception No. 2: Where two to six service disconnecting means in separate enclosures are grouped at one location and supply separate loads from one service drop, set of overhead service conductors, set of underground service conductors, or service lateral, one set of service-entrance conductors shall be permitted to supply each or several such service equipment enclosures.

Exception No. 3: A one-family dwelling unit and its accessory structures shall be permitted to have one set of service-entrance conductors run to each from a single service drop, set of overhead service conductors, set of underground service conductors, or service lateral.

Exception No. 4: Two-family dwellings, multifamily dwellings, and multiple occupancy buildings shall be permitted to have one set of service-entrance conductors installed to supply the circuits covered in 210.25.

Exception No. 5: One set of service-entrance conductors connected to the supply side of the normal service disconnecting means shall be permitted to supply each or several systems covered by 230.82(5) or 230.82(6).

230.41 Insulation of Service-Entrance Conductors. Service-entrance conductors entering or on the exterior of buildings or other structures shall be insulated.

Exception: A grounded conductor shall be permitted to be uninsulated as follows:

(1) Bare copper used in a raceway or part of a service cable assembly
(2) Bare copper for direct burial where bare copper is approved for the soil conditions
(3) Bare copper for direct burial without regard to soil conditions where part of a cable assembly identified for underground use
(4) Aluminum or copper-clad aluminum without individual insulation or covering where part of a cable assembly or identified for underground use in a raceway, or for direct burial
(5) Bare conductors used in an auxiliary gutter

230.42 Minimum Size and Ampacity.

(A) General. Service-entrance conductors shall have an ampacity of not less than the maximum load to be served. Conductors shall be sized not less than the largest of 230.42(A)(1) or (A)(2). Loads shall be determined in accordance with Part III, IV, or V of Article 220, as applicable. Ampacity shall be determined from 310.14 and shall comply with 110.14(C). The maximum current of busways shall be that value for which the busway has been listed or labeled.

Informational Note: For information on busways, see UL 857, *Standard for Safety for Busways*.

(1) Where the service-entrance conductors supply continuous loads or any combination of noncontinuous and continuous loads, the minimum service-entrance conductor size shall have an ampacity not less than the sum of the noncontinuous loads plus 125 percent of continuous loads.

Exception No. 1: Grounded conductors that are not connected to an overcurrent device shall be permitted to be sized at 100 percent of the sum of the continuous and noncontinuous load.

Exception No. 2: The sum of the noncontinuous load and the continuous load if the service-entrance conductors terminate in an overcurrent device where both the overcurrent device and its assembly are listed for operation at 100 percent of their rating shall be permitted.

(2) The minimum service-entrance conductor size shall have an ampacity not less than the maximum load to be served after the application of any adjustment or correction factors.

(B) Specific Installations. In addition to the requirements of 230.42(A), the minimum ampacity for ungrounded conductors for specific installations shall not be less than the rating of the service disconnecting means specified in 230.79(A) through (D).

(C) Grounded Conductors. The grounded conductor shall not be smaller than the minimum size as required by 250.24(C).

230.43 Wiring Methods for 1000 Volts, Nominal, or Less.
Service-entrance conductors shall be installed in accordance with the applicable

Services 230.46

requirements of this *Code* covering the type of wiring method used and shall be limited to the following methods:

(1) Open wiring on insulators
(2) Type IGS cable
(3) Rigid metal conduit (RMC)
(4) Intermediate metal conduit (IMC)
(5) Electrical metallic tubing (EMT)
(6) Electrical nonmetallic tubing
(7) Service-entrance cables
(8) Wireways
(9) Busways
(10) Auxiliary gutters
(11) Rigid polyvinyl chloride conduit (PVC)
(12) Cablebus
(13) Type MC cable
(14) Mineral-insulated, metal-sheathed cable, Type MI
(15) Flexible metal conduit (FMC) not over 1.8 m (6 ft) long or liquidtight flexible metal conduit (LFMC) not over 1.8 m (6 ft) long between a raceway, or between a raceway and service equipment, with a supply-side bonding jumper routed with the flexible metal conduit (FMC) or the liquidtight flexible metal conduit (LFMC) according to 250.102(A), (B), (C), and (E)
(16) Liquidtight flexible nonmetallic conduit (LFNC)
(17) High density polyethylene conduit (HDPE)
(18) Nonmetallic underground conduit with conductors (NUCC)
(19) Reinforced thermosetting resin conduit (RTRC)
(20) Type TC-ER cable

230.46 Spliced and Tapped Conductors. Service-entrance conductors shall be permitted to be spliced or tapped in accordance with 110.14, 300.5(E), 300.13, and 300.15. Power distribution blocks, pressure connectors, and devices for splices and taps shall be listed. Power distribution blocks installed on service conductors shall be marked "suitable for use on the line side of the service equipment" or equivalent.

Effective January 1, 2023, pressure connectors and devices for splices and taps installed on service conductors shall be marked "suitable for use on the line side of the service equipment" or equivalent.

230.50 Protection Against Physical Damage.

(A) Underground Service-Entrance Conductors. Underground service-entrance conductors shall be protected against physical damage in accordance with 300.5.

(B) All Other Service-Entrance Conductors. All other service-entrance conductors, other than underground service entrance conductors, shall be protected against physical damage as specified in 230.50(B)(1) or (B)(2).

(1) Service-Entrance Cables. Service-entrance cables, where subject to physical damage, shall be protected by any of the following:

(1) Rigid metal conduit (RMC)
(2) Intermediate metal conduit (IMC)
(3) Schedule 80 PVC conduit
(4) Electrical metallic tubing (EMT)
(5) Reinforced thermosetting resin conduit (RTRC)
(6) Other approved means

230.51 Mounting Supports. Service-entrance cables or individual open service-entrance conductors shall be supported as specified in 230.51(A), (B), or (C).

(A) Service-Entrance Cables. Service-entrance cables shall be supported by straps or other approved means within 300 mm (12 in.) of every service head, gooseneck, or connection to a raceway or enclosure and at intervals not exceeding 750 mm (30 in.).

230.53 Raceways to Drain. Where exposed to the weather, raceways enclosing service-entrance conductors shall be listed or approved for use in wet locations and arranged to drain. Where embedded in masonry, raceways shall be arranged to drain.

230.54 Overhead Service Locations.

(A) Service Head. Service raceways shall be equipped with a service head at the point of connection to service-drop or overhead service conductors. The service head shall be listed for use in wet locations.

(B) Service-Entrance Cables Equipped with Service Head or Gooseneck. Service-entrance cables shall be equipped with a service head. The service head shall be listed for use in wet locations.

Exception: Type SE cable shall be permitted to be formed in a gooseneck and taped with a self-sealing weather-resistant thermoplastic.

(C) Service Heads and Goosenecks Above Service-Drop or Overhead Service Attachment. Service heads on raceways or service-entrance cables and goosenecks in service-entrance cables shall be located above the point of attachment of the service-drop or overhead service conductors to the building or other structure.

Exception: Where it is impracticable to locate the service head or gooseneck above the point of attachment, the service head or gooseneck location shall be permitted not farther than 600 mm (24 in.) from the point of attachment.

(D) Secured. Service-entrance cables shall be held securely in place.

(E) Separately Bushed Openings. Service heads shall have conductors of different potential brought out through separately bushed openings.

Exception: For jacketed multiconductor service-entrance cable without splice.

(F) Drip Loops. Drip loops shall be formed on individual conductors. To prevent the entrance of moisture, service-entrance conductors shall be connected to the service-drop or overhead service conductors either (1) below the level of the service head or (2) below the level of the termination of the service-entrance cable sheath.

(G) Arranged That Water Will Not Enter Service Raceway or Equipment. Service-entrance and overhead service conductors shall be arranged so that water will not enter service raceway or equipment.

230.56 Service Conductor with the Higher Voltage to Ground. On a 4-wire, delta-connected service where the midpoint of one phase winding is grounded, the service conductor having the higher phase voltage to ground shall be durably and permanently marked by an outer finish that is orange in color, or by other effective means, at each termination or junction point.

PART V. SERVICE EQUIPMENT — GENERAL

230.62 Service Equipment — Enclosed or Guarded. Energized parts of service equipment shall be enclosed as specified in 230.62(A) or guarded as specified in 230.62(B).

(A) Enclosed. Energized parts shall be enclosed so that they will not be exposed to accidental contact or shall be guarded as in 230.62(B).

(B) Guarded. Energized parts that are not enclosed shall be installed on a switchboard, panelboard, or control board and guarded in accordance with 110.18 and 110.27. Where energized parts are guarded as provided in 110.27(A)(1) and (A)(2), a means for locking or sealing doors providing access to energized parts shall be provided.

230.66 Marking.

(A) General. Service equipment rated at 1000 volts or less shall be marked to identify it as being suitable for use as service equipment. All service equipment shall be listed or field evaluated.

(B) Meter Sockets. Meter sockets shall not be considered service equipment but shall be listed and rated for the voltage and current rating of the service.

Exception: Meter sockets supplied by and under the exclusive control of an electric utility shall not be required to be listed.

230.67 Surge Protection.

(A) Surge-Protective Device. All services supplying dwelling units shall be provided with a surge-protective device (SPD).

(B) Location. The SPD shall be an integral part of the service equipment or shall be located immediately adjacent thereto.

Exception: The SPD shall not be required to be located in the service equipment as required in (B) if located at each next level distribution equipment downstream toward the load.

(C) Type. The SPD shall be a Type 1 or Type 2 SPD.

(D) Replacement. Where service equipment is replaced, all of the requirements of this section shall apply.

PART VI. SERVICE EQUIPMENT — DISCONNECTING MEANS

230.70 General. Means shall be provided to disconnect all ungrounded conductors in a building or other structure from the service conductors.

(A) Location. The service disconnecting means shall be installed in accordance with 230.70(A)(1), (A)(2), and (A)(3).

Services 230.71

(1) Readily Accessible Location. The service disconnecting means shall be installed at a readily accessible location either outside of a building or structure or inside nearest the point of entrance of the service conductors.

(2) Bathrooms. Service disconnecting means shall not be installed in bathrooms.

(B) Marking. Each service disconnect shall be permanently marked to identify it as a service disconnect.

(C) Suitable for Use. Each service disconnecting means shall be suitable for the prevailing conditions. Service equipment installed in hazardous (classified) locations shall comply with the requirements of Articles 500 through 517.

230.71 Maximum Number of Disconnects. Each service shall have only one disconnecting means unless the requirements of 230.71(B) are met.

(A) General. For the purpose of this section, disconnecting means installed as part of listed equipment and used solely for the following shall not be considered a service disconnecting means:

(1) Power monitoring equipment
(2) Surge-protective device(s)
(3) Control circuit of the ground-fault protection system
(4) Power-operable service disconnecting means

(B) Two to Six Service Disconnecting Means. Two to six service disconnects shall be permitted for each service permitted by 230.2 or for each set of service-entrance conductors permitted by 230.40, Exception No. 1, 3, 4, or 5. The two to six service disconnecting means shall be permitted to consist of a combination of any of the following:

(1) Separate enclosures with a main service disconnecting means in each enclosure
(2) Panelboards with a main service disconnecting means in each panelboard enclosure
(3) Switchboard(s) where there is only one service disconnect in each separate vertical section where there are barriers separating each vertical section

(4) Service disconnects in switchgear or metering centers where each disconnect is located in a separate compartment

Informational Note No. 1: Metering centers are addressed in UL 67, *Standard for Panelboards*.

Informational Note No. 2: Examples of separate enclosures with a main service disconnecting means in each enclosure include but are not limited to motor control centers, fused disconnects, circuit breaker enclosures, and transfer switches that are suitable for use as service equipment.

230.72 Grouping of Disconnects.

(A) General. The two to six disconnects, if permitted in 230.71, shall be grouped. Each disconnect shall be marked to indicate the load served.

(C) Access to Occupants. In a multiple-occupancy building, each occupant shall have access to the occupant's service disconnecting means.

Exception: In a multiple-occupancy building where electric service and electrical maintenance are provided by the building management and where these are under continuous building management supervision, the service disconnecting means supplying more than one occupancy shall be permitted to be accessible to authorized management personnel only.

230.74 Simultaneous Opening of Poles.
Each service disconnect shall simultaneously disconnect all ungrounded service conductors that it controls from the premises wiring system.

230.75 Disconnection of Grounded Conductor.
Where the service disconnecting means does not disconnect the grounded conductor from the premises wiring, other means shall be provided for this purpose in the service equipment. A terminal or bus to which all grounded conductors can be attached by means of pressure connectors shall be permitted for this purpose. In a multisection switchboard or switchgear, disconnects for the grounded conductor shall be permitted to be in any section of the switchboard or switchgear.

Services

230.76 Manually or Power Operable. The service disconnecting means for ungrounded service conductors shall consist of one of the following:

(1) A manually operable switch or circuit breaker equipped with a handle or other suitable operating means
(2) A power-operated switch or circuit breaker, provided the switch or circuit breaker can be opened by hand in the event of a power supply failure

230.77 Indicating. The service disconnecting means shall plainly indicate whether it is in the open (off) or closed (on) position.

230.79 Rating of Service Disconnecting Means. The service disconnecting means shall have a rating not less than the calculated load to be carried, determined in accordance with Part III, IV, or V of Article 220, as applicable. In no case shall the rating be lower than specified in 230.79(A), (B), (C), or (D).

(A) One-Circuit Installations. For installations to supply only limited loads of a single branch circuit, the service disconnecting means shall have a rating of not less than 15 amperes.

(B) Two-Circuit Installations. For installations consisting of not more than two 2-wire branch circuits, the service disconnecting means shall have a rating of not less than 30 amperes.

(C) One-Family Dwellings. For a one-family dwelling, the service disconnecting means shall have a rating of not less than 100 amperes, 3-wire.

(D) All Others. For all other installations, the service disconnecting means shall have a rating of not less than 60 amperes.

230.80 Combined Rating of Disconnects. Where the service disconnecting means consists of more than one switch or circuit breaker, as permitted by 230.71, the combined ratings of all the switches or circuit breakers used shall not be less than the rating required by 230.79.

230.81 Connection to Terminals. The service conductors shall be connected to the service disconnecting means by pressure connectors, clamps, or other approved means. Connections that depend on solder shall not be used.

230.82 Equipment Connected to the Supply Side of Service Disconnect. Only the following equipment shall be permitted to be connected to the supply side of the service disconnecting means:

(1) Cable limiters.
(2) Meters and meter sockets nominally rated not in excess of 1000 volts, if all metal housings and service enclosures are grounded in accordance with Part VII and bonded in accordance with Part V of Article 250.
(3) Meter disconnect switches nominally rated not in excess of 1000 volts that have a short-circuit current rating equal to or greater than the available fault current, if all metal housings and service enclosures are grounded in accordance with Part VII and bonded in accordance with Part V of Article 250. A meter disconnect switch shall be capable of interrupting the load served. A meter disconnect shall be legibly field marked on its exterior in a manner suitable for the environment as follows:

METER DISCONNECT
NOT SERVICE EQUIPMENT

(4) Instrument transformers (current and voltage), impedance shunts, load management devices, surge arresters, and Type 1 surge-protective devices.
(5) Conductors used to supply load management devices, circuits for standby power systems, fire pump equipment, and fire and sprinkler alarms, if provided with service equipment and installed in accordance with requirements for service-entrance conductors.
(6) Solar photovoltaic systems, fuel cell systems, wind electric systems, energy storage systems, or interconnected electric power production sources, if provided with a disconnecting means listed as suitable for use as service equipment, and overcurrent protection as specified in Part VII of Article 230.
(7) Control circuits for power-operable service disconnecting means, if suitable overcurrent protection and disconnecting means are provided.
(8) Ground-fault protection systems or Type 2 surge-protective devices, where installed as part of listed equipment, if suitable overcurrent protection and disconnecting means are provided.

Services 230.85

(9) Connections used only to supply listed communications equipment under the exclusive control of the serving electric utility, if suitable overcurrent protection and disconnecting means are provided. For installations of equipment by the serving electric utility, a disconnecting means is not required if the supply is installed as part of a meter socket, such that access can only be gained with the meter removed.

(10) Emergency disconnects in accordance with 230.85, if all metal housings and service enclosures are grounded in accordance with Part VII and bonded in accordance with Part V of Article 250.

(11) Meter-mounted transfer switches nominally rated not in excess of 1000 volts that have a short-circuit current rating equal to or greater than the available fault current. A meter-mounted transfer switch shall be listed and be capable of transferring the load served. A meter-mounted transfer switch shall be marked on its exterior with both of the following:

   a. Meter-mounted transfer switch
   b. Not service equipment

230.85 Emergency Disconnects. For one- and two-family dwelling units, all service conductors shall terminate in disconnecting means having a short-circuit current rating equal to or greater than the available fault current, installed in a readily accessible outdoor location. If more than one disconnect is provided, they shall be grouped. Each disconnect shall be one of the following:

(1) Service disconnects marked as follows: EMERGENCY DISCONNECT, SERVICE DISCONNECT
(2) Meter disconnects installed per 230.82(3) and marked as follows: EMERGENCY DISCONNECT, METER DISCONNECT, NOT SERVICE EQUIPMENT
(3) Other listed disconnect switches or circuit breakers on the supply side of each service disconnect that are suitable for use as service equipment and marked as follows: EMERGENCY DISCONNECT, NOT SERVICE EQUIPMENT

Markings shall comply with 110.21(B).

PART VII. SERVICE EQUIPMENT — OVERCURRENT PROTECTION

230.90 Where Required. Each ungrounded service conductor shall have overload protection.

(A) Ungrounded Conductor. Such protection shall be provided by an overcurrent device in series with each ungrounded service conductor that has a rating or setting not higher than the ampacity of the conductor. A set of fuses shall be considered all the fuses required to protect all the ungrounded conductors of a circuit. Single-pole circuit breakers, grouped in accordance with 230.71(B), shall be considered as one protective device.

Exception No. 1: For motor-starting currents, ratings that comply with 430.52, 430.62, and 430.63 shall be permitted.

Exception No. 2: Fuses and circuit breakers with a rating or setting that complies with 240.4(B) or (C) and 240.6 shall be permitted.

Exception No. 3: Two to six circuit breakers or sets of fuses shall be permitted as the overcurrent device to provide the overload protection. The sum of the ratings of the circuit breakers or fuses shall be permitted to exceed the ampacity of the service conductors, provided the calculated load does not exceed the ampacity of the service conductors.

Exception No. 5: Overload protection for 120/240-volt, 3-wire, single-phase dwelling services shall be permitted in accordance with the requirements of 310.12.

(B) Not in Grounded Conductor. No overcurrent device shall be inserted in a grounded service conductor except a circuit breaker that simultaneously opens all conductors of the circuit.

230.91 Location. The service overcurrent device shall be an integral part of the service disconnecting means or shall be located immediately adjacent thereto. Where fuses are used as the service overcurrent device, the disconnecting means shall be located ahead of the supply side of the fuses.

230.92 Locked Service Overcurrent Devices. Where the service overcurrent devices are locked or sealed or are not readily accessible to the occupant, branch-circuit or feeder overcurrent devices shall be installed on the load side, shall be mounted in a readily accessible location, and shall be of lower ampere rating than the service overcurrent device.

230.93 Protection of Specific Circuits. Where necessary to prevent tampering, an automatic overcurrent device that protects service conductors supplying only a specific load, such as a water heater, shall be permitted to be locked or sealed where located so as to be accessible.

230.94 Relative Location of Overcurrent Device and Other Service Equipment. The overcurrent device shall protect all circuits and devices.

CHAPTER 7
FEEDERS

INTRODUCTION

This chapter contains requirements from Article 215 — Feeders. This article covers installation, overcurrent protection, minimum size, and ampacity of conductors for feeders that supply other feeders and/or branch-circuit loads. Feeder loads must be calculated in accordance with Article 220. A feeder consists of all circuit conductors located between the service equipment, the source of a separately derived system, or other power supply source and the final branch-circuit overcurrent device. Feeder conductors supply power to feeder panelboards or other distribution equipment from which branch circuits are supplied. Feeder panelboards are often referred to in the industry as subpanels and are sometimes misidentified as service equipment. Service equipment is supplied only by service conductors, never by a feeder. Feeder conductors typically are provided with the full range of overcurrent protection meeting the applicable requirements of Article 240 at the point they receive their supply. There are allowances within Article 240 to provide one level of overcurrent protection at the point the feeder receives its supply, with the overload protection provided at the point the feeder conductor terminates. The conductors used in this type of application are called "tap" conductors.

77

ARTICLE 215

Feeders

215.2 Minimum Rating and Size.

(A) Feeders Not More Than 1000 Volts.

(1) General. Feeder conductors shall have an ampacity not less than the larger of 215.2(A)(1)(a) or (A)(1)(b) and shall comply with 110.14(C).

(a) Where a feeder supplies continuous loads or any combination of continuous and noncontinuous loads, the minimum feeder conductor size shall have an ampacity not less than the noncontinuous load plus 125 percent of the continuous load.

(b) The minimum feeder conductor size shall have an ampacity not less than the maximum load to be served after the application of any adjustment or correction factors in accordance with 310.14.

> Informational Note No. 2: Conductors for feeders, as defined in Article 100, sized to prevent a voltage drop exceeding 3 percent at the farthest outlet of power, heating, and lighting loads, or combinations of such loads, and where the maximum total voltage drop on both feeders and branch circuits to the farthest outlet does not exceed 5 percent, will provide reasonable efficiency of operation.
>
> Informational Note No. 3: See 210.19(A), Informational Note No. 4, for voltage drop for branch circuits.

(2) Grounded Conductor. The size of the feeder circuit grounded conductor shall not be smaller than that required by 250.122, except that 250.122(F) shall not apply where grounded conductors are run in parallel.

Additional minimum sizes shall be as specified in 215.2(A)(3) under the conditions stipulated.

(3) Ampacity Relative to Service Conductors. The feeder conductor ampacity shall not be less than that of the service conductors where the feeder conductors carry the total load supplied by service conductors with an ampacity of 55 amperes or less.

215.3 Overcurrent Protection. Feeders shall be protected against overcurrent in accordance with Part I of Article 240. Where a feeder supplies continuous loads or any combination of continuous and noncontinuous loads, the rating of the overcurrent device shall not be less than the noncontinuous load plus 125 percent of the continuous load.

215.4 Feeders with Common Neutral Conductor.

(A) Feeders with Common Neutral. Up to three sets of 3-wire feeders or two sets of 4-wire or 5-wire feeders shall be permitted to utilize a common neutral.

(B) In Metal Raceway or Enclosure. Where installed in a metal raceway or other metal enclosure, all conductors of all feeders using a common neutral conductor shall be enclosed within the same raceway or other enclosure as required in 300.20.

215.5 Diagrams of Feeders.
If required by the authority having jurisdiction, a diagram showing feeder details shall be provided prior to the installation of the feeders. Such a diagram shall show the area in square feet of the building or other structure supplied by each feeder, the total calculated load before applying demand factors, the demand factors used, the calculated load after applying demand factors, and the size and type of conductors to be used.

215.6 Feeder Equipment Grounding Conductor.
Where a feeder supplies branch circuits in which equipment grounding conductors are required, the feeder shall include or provide an equipment grounding conductor, to which the equipment grounding conductors of the branch circuits shall be connected. Where the feeder supplies a separate building or structure, the requirements of 250.32 shall apply.

215.9 Ground-Fault Circuit-Interrupter Protection for Personnel.
Feeders shall be permitted to be protected by a ground-fault circuit interrupter installed in a readily accessible location in lieu of the provisions for such interrupters as specified in 210.8 and 590.6(A).

CHAPTER 8
PANELBOARDS AND DISCONNECTS

INTRODUCTION

This chapter contains requirements from Article 408 — Switchboards, Switchgear, and Panelboards, and Article 404 — Switches. Article 408 covers switchboards, switchgear, and panelboards installed for the control of light and power circuits. A switchboard is a large single panel, frame, or assembly of panels on which are mounted, on the front or back (or both), switches, overcurrent and other protective devices, buses, and usually instruments. Switchboards are generally accessible from both the front and the back and are not intended to be installed in cabinets. Switchboards and switchgear are similar in construction in that both types of equipment are freestanding and fully enclosed (other than ventilating openings) and can be an assembly of sections. Two factors distinguish low voltage (600 volts and below) switchgear from switchboards — internal construction differences, such as full compartmentalization for overcurrent protective devices and for bus and terminal connections, and the ability to withdraw circuit breakers with enclosure doors closed. Additionally, switchgear is constructed for installation in systems rated over 600 volts and is commonly used in medium voltage systems.

A panelboard consists of one (or more) panel units designed for assembly in the form of a single panel. It includes buses and automatic overcurrent devices and may contain switches for the control of light, heat, or power circuits. Panelboards are accessible only from the front and are designed to be enclosed within cabinets or cutout boxes. Switchboards and panelboards that are used as service equipment must be identified specifically for such applications per 230.66.

Requirements from Article 404 are in two chapters of this *Pocket Guide,* Chapters 8 and 18. This chapter covers switches and circuit breakers used as a switch or disconnecting means. The switches discussed here are more commonly called disconnect switches or safety switches. Disconnects may contain overcurrent protective devices (fuses or circuit breakers), or they may be nonfusible or molded-case switches.

ARTICLE 408
Switchboards, Switchgear, and Panelboards

PART I. GENERAL

408.3 Support and Arrangement of Busbars and Conductors.

(C) Used as Service Equipment. Each switchboard, switchgear, or panelboard, if used as service equipment, shall be provided with a main bonding jumper sized in accordance with 250.28(D) or the equivalent placed within the panelboard or one of the sections of the switchboard or switchgear for connecting the grounded service conductor on its supply side to the switchboard, switchgear, or panelboard frame. All sections of a switchboard or switchgear shall be bonded together using an equipment-bonding jumper or a supply-side bonding jumper sized in accordance with 250.122 or 250.102(C)(1) as applicable.

408.4 Field Identification Required.

(A) Circuit Directory or Circuit Identification. Every circuit and circuit modification shall be legibly identified as to its clear, evident, and specific purpose or use. The identification shall include an approved degree of detail that allows each circuit to be distinguished from all others. Spare positions that contain unused overcurrent devices or switches shall be described accordingly. The identification shall be included in a circuit directory that is located on the face, inside of, or in an approved location adjacent to the panel door in the case of a panelboard and at each switch or circuit breaker in a switchboard or switchgear. No circuit shall be described in a manner that depends on transient conditions of occupancy.

(B) Source of Supply. All switchboards, switchgear, and panelboards supplied by a feeder(s) in other than one- or two-family dwellings shall be permanently marked to indicate each device or equipment where the power originates. The label shall be permanently affixed, of sufficient durability to withstand the environment involved, and not handwritten.

408.7 Unused Openings.
Unused openings for circuit breakers and switches shall be closed using identified closures, or other approved means that provide protection substantially equivalent to the wall of the enclosure.

408.8 Reconditioning of Equipment. Reconditioning of equipment within the scope of this article shall be limited as described in 408.8(A) and (B). The reconditioning process shall use design qualified parts verified under applicable standards and be performed in accordance with any instructions provided by the manufacturer. If equipment has been damaged by fire, products of combustion, or water, it shall be specifically evaluated by its manufacturer or a qualified testing laboratory prior to being returned to service.

(A) Panelboards. Panelboards shall not be permitted to be reconditioned. This shall not prevent the replacement of a panelboard within an enclosure. In the event the replacement has not been listed for the specific enclosure and the available fault current is greater than 10,000 amperes, the completed work shall be field labeled, and any previously applied listing marks on the cabinet that pertain to the panelboard shall be removed.

PART III. PANELBOARDS

408.30 General. All panelboards shall have a rating not less than the minimum feeder capacity required for the load calculated in accordance with Part III, IV, or V of Article 220, as applicable.

408.36 Overcurrent Protection. In addition to the requirement of 408.30, a panelboard shall be protected by an overcurrent protective device having a rating not greater than that of the panelboard. This overcurrent protective device shall be located within or at any point on the supply side of the panelboard.

Exception No. 1: Individual protection shall not be required for a panelboard protected by two main circuit breakers or two sets of fuses in other than service equipment, having a combined rating not greater than that of the panelboard. A panelboard constructed or wired under this exception shall not contain more than 42 overcurrent devices. For the purposes of determining the maximum of 42 overcurrent devices, a 2-pole or a 3-pole circuit breaker shall be considered as two or three overcurrent devices, respectively.

Exception No. 2: For existing panelboards, individual protection shall not be required for a panelboard used as service equipment for an individual residential occupancy.

(A) Snap Switches Rated at 30 Amperes or Less. Panelboards equipped with snap switches rated at 30 amperes or less shall have overcurrent protection of 200 amperes or less.

(D) Back-Fed Devices. Plug-in-type overcurrent protection devices or plug-in type main lug assemblies that are backfed and used to terminate field-installed ungrounded supply conductors shall be secured in place by an additional fastener that requires other than a pull to release the device from the mounting means on the panel.

408.37 Panelboards in Damp or Wet Locations. Panelboards in damp or wet locations shall be installed to comply with 312.2.

> [312.2 **Damp and Wet Locations.** In damp or wet locations, surface-type enclosures within the scope of this article shall be placed or equipped so as to prevent moisture or water from entering and accumulating within the cabinet or cutout box, and shall be mounted so that there is at least 6-mm (¼-in.) airspace between the enclosure and the wall or other supporting surface. Enclosures installed in wet locations shall be weatherproof. For enclosures in wet locations, raceways or cables entering above the level of uninsulated live parts shall use fittings listed for wet locations.
>
> *Exception: Nonmetallic enclosures shall be permitted to be installed without the airspace on a concrete, masonry, tile, or similar surface.*
>
> Informational Note: For protection against corrosion, see 300.6.]

408.40 Grounding of Panelboards. Panelboard cabinets and panelboard frames, if of metal, shall be in physical contact with each other and shall be connected to an equipment grounding conductor. Where the panelboard is used with nonmetallic raceway or cable or where separate equipment grounding conductors are provided, a terminal bar for the equipment grounding conductors shall be secured inside the cabinet. The terminal bar shall be bonded to the cabinet and panelboard frame, if of metal; otherwise it shall be connected to the equipment grounding conductor that is run with the conductors feeding the panelboard.

Exception: Where an isolated equipment grounding conductor for a branch circuit or a feeder is provided as permitted by 250.146(D), the insulated equipment grounding conductor that is run with the circuit conductors shall be permitted to pass through the panelboard without being connected to the panelboard's equipment grounding terminal bar.

Panelboards and Disconnects

Equipment grounding conductors shall not be connected to a terminal bar provided for grounded conductors or neutral conductors unless the bar is identified for the purpose and is located where interconnection between equipment grounding conductors and grounded circuit conductors is permitted or required by Article 250.

408.41 Grounded Conductor Terminations. Each grounded conductor shall terminate within the panelboard in an individual terminal that is not also used for another conductor.

Exception: Grounded conductors of circuits with parallel conductors shall be permitted to terminate in a single terminal if the terminal is identified for connection of more than one conductor.

408.43 Panelboard Orientation. Panelboards shall not be installed in the face-up position.

ARTICLE 404
Switches

404.3 Enclosure.

(A) General. Switches and circuit breakers shall be of the externally operable type mounted in an enclosure listed for the intended use. The minimum wire-bending space at terminals and minimum gutter space provided in switch enclosures shall be as required in 312.6.

(B) Used as a Raceway. Enclosures shall not be used as junction boxes, auxiliary gutters, or raceways for conductors feeding through or tapping off to other switches or overcurrent devices, unless the enclosure complies with 312.8.

404.4 Damp or Wet Locations.

(A) Surface-Mounted Switch or Circuit Breaker. A surface-mounted switch or circuit breaker shall be enclosed in a weatherproof enclosure or cabinet that complies with 312.2.

(B) Flush-Mounted Switch or Circuit Breaker. A flush-mounted switch or circuit breaker shall be equipped with a weatherproof cover.

(C) Switches in Tub or Shower Spaces. Switches shall not be installed within tub or shower spaces unless installed as part of a listed tub or shower assembly.

404.6 Position and Connection of Switches.

(A) Single-Throw Knife Switches. Single-throw knife switches shall be placed so that gravity will not tend to close them. Single-throw knife switches, approved for use in the inverted position, shall be provided with an integral mechanical means that ensures that the blades remain in the open position when so set.

404.7 Indicating. General-use and motor-circuit switches, circuit breakers, and molded case switches, where mounted in an enclosure as described in 404.3, shall indicate, in a location that is visible when accessing the external operating means, whether they are in the open (off) or closed (on) position.

Where these switch or circuit breaker handles are operated vertically rather than rotationally or horizontally, the up position of the handle shall be the closed (on) position.

Exception No. 1: Vertically operated double-throw switches shall be permitted to be in the closed (on) position with the handle in either the up or down position.

404.8 Accessibility and Grouping.

(A) Location. All switches and circuit breakers used as switches shall be located so that they may be operated from a readily accessible place. They shall be installed such that the center of the grip of the operating handle of the switch or circuit breaker, when in its highest position, is not more than 2.0 m (6 ft 7 in.) above the floor or working platform.

404.11 Circuit Breakers as Switches. A hand-operable circuit breaker equipped with a lever or handle, or a power-operated circuit breaker capable of being opened by hand in the event of a power failure, shall be permitted to serve as a switch if it has the required number of poles.

Informational Note: See the provisions contained in 240.81 and 240.83.

CHAPTER 9

OVERCURRENT PROTECTION

INTRODUCTION

This chapter contains requirements from Article 240 — Overcurrent Protection. Article 240 provides general requirements for overcurrent protection of conductors and equipment and for overcurrent protective devices. Generally for equipment rated 1000 volts and less, overcurrent protective devices are plug fuses, cartridge fuses, and circuit breakers. Overcurrent protection for conductors and equipment is provided to open the circuit if the current reaches a value that will cause an excessive or dangerous temperature in conductors or conductor insulation.

As a general rule, overcurrent protection of conductors is required to be provided at the point a conductor receives its supply. Using this approach, overload, short-circuit, and line to ground fault protection is provided using a single overcurrent protective device. However, this approach is not always attainable, as in the case of conductors connected to a transformer secondary or to the output terminals of a generator. Alternative protection arrangements are provided by 240.21 that meet the overall objective to provide the full range of overcurrent protection for conductors.

Contained in this *Pocket Guide* are six of the nine parts that are included in Article 240: Part I General; Part II Location; Part III Enclosures; Part IV Disconnecting and Guarding; Part V Plug Fuses, Fuseholders, and Adapters; and Part VII Circuit Breakers.

ARTICLE 240
Overcurrent Protection

PART I. GENERAL

240.4 Protection of Conductors. Conductors, other than flexible cords, flexible cables, and fixture wires, shall be protected against overcurrent in accordance with their ampacities specified in 310.14, unless otherwise permitted or required in 240.4(A) through (G).

Informational Note: See ICEA P-32-382-2007 (R2013), *Short Circuit Characteristics of Insulated Cables,* for information on allowable short-circuit currents for insulated copper and aluminum conductors.

(B) Overcurrent Devices Rated 800 Amperes or Less. The next higher standard overcurrent device rating (above the ampacity of the conductors being protected) shall be permitted to be used, provided all of the following conditions are met:

(1) The conductors being protected are not part of a branch circuit supplying more than one receptacle for cord-and-plug-connected portable loads.
(2) The ampacity of the conductors does not correspond with the standard ampere rating of a fuse or a circuit breaker without overload trip adjustments above its rating (but that shall be permitted to have other trip or rating adjustments).
(3) The next higher standard rating selected does not exceed 800 amperes.

(C) Overcurrent Devices Rated over 800 Amperes. Where the overcurrent device is rated over 800 amperes, the ampacity of the conductors it protects shall be equal to or greater than the rating of the overcurrent device defined in 240.6.

(D) Small Conductors. Unless specifically permitted in 240.4(E) or (G), the overcurrent protection shall not exceed that required by (D)(1) through (D)(7) after any correction factors for ambient temperature and number of conductors have been applied.

Overcurrent Protection — 240.4

(1) 18 AWG Copper. 7 amperes, provided all the following conditions are met:

(1) Continuous loads do not exceed 5.6 amperes.
(2) Overcurrent protection is provided by one of the following:

   a. Branch-circuit-rated circuit breakers listed and marked for use with 18 AWG copper wire
   b. Branch-circuit-rated fuses listed and marked for use with 18 AWG copper wire
   c. Class CC, Class J, or Class T fuses

Table 240.4(G) Specific Conductor Applications

Conductor	Article	Section
Air-conditioning and refrigeration equipment circuit conductors	440, Parts III, VI	
Capacitor circuit conductors	460	460.8(B) and 460.25
Control and instrumentation circuit conductors (Type ITC)	727	727.9
Electric welder circuit conductors	630	630.12 and 630.32
Fire alarm system circuit conductors	760	760.43, 760.45, 760.121, and Chapter 9, Tables 12(A) and 12(B)
Motor-operated appliance circuit conductors	422, Part II	
Motor and motor-control circuit conductors	430, Parts II, III, IV, V, VI, VII	
Phase converter supply conductors	455	455.7
Remote-control, signaling, and power-limited circuit conductors	725	725.43, 725.45, 725.121, and Chapter 9, Tables 11(A) and 11(B)
Secondary tie conductors	450	450.6

(2) 16 AWG Copper. 10 amperes, provided all the following conditions are met:

(1) Continuous loads do not exceed 8 amperes.
(2) Overcurrent protection is provided by one of the following:
   a. Branch-circuit-rated circuit breakers listed and marked for use with 16 AWG copper wire
   b. Branch-circuit-rated fuses listed and marked for use with 16 AWG copper wire
   c. Class CC, Class J, or Class T fuses

(3) 14 AWG Copper. 15 amperes

(4) 12 AWG Aluminum and Copper-Clad Aluminum. 15 amperes

(5) 12 AWG Copper. 20 amperes

(6) 10 AWG Aluminum and Copper-Clad Aluminum. 25 amperes

(7) 10 AWG Copper. 30 amperes

(E) Tap Conductors. Tap conductors shall be permitted to be protected against overcurrent in accordance with the following:

(1) 210.19(A)(3) and (A)(4), Household Ranges and Cooking Appliances and Other Loads
(3) 240.21, Location in Circuit

240.6 Standard Ampere Ratings.

(A) Fuses and Fixed-Trip Circuit Breakers. The standard ampere ratings for fuses and inverse time circuit breakers shall be considered as shown in Table 240.6(A). Additional standard ampere ratings for fuses shall be 1, 3, 6, 10, and 601. The use of fuses and inverse time circuit breakers with nonstandard ampere ratings shall be permitted.

240.10 Supplementary Overcurrent Protection. Where supplementary overcurrent protection is used for luminaires, appliances, and other equipment or for internal circuits and components of equipment, it shall not be used as a substitute for required branch-circuit overcurrent devices or in place of the required branch-circuit protection. Supplementary overcurrent devices shall not be required to be readily accessible.

240.15 Ungrounded Conductors.

(A) Overcurrent Device Required. A fuse or an overcurrent trip unit of a circuit breaker shall be connected in series with each ungrounded

Overcurrent Protection

Table 240.6(A) Standard Ampere Ratings for Fuses and Inverse Time Circuit Breakers

Standard Ampere Ratings				
15	20	25	30	35
40	45	50	60	70
80	90	100	110	125
150	175	200	225	250
300	350	400	450	500
600	700	800	1000	1200
1600	2000	2500	3000	4000
5000	6000	—	—	—

conductor. A combination of a current transformer and overcurrent relay shall be considered equivalent to an overcurrent trip unit.

Informational Note: For motor circuits, see Parts III, IV, V, and XI of Article 430.

(B) Circuit Breaker as Overcurrent Device. Circuit breakers shall open all ungrounded conductors of the circuit both manually and automatically unless otherwise permitted in 240.15(B)(1), (B)(2), (B)(3), and (B)(4).

(1) Multiwire Branch Circuits. Individual single-pole circuit breakers, with identified handle ties, shall be permitted as the protection for each ungrounded conductor of multiwire branch circuits that serve only single-phase line-to-neutral loads.

(2) Grounded Single-Phase Alternating-Current Circuits. In grounded systems, individual single-pole circuit breakers rated 120/240 volts ac, with identified handle ties, shall be permitted as the protection for each ungrounded conductor for line-to-line connected loads for single-phase circuits.

PART II. LOCATION

240.21 Location in Circuit. Overcurrent protection shall be provided in each ungrounded circuit conductor and shall be located at the point where the conductors receive their supply except as specified in

240.21 240.21(A) through (H). Conductors supplied under 240.21(A) through (H) shall not supply another conductor except through an overcurrent protective device meeting the requirements of 240.4.

(A) Branch-Circuit Conductors. Branch-circuit tap conductors meeting the requirements specified in 210.19 shall be permitted to have overcurrent protection as specified in 210.20.

(D) Service Conductors. Service conductors shall be permitted to be protected by overcurrent devices in accordance with 230.91.

240.24 Location in or on Premises.

(A) Accessibility. Circuit breakers and switches containing fuses shall be readily accessible and installed so that the center of the grip of the operating handle of the switch or circuit breaker, when in its highest position, is not more than 2.0 m (6 ft 7 in.) above the floor or working platform, unless one of the following applies:

(2) For supplementary overcurrent protection, as described in 240.10.
(3) For overcurrent devices, as described in 225.40 and 230.92.
(4) For overcurrent devices adjacent to utilization equipment that they supply, access shall be permitted to be by portable means.

(B) Occupancy. Each occupant shall have ready access to all overcurrent devices protecting the conductors supplying that occupancy, unless otherwise permitted in 240.24(B)(1) and (B)(2).

(1) Service and Feeder Overcurrent Devices. Where electric service and electrical maintenance are provided by the building management and where these are under continuous building management supervision, the service overcurrent devices and feeder overcurrent devices supplying more than one occupancy shall be permitted to be accessible only to authorized management personnel in the following:

(1) Multiple-occupancy buildings
(2) Guest rooms or guest suites

(2) Branch-Circuit Overcurrent Devices. Where electric service and electrical maintenance are provided by the building management and where these are under continuous building management supervision, the branch-circuit overcurrent devices supplying any guest rooms or guest suites without permanent provisions for cooking shall be permitted to be accessible only to authorized management personnel.

Overcurrent Protection 240.40

(C) Not Exposed to Physical Damage. Overcurrent devices shall be located where they will not be exposed to physical damage.

Informational Note: See 110.11, Deteriorating Agents.

(D) Not in Vicinity of Easily Ignitible Material. Overcurrent devices shall not be located in the vicinity of easily ignitible material, such as in clothes closets.

(E) Not Located in Bathrooms. In dwelling units, dormitory units, and guest rooms or guest suites, overcurrent devices, other than supplementary overcurrent protection, shall not be located in bathrooms.

(F) Not Located over Steps. Overcurrent devices shall not be located over steps of a stairway.

PART III. ENCLOSURES

240.32 Damp or Wet Locations. Enclosures for overcurrent devices in damp or wet locations shall comply with 312.2.

240.33 Vertical Position. Enclosures for overcurrent devices shall be mounted in a vertical position. Circuit breaker enclosures shall be permitted to be installed horizontally where the circuit breaker is installed in accordance with 240.81. Listed busway plug-in units shall be permitted to be mounted in orientations corresponding to the busway mounting position.

PART IV. DISCONNECTING AND GUARDING

240.40 Disconnecting Means for Fuses. Cartridge fuses in circuits of any voltage, and all fuses in circuits over 150 volts to ground, shall be provided with a disconnecting means on their supply side so that each circuit containing fuses can be independently disconnected from the source of power. A cable limiter without a disconnecting means shall be permitted on the supply side of the service disconnecting means as permitted by 230.82. A single disconnecting means shall be permitted on the supply side of more than one set of fuses as permitted by 430.112, Exception, for group operation of motors, 424.22(C) for fixed electric space-heating equipment, and 425.22(C) for fixed resistance and electrode industrial process heating equipment, or where specifically permitted elsewhere in this *Code*.

PART V. PLUG FUSES, FUSEHOLDERS, AND ADAPTERS

240.50 General.

(A) Maximum Voltage. Plug fuses shall be permitted to be used in the following circuits:

(1) Circuits not exceeding 125 volts between conductors
(2) Circuits supplied by a system having a grounded neutral point where the line-to-neutral voltage does not exceed 150 volts

(B) Marking. Each fuse, fuseholder, and adapter shall be marked with its ampere rating.

(C) Hexagonal Configuration. Plug fuses of 15-ampere and lower rating shall be identified by a hexagonal configuration of the window, cap, or other prominent part to distinguish them from fuses of higher ampere ratings.

(D) No Energized Parts. Plug fuses, fuseholders, and adapters shall have no exposed energized parts after fuses or fuses and adapters have been installed.

(E) Screw Shell. The screw shell of a plug-type fuseholder shall be connected to the load side of the circuit.

240.51 Edison-Base Fuses.

(A) Classification. Plug fuses of the Edison-base type shall be classified at not over 125 volts and 30 amperes and below.

(B) Replacement Only. Plug fuses of the Edison-base type shall be used only for replacements in existing installations where there is no evidence of overfusing or tampering.

240.52 Edison-Base Fuseholders.
Fuseholders of the Edison-base type shall be installed only where they are made to accept Type S fuses by the use of adapters.

240.53 Type S Fuses.
Type S fuses shall be of the plug type and shall comply with 240.53(A) and (B).

(A) Classification. Type S fuses shall be classified at not over 125 volts and 0 to 15 amperes, 16 to 20 amperes, and 21 to 30 amperes.

(B) Noninterchangeable. Type S fuses of an ampere classification as specified in 240.53(A) shall not be interchangeable with a lower ampere

240.54 Type S Fuses, Adapters, and Fuseholders.

(A) To Fit Edison-Base Fuseholders. Type S adapters shall fit Edison-base fuseholders.

(B) To Fit Type S Fuses Only. Type S fuseholders and adapters shall be designed so that either the fuseholder itself or the fuseholder with a Type S adapter inserted cannot be used for any fuse other than a Type S fuse.

(C) Nonremovable. Type S adapters shall be designed so that once inserted in a fuseholder, they cannot be removed.

(D) Nontamperable. Type S fuses, fuseholders, and adapters shall be designed so that tampering or shunting (bridging) would be difficult.

(E) Interchangeability. Dimensions of Type S fuses, fuseholders, and adapters shall be standardized to permit interchangeability regardless of the manufacturer.

PART VII. CIRCUIT BREAKERS

240.80 Method of Operation. Circuit breakers shall be trip free and capable of being closed and opened by manual operation. Their normal method of operation by other than manual means, such as electrical or pneumatic, shall be permitted if means for manual operation are also provided.

240.81 Indicating. Circuit breakers shall clearly indicate whether they are in the open "off" or closed "on" position.

Where circuit breaker handles are operated vertically rather than rotationally or horizontally, the "up" position of the handle shall be the "on" position.

240.83 Marking.

(A) Durable and Visible. Circuit breakers shall be marked with their ampere rating in a manner that will be durable and visible after installation. Such marking shall be permitted to be made visible by removal of a trim or cover.

(B) Location. Circuit breakers rated at 100 amperes or less and 1000 volts or less shall have the ampere rating molded, stamped, etched, or similarly marked into their handles or escutcheon areas.

(C) Interrupting Rating. Every circuit breaker having an interrupting rating other than 5000 amperes shall have its interrupting rating shown on the circuit breaker. The interrupting rating shall not be required to be marked on circuit breakers used for supplementary protection.

CHAPTER 10
GROUNDED (NEUTRAL) CONDUCTORS

INTRODUCTION

This chapter contains requirements from Article 200 — Use and Identification of Grounded Conductors. Article 200 provides the general requirements on the installation of grounded conductors in premises wiring systems and specifies identification methods for these conductors and their terminals. A grounded conductor is a system or circuit conductor that is intentionally grounded (connected to earth). The grounded conductor in many electrical systems meets the Article 100 definition of *neutral conductor*, but not all grounded conductors are neutral conductors. For instance, a 120-volt, two-wire system has a conductor that is intentionally grounded; it is indeed a grounded conductor, but it is not a neutral conductor. The use and identification requirements of Article 200 apply to all electrical systems conductors that are intentionally grounded, regardless of what the conductor is called on the job site. Grounded conductors are generally identified by a continuous white or gray outer finish or by three continuous white or gray stripes on other than green insulation along the entire length. Other methods of identification are permitted for certain wiring methods and for flexible cords and cables. Although Article 200 is a short article, the proper use and identification of grounded conductors is extremely important for ensuring the safety of those who work on electrical systems. Other requirements covering the use and installation of grounded circuit conductors are located in Article 250 of the *Code*.

ARTICLE 200
Use and Identification of Grounded Conductors

200.1 Scope. This article provides requirements for the following:

(1) Identification of terminals
(2) Grounded conductors in premises wiring systems
(3) Identification of grounded conductors

Informational Note: See Article 100 for definitions of *Grounded Conductor, Equipment Grounding Conductor,* and *Grounding Electrode Conductor.*

200.2 General. Grounded conductors shall comply with 200.2(A) and (B).

(A) Insulation. The grounded conductor, if insulated, shall have insulation that is (1) suitable, other than color, for any ungrounded conductor of the same circuit for systems of 1000 volts or less, or impedance grounded neutral systems of over 1000 volts, or (2) rated not less than 600 volts for solidly grounded neutral systems of over 1000 volts as described in 250.184(A).

(B) Continuity. The continuity of a grounded conductor shall not depend on a connection to a metallic enclosure, raceway, or cable armor.

200.3 Connection to Grounded System. Grounded conductors of premises wiring systems shall be electrically connected to the supply system grounded conductor to ensure a common, continuous grounded system. For the purpose of this section, *electrically connected* shall mean making a direct electrical connection capable of carrying current, as distinguished from induced currents.

200.4 Neutral Conductors. Neutral conductors shall be installed in accordance with 200.4(A) and (B).

(A) Installation. Neutral conductors shall not be used for more than one branch circuit, for more than one multiwire branch circuit, or for more than one set of ungrounded feeder conductors unless specifically permitted elsewhere in this *Code*.

(B) Multiple Circuits. Where more than one neutral conductor associated with different circuits is in an enclosure, grounded circuit conductors of each circuit shall be identified or grouped to correspond with the ungrounded circuit conductor(s) by wire markers, cable ties, or similar means in at least one location within the enclosure.

Grounded (Neutral) Conductors

Exception No. 1: The requirement for grouping or identifying shall not apply if the branch-circuit or feeder conductors enter from a cable or a raceway unique to the circuit that makes the grouping obvious.

Exception No. 2: The requirement for grouping or identifying shall not apply where branch-circuit conductors pass through a box or conduit body without a loop as described in 314.16(B)(1) or without a splice or termination.

200.6 Means of Identifying Grounded Conductors.

(A) Sizes 6 AWG or Smaller. An insulated grounded conductor of 6 AWG or smaller shall be identified by one of the following means:

(1) The insulated conductor shall have a continuous white outer finish.
(2) The insulated conductor shall have a continuous gray outer finish.
(3) The insulated conductor shall have three continuous white or gray stripes along the conductor's entire length on other than green insulation.
(4) Insulated conductors that have their outer covering finished to show a white or gray color but have colored tracer threads in the braid identifying the source of manufacture are acceptable means of identification.
(5) A single-conductor, sunlight-resistant, outdoor-rated cable used as a solidly grounded conductor in photovoltaic power systems, as permitted by 690.41, shall be identified at the time of installation by markings at terminations in accordance with 200.6(A)(1) through (A)(4).
(6) The grounded conductor of a mineral-insulated, metal-sheathed cable (Type MI) shall be identified at the time of installation by distinctive marking at its terminations.
(7) Fixture wire shall comply with the requirements for grounded conductor identification as specified in 402.8.
(8) For aerial cable, the identification shall comply with one of the methods in 200.6(A)(1) through (A)(5), or by means of a ridge located on the exterior of the cable so as to identify it.

(B) Sizes 4 AWG or Larger. An insulated grounded conductor 4 AWG or larger shall be identified by one of the following means:

(1) A continuous white outer finish.
(2) A continuous gray outer finish.
(3) Three continuous white or gray stripes along the conductor's entire length on other than green insulation.

(4) At the time of installation, by a distinctive white or gray marking at its terminations. This marking shall encircle the conductor or insulation.

(C) Flexible Cords. An insulated conductor that is intended for use as a grounded conductor, where contained within a flexible cord, shall be identified by a white or gray outer finish or by methods permitted by 400.22.

(E) Grounded Conductors of Multiconductor Cables. The insulated grounded conductor(s) in a multiconductor cable shall be identified by a continuous white or gray outer finish or by three continuous white or gray stripes on other than green insulation along its entire length. For conductors that are 4 AWG or larger in cables, identification of the grounded conductor shall be permitted to comply with 200.6(B). For multiconductor flat cable with conductors that are 4 AWG or larger, an external ridge shall be permitted to identify the grounded conductor.

200.7 Use of Insulation of a White or Gray Color or with Three Continuous White or Gray Stripes.

(A) General. The following shall be used only for the grounded circuit conductor, unless otherwise permitted in 200.7(B) and (C):

(1) A conductor with continuous white or gray covering
(2) A conductor with three continuous white or gray stripes on other than green insulation
(3) A marking of white or gray color at the termination

(B) Circuits of Less Than 50 Volts. A conductor with white or gray color insulation or three continuous white stripes or having a marking of white or gray at the termination for circuits of less than 50 volts shall be required to be grounded only as required by 250.20(A).

(C) Circuits of 50 Volts or More. The use of insulation that is white or gray or that has three continuous white or gray stripes for other than a grounded conductor for circuits of 50 volts or more shall be permitted only as in (1) and (2).

(1) If part of a cable assembly that has the insulation permanently reidentified to indicate its use as an ungrounded conductor by marking tape, painting, or other effective means at its termination and at each location where the conductor is visible and accessible. Identification shall encircle the insulation and shall be a color other than white, gray, or green. If used for single-pole, 3-way or 4-way switch loops, the reidentified conductor with white or gray insulation or three

Grounded (Neutral) Conductors

continuous white or gray stripes shall be used only for the supply to the switch, but not as a return conductor from the switch to the outlet.

(2) A flexible cord having one conductor identified by a white or gray outer finish or three continuous white or gray stripes, or by any other means permitted by 400.22, that is used for connecting an appliance or equipment permitted by 400.10. This shall apply to flexible cords connected to outlets whether or not the outlet is supplied by a circuit that has a grounded conductor.

200.9 Means of Identification of Terminals. In devices or utilization equipment with polarized connections, identification of terminals to which a grounded conductor is to be connected shall be substantially white or silver in color. The identification of other terminals shall be of a readily distinguishable different color.

CHAPTER 11

GROUNDING AND BONDING

INTRODUCTION

Chapter 11 contains requirements from Article 250 — Grounding and Bonding. Article 250 contains requirements for grounding and bonding of systems, circuits and equipment. *Grounded* (grounding) is defined in Article 100 as connected (connecting) to ground or to a conductive body that extends the ground connection. Connecting to ground simply means connecting to the earth. A system or circuit conductor that is intentionally grounded is a grounded conductor. Bonding is connecting parts of an electrical system together into a conductive path. There are bonding requirements for circuit components that are normally current-carrying (e.g., main and system bonding jumpers), and there are requirements to bond normally noncurrent carrying conductive parts together to form a continuous electrical conductor. Bonded conductors and equipment are not grounded until an intentional connection to earth is completed.

Understanding the terminology in Article 250 enables *Code* users to better understand and apply the requirements in Article 250. Some of these terms are defined in Article 250, but many are used in other articles of the *Code* in addition to Article 250 and are defined in Article 100. While Article 250 contains the majority of requirements on grounding and bonding, there are requirements elsewhere in the *Code* that have a unique application to the equipment or occupancy covered in a particular article. Table 250.3 provides a compilation of those special requirements.

Contained in this *Pocket Guide* are seven of the ten parts that are included in Article 250: Part I General; Part II Circuit System Grounding; Part III Grounding Electrode System and Grounding Electrode Conductor; Part IV Enclosure, Raceway, and Service Cable Grounding; Part V Bonding; Part VI Equipment Grounding and Equipment Grounding Conductors; and Part VII Methods of Equipment Grounding.

ARTICLE 250
Grounding and Bonding

PART I. GENERAL

250.1 Scope. This article covers general requirements for grounding and bonding of electrical installations, and the specific requirements in (1) through (6).

(1) Systems, circuits, and equipment required, permitted, or not permitted to be grounded
(2) Circuit conductor to be grounded on grounded systems
(3) Location of grounding connections
(4) Types and sizes of grounding and bonding conductors and electrodes
(5) Methods of grounding and bonding
(6) Conditions under which guards, isolation, or insulation may be substituted for grounding

250.4 General Requirements for Grounding and Bonding. The following general requirements identify what grounding and bonding of electrical systems are required to accomplish. The prescriptive methods contained in Article 250 shall be followed to comply with the performance requirements of this section.

(A) Grounded Systems.

(1) Electrical System Grounding. Electrical systems that are grounded shall be connected to earth in a manner that will limit the voltage imposed by lightning, line surges, or unintentional contact with higher-voltage lines and that will stabilize the voltage to earth during normal operation.

(2) Grounding of Electrical Equipment. Normally non–current-carrying conductive materials enclosing electrical conductors or equipment, or forming part of such equipment, shall be connected to earth so as to limit the voltage to ground on these materials.

(3) Bonding of Electrical Equipment. Normally non–current-carrying conductive materials enclosing electrical conductors or equipment, or forming part of such equipment, shall be connected together and to the electrical supply source in a manner that establishes an effective ground-fault current path.

(4) Bonding of Electrically Conductive Materials and Other Equipment. Normally non–current-carrying electrically conductive materials

Grounding and Bonding 250.8

that are likely to become energized shall be connected together and to the electrical supply source in a manner that establishes an effective ground-fault current path.

(5) Effective Ground-Fault Current Path. Electrical equipment and wiring and other electrically conductive material likely to become energized shall be installed in a manner that creates a low-impedance circuit facilitating the operation of the overcurrent device or ground detector for high-impedance grounded systems. It shall be capable of safely carrying the maximum ground-fault current likely to be imposed on it from any point on the wiring system where a ground fault may occur to the electrical supply source. The earth shall not be considered as an effective ground-fault current path.

250.6 Objectionable Current.

(A) Arrangement to Prevent Objectionable Current. The grounding of electrical systems, circuit conductors, surge arresters, surge-protective devices, and conductive normally non–current-carrying metal parts of equipment shall be installed and arranged in a manner that will prevent objectionable current.

(B) Alterations to Stop Objectionable Current. If the use of multiple grounding connections results in objectionable current and the requirements of 250.4(A)(5) or (B)(4) are met, one or more of the following alterations shall be permitted:

(1) Discontinue one or more but not all of such grounding connections.
(2) Change the locations of the grounding connections.
(3) Interrupt the continuity of the conductor or conductive path causing the objectionable current.
(4) Take other suitable remedial and approved action.

250.8 Connection of Grounding and Bonding Equipment.

(A) Permitted Methods. Equipment grounding conductors, grounding electrode conductors, and bonding jumpers shall be connected by one or more of the following means:

(1) Listed pressure connectors
(2) Terminal bars
(3) Pressure connectors listed as grounding and bonding equipment
(4) Exothermic welding process
(5) Machine screw-type fasteners that engage not less than two threads or are secured with a nut

(6) Thread-forming machine screws that engage not less than two threads in the enclosure
(7) Connections that are part of a listed assembly
(8) Other listed means

(B) Methods Not Permitted. Connection devices or fittings that depend solely on solder shall not be used.

250.10 Protection of Ground Clamps and Fittings. Ground clamps or other fittings exposed to physical damage shall be enclosed in metal, wood, or equivalent protective covering.

250.12 Clean Surfaces. Nonconductive coatings (such as paint, lacquer, and enamel) on equipment to be grounded or bonded shall be removed from threads and other contact surfaces to ensure good electrical continuity or shall be connected by means of fittings designed so as to make such removal unnecessary.

PART II. SYSTEM GROUNDING

250.20 Alternating-Current Systems to Be Grounded. Alternating-current systems shall be grounded as provided for in 250.20(A), (B), (C), or (D). Other systems shall be permitted to be grounded. If such systems are grounded, they shall comply with the applicable provisions of this article.

(A) Alternating-Current Systems of Less Than 50 Volts. Alternating-current systems of less than 50 volts shall be grounded under any of the following conditions:

(1) Where supplied by transformers, if the transformer supply system exceeds 150 volts to ground
(2) Where supplied by transformers, if the transformer supply system is ungrounded
(3) Where installed outside as overhead conductors

(B) Alternating-Current Systems of 50 Volts to 1000 Volts. Alternating-current systems of 50 volts to 1000 volts that supply premises wiring and premises wiring systems shall be grounded under any of the following conditions:

(1) Where the system can be grounded so that the maximum voltage to ground on the ungrounded conductors does not exceed 150 volts
(2) Where the system is 3-phase, 4-wire, wye connected in which the neutral conductor is used as a circuit conductor

Grounding and Bonding 250.24

(3) Where the system is 3-phase, 4-wire, delta connected in which the midpoint of one phase winding is used as a circuit conductor

250.24 Grounding of Service-Supplied Alternating-Current Systems.

(A) System Grounding Connections. A premises wiring system supplied by a grounded ac service shall have a grounding electrode conductor connected to the grounded service conductor, at each service, in accordance with 250.24(A)(1) through (A)(5).

(1) General. The grounding electrode conductor connection shall be made at any accessible point from the load end of the overhead service conductors, service drop, underground service conductors, or service lateral to, including the terminal or bus to which the grounded service conductor is connected at the service disconnecting means.

> Informational Note: See definitions of *Service Conductors, Overhead; Service Conductors, Underground; Service Drop;* and *Service Lateral* in Article 100.

(4) Main Bonding Jumper as Wire or Busbar. Where the main bonding jumper specified in 250.28 is a wire or busbar and is installed from the grounded conductor terminal bar or bus to the equipment grounding terminal bar or bus in the service equipment, the grounding electrode conductor shall be permitted to be connected to the equipment grounding terminal, bar, or bus to which the main bonding jumper is connected.

(5) Load-Side Grounding Connections. A grounded conductor shall not be connected to normally non–current-carrying metal parts of equipment, to equipment grounding conductor(s), or be reconnected to ground on the load side of the service disconnecting means except as otherwise permitted in this article.

(B) Main Bonding Jumper. For a grounded system, an unspliced main bonding jumper shall be used to connect the equipment grounding conductor(s) and the service-disconnect enclosure to the grounded conductor within the enclosure for each service disconnect in accordance with 250.28.

(C) Grounded Conductor Brought to Service Equipment. Where an ac system operating at 1000 volts or less is grounded at any point, the grounded conductor(s) shall be routed with the ungrounded conductors to each service disconnecting means and shall be connected to each disconnecting means grounded conductor(s) terminal or bus. A main bonding jumper shall connect the grounded conductor(s) to each service disconnecting means enclosure. The grounded conductor(s) shall be installed in accordance with 250.24(C)(1) through (C)(4).

(1) Sizing for a Single Raceway or Cable. The grounded conductor shall not be smaller than specified in Table 250.102(C)(1).

(2) Parallel Conductors in Two or More Raceways or Cables. If the ungrounded service-entrance conductors are installed in parallel in two or more raceways or cables, the grounded conductor shall also be installed in parallel. The size of the grounded conductor in each raceway or cable shall be based on the total circular mil area of the parallel ungrounded conductors in the raceway or cable, as indicated in 250.24(C)(1), but not smaller than 1/0 AWG.

(D) Grounding Electrode Conductor. A grounding electrode conductor shall be used to connect the equipment grounding conductors, the service-equipment enclosures, and, where the system is grounded, the grounded service conductor to the grounding electrode(s) required by Part III of this article. This conductor shall be sized in accordance with 250.66.

High-impedance grounded neutral system connections shall be made as covered in 250.36.

250.25 Grounding Systems Permitted to Be Connected on the Supply Side of the Disconnect. The grounding of systems connected on the supply side of the service disconnect, as permitted in 230.82, that are in enclosures separate from the service equipment enclosure shall comply with 250.25(A) or (B).

(A) Grounded System. If the utility supply system is grounded, the grounding of systems permitted to be connected on the supply side of the service disconnect and are installed in one or more separate enclosures from the service equipment enclosure shall comply with the requirements of 250.24(A) through (D).

250.26 Conductor to Be Grounded — Alternating-Current Systems. For grounded ac premises wiring systems, the conductor to be grounded shall be as specified in the following:

(1) Single-phase, 2-wire — one conductor
(2) Single-phase, 3-wire — the neutral conductor
(3) Multiphase systems having one wire common to all phases — the neutral conductor
(4) Multiphase systems where one phase is grounded — that phase conductor

Grounding and Bonding 250.50

(5) Multiphase systems in which one phase is used as in (2) — the neutral conductor

250.28 Main Bonding Jumper and System Bonding Jumper. For a grounded system, main bonding jumpers and system bonding jumpers shall be installed as follows:

(A) Material. Main bonding jumpers and system bonding jumpers shall be of copper, aluminum, copper-clad aluminum, or other corrosion-resistant material. A main bonding jumper and a system bonding jumper shall be a wire, bus, screw, or similar suitable conductor.

(B) Construction. Where a main bonding jumper or a system bonding jumper is a screw only, the screw shall be identified with a green finish that shall be visible with the screw installed.

(C) Attachment. Main bonding jumpers and system bonding jumpers shall be connected in the manner specified in 250.8.

(D) Size. Main bonding jumpers and system bonding jumpers shall be sized in accordance with 250.28(D)(1) through (D)(3).

(1) General. Main bonding jumpers and system bonding jumpers shall not be smaller than specified in Table 250.102(C)(1).

(2) Main Bonding Jumper for Service with More Than One Enclosure. If a service consists of more than a single enclosure as permitted in 230.71(B), the main bonding jumper for each enclosure shall be sized in accordance with 250.28(D)(1) based on the largest ungrounded service conductor serving that enclosure.

250.32 Buildings or Structures Supplied by a Feeder(s) or Branch Circuit(s). See 250.32 in Chapter 24.

PART III. GROUNDING ELECTRODE SYSTEM AND GROUNDING ELECTRODE CONDUCTOR

250.50 Grounding Electrode System. All grounding electrodes as described in 250.52(A)(1) through (A)(7) that are present at each building or structure served shall be bonded together to form the grounding electrode system. Where none of these grounding electrodes exist, one or more of the grounding electrodes specified in 250.52(A)(4) through (A)(8) shall be installed and used.

Exception: Concrete-encased electrodes of existing buildings or structures shall not be required to be part of the grounding electrode system where the steel reinforcing bars or rods are not accessible for use without disturbing the concrete.

250.52 Grounding Electrodes.

(A) Electrodes Permitted for Grounding.

(1) Metal Underground Water Pipe. A metal underground water pipe in direct contact with the earth for 3.0 m (10 ft) or more (including any metal well casing bonded to the pipe) and electrically continuous (or made electrically continuous by bonding around insulating joints or insulating pipe) to the points of connection of the grounding electrode conductor and the bonding conductor(s) or jumper(s), if installed.

(2) Metal In-ground Support Structure(s). One or more metal in-ground support structure(s) in direct contact with the earth vertically for 3.0 m (10 ft) or more, with or without concrete encasement. If multiple metal in-ground support structures are present at a building or a structure, it shall be permissible to bond only one into the grounding electrode system.

Informational Note: Metal in-ground support structures include, but are not limited to, pilings, casings, and other structural metal.

(3) Concrete-Encased Electrode. A concrete-encased electrode shall consist of at least 6.0 m (20 ft) of either (1) or (2):

(1) One or more bare or zinc galvanized or other electrically conductive coated steel reinforcing bars or rods of not less than 13 mm (½ in.) in diameter, installed in one continuous 6.0 m (20 ft) length, or if in multiple pieces connected together by the usual steel tie wires, exothermic welding, welding, or other effective means to create a 6.0 m (20 ft) or greater length; or
(2) Bare copper conductor not smaller than 4 AWG

Metallic components shall be encased by at least 50 mm (2 in.) of concrete and shall be located horizontally within that portion of a concrete foundation or footing that is in direct contact with the earth or within vertical foundations or structural components or members that are in direct contact with the earth. If multiple concrete-encased electrodes are present at a building or structure, it shall be permissible to bond only one into the grounding electrode system.

Grounding and Bonding 250.52

Informational Note: Concrete installed with insulation, vapor barriers, films or similar items separating the concrete from the earth is not considered to be in "direct contact" with the earth.

(4) Ground Ring. A ground ring encircling the building or structure, in direct contact with the earth, consisting of at least 6.0 m (20 ft) of bare copper conductor not smaller than 2 AWG.

(5) Rod and Pipe Electrodes. Rod and pipe electrodes shall not be less than 2.44 m (8 ft) in length and shall consist of the following materials.

(a) Grounding electrodes of pipe or conduit shall not be smaller than metric designator 21 (trade size ¾) and, where of steel, shall have the outer surface galvanized or otherwise metal-coated for corrosion protection.

(b) Rod-type grounding electrodes of stainless steel and copper or zinc coated steel shall be at least 15.87 mm (⅝ in.) in diameter, unless listed.

(6) Other Listed Electrodes. Other listed grounding electrodes shall be permitted.

(7) Plate Electrodes. Each plate electrode shall expose not less than 0.186 m² (2 ft²) of surface to exterior soil. Electrodes of bare or electrically conductive coated iron or steel plates shall be at least 6.4 mm (¼ in.) in thickness. Solid, uncoated electrodes of nonferrous metal shall be at least 1.5 mm (0.06 in.) in thickness.

(8) Other Local Metal Underground Systems or Structures. Other local metal underground systems or structures such as piping systems, underground tanks, and underground metal well casings that are not bonded to a metal water pipe.

(B) Not Permitted for Use as Grounding Electrodes. The following systems and materials shall not be used as grounding electrodes:

(1) Metal underground gas piping systems
(2) Aluminum
(3) The structures and structural reinforcing steel described in 680.26(B)(1) and (B)(2)

Informational Note: See 250.104(B) for bonding requirements of gas piping.

250.53 Grounding Electrode System Installation.

(A) Rod, Pipe, and Plate Electrodes. Rod, pipe, and plate electrodes shall meet the requirements of 250.53(A)(1) through (A)(3).

(1) Below Permanent Moisture Level. If practicable, rod, pipe, and plate electrodes shall be embedded below permanent moisture level. Rod, pipe, and plate electrodes shall be free from nonconductive coatings such as paint or enamel.

(2) Supplemental Electrode Required. A single rod, pipe, or plate electrode shall be supplemented by an additional electrode of a type specified in 250.52(A)(2) through (A)(8). The supplemental electrode shall be permitted to be bonded to one of the following:

(1) Rod, pipe, or plate electrode
(2) Grounding electrode conductor
(3) Grounded service-entrance conductor
(4) Nonflexible grounded service raceway
(5) Any grounded service enclosure

Exception: If a single rod, pipe, or plate grounding electrode has a resistance to earth of 25 ohms or less, the supplemental electrode shall not be required.

(3) Supplemental Electrode. f multiple rod, pipe, or plate electrodes are installed to meet the requirements of this section, they shall not be less than 1.8 m (6 ft) apart.

> Informational Note: The paralleling efficiency of rods is increased by spacing them twice the length of the longest rod.

(B) Electrode Spacing. Where more than one of the electrodes of the type specified in 250.52(A)(5) or (A)(7) are used, each electrode of one grounding system (including that used for strike termination devices) shall not be less than 1.83 m (6 ft) from any other electrode of another grounding system. Two or more grounding electrodes that are bonded together shall be considered a single grounding electrode system.

(C) Bonding Jumper. The bonding jumper(s) used to connect the grounding electrodes together to form the grounding electrode system shall be installed in accordance with 250.64(A), (B), and (E), shall be sized in accordance with 250.66, and shall be connected in the manner specified in 250.70. Rebar shall not be used as a conductor to interconnect the electrodes of grounding electrode systems.

Grounding and Bonding 250.58

(D) Metal Underground Water Pipe. If used as a grounding electrode, metal underground water pipe shall meet the requirements of 250.53 (D)(1) and (D)(2).

(1) Continuity. Continuity of the grounding path or the bonding connection to interior piping shall not rely on water meters or filtering devices and similar equipment.

(2) Supplemental Electrode Required. A metal underground water pipe shall be supplemented by an additional electrode of a type specified in 250.52(A)(2) through (A)(8). If the supplemental electrode is of the rod, pipe, or plate type, it shall comply with 250.53(A). The supplemental electrode shall be bonded to one of the following:

(1) Grounding electrode conductor
(2) Grounded service-entrance conductor
(3) Nonflexible grounded service raceway
(4) Any grounded service enclosure
(5) As provided by 250.32(B)

Exception: The supplemental electrode shall be permitted to be bonded to the interior metal water piping as specified in 250.68(C)(1).

(E) Supplemental Electrode Bonding Connection Size. Where the supplemental electrode is a rod, pipe, or plate electrode, that portion of the bonding jumper that is the sole connection to the supplemental grounding electrode shall not be required to be larger than 6 AWG copper wire or 4 AWG aluminum wire.

(F) Ground Ring. The ground ring shall be installed not less than 750 mm (30 in.) below the surface of the earth.

250.54 Auxiliary Grounding Electrodes. One or more grounding electrodes shall be permitted to be connected to the equipment grounding conductors specified in 250.118 and shall not be required to comply with the electrode bonding requirements of 250.50 or 250.53(C) or the resistance requirements of 250.53(A)(2) Exception, but the earth shall not be used as an effective ground-fault current path as specified in 250.4(A)(5) and 250.4(B)(4).

250.58 Common Grounding Electrode. Where an ac system is connected to a grounding electrode in or at a building or structure, the same electrode shall be used to ground conductor enclosures and equipment in or on that building or structure. Where separate services, feeders, or branch circuits supply a building and are required to be connected to a grounding electrode(s), the same grounding electrode(s) shall be used.

Two or more grounding electrodes that are bonded together shall be considered as a single grounding electrode system in this sense.

250.60 Use of Strike Termination Devices. Conductors and driven pipes, rods, or plate electrodes used for grounding strike termination devices shall not be used in lieu of the grounding electrodes required by 250.50 for grounding wiring systems and equipment. This provision shall not prohibit the required bonding together of grounding electrodes of different systems.

> Informational Note No. 1: See 250.106 for the bonding requirement of the lightning protection system components to the building or structure grounding electrode system.
>
> Informational Note No. 2: Bonding together of all separate grounding electrodes will limit voltage differences between them and between their associated wiring systems.

250.62 Grounding Electrode Conductor Material. The grounding electrode conductor shall be of copper, aluminum, copper-clad aluminum, or the items as permitted in 250.68(C). The material selected shall be resistant to any corrosive condition existing at the installation or shall be protected against corrosion. Conductors of the wire type shall be solid or stranded, insulated, covered, or bare.

250.64 Grounding Electrode Conductor Installation. Grounding electrode conductors at the service, at each building or structure where supplied by a feeder(s) or branch circuit(s), or at a separately derived system shall be installed as specified in 250.64(A) through (F).

(A) Aluminum or Copper-Clad Aluminum Conductors. Grounding electrode conductors of bare, covered, or insulated aluminum or copper-clad aluminum shall comply with the following:

(1) Bare or covered conductors without an extruded polymeric covering shall not be installed where subject to corrosive conditions or be installed in direct contact with concrete.
(2) Terminations made within outdoor enclosures that are listed and identified for the environment shall be permitted within 450 mm (18 in.) of the bottom of the enclosure.
(3) Aluminum or copper-clad aluminum conductors external to buildings or equipment enclosures shall not be terminated within 450 mm (18 in.) of the earth.

Grounding and Bonding 250.64

(B) Securing and Protection Against Physical Damage. Where exposed, a grounding electrode conductor or its enclosure shall be securely fastened to the surface on which it is carried. Grounding electrode conductors shall be permitted to be installed on or through framing members.

(1) Not Exposed to Physical Damage. A 6 AWG or larger copper or aluminum grounding electrode conductor not exposed to physical damage shall be permitted to be run along the surface of the building construction without metal covering or protection.

(2) Exposed to Physical Damage. A 6 AWG or larger copper or aluminum grounding electrode conductor exposed to physical damage shall be protected in rigid metal conduit (RMC), intermediate metal conduit (IMC), Schedule 80 rigid polyvinyl chloride conduit (PVC), reinforced thermosetting resin conduit Type XW (RTRC-XW), electrical metallic tubing (EMT), or cable armor.

(3) Smaller Than 6 AWG. Grounding electrode conductors smaller than 6 AWG shall be protected in RMC, IMC, Schedule 80 PVC, RTRC-XW, EMT, or cable armor.

(4) In Contact with the Earth. Grounding electrode conductors and grounding electrode bonding jumpers in contact with the earth shall not be required to comply with 300.5, but shall be buried or otherwise protected if subject to physical damage.

(C) Continuous. Except as provided in 250.30(A)(5) and (A)(6), 250.30(B)(1), and 250.68(C), grounding electrode conductor(s) shall be installed in one continuous length without a splice or joint. If necessary, splices or connections shall be made as permitted in (1) through (4):

(1) Splicing of the wire-type grounding electrode conductor shall be permitted only by irreversible compression-type connectors listed as grounding and bonding equipment or by the exothermic welding process.
(2) Sections of busbars shall be permitted to be connected together to form a grounding electrode conductor.
(3) Bolted, riveted, or welded connections of structural metal frames of buildings or structures.
(4) Threaded, welded, brazed, soldered or bolted-flange connections of metal water piping.

(D) Building or Structure with Multiple Disconnecting Means in Separate Enclosures. If a building or structure is supplied by a service or feeder with two or more disconnecting means in separate enclosures, the grounding electrode connections shall be made in accordance with 250.64(D)(1), 250.64(D)(2), or 250.64(D)(3).

(1) Common Grounding Electrode Conductor and Taps. A common grounding electrode conductor and grounding electrode conductor taps shall be installed. The common grounding electrode conductor shall be sized in accordance with 250.66, based on the sum of the circular mil area of the largest ungrounded conductor(s) of each set of conductors that supplies the disconnecting means. If the service-entrance conductors connect directly to the overhead service conductors, service drop, underground service conductors, or service lateral, the common grounding electrode conductor shall be sized in accordance with Table 250.66, note 1.

A grounding electrode conductor tap shall extend to the inside of each disconnecting means enclosure. The grounding electrode conductor taps shall be sized in accordance with 250.66 for the largest service-entrance or feeder conductor serving the individual enclosure. The tap conductors shall be connected to the common grounding electrode conductor by one of the following methods in such a manner that the common grounding electrode conductor remains without a splice or joint:

(1) Exothermic welding.
(2) Connectors listed as grounding and bonding equipment.
(3) Connections to an aluminum or copper busbar not less than 6 mm thick × 50 mm wide (¼ in. thick × 2 in. wide) and of sufficient length to accommodate the number of terminations necessary for the installation. The busbar shall be securely fastened and shall be installed in an accessible location. Connections shall be made by a listed connector or by the exothermic welding process. If aluminum busbars are used, the installation shall comply with 250.64(A).

(2) Individual Grounding Electrode Conductors. A grounding electrode conductor shall be connected between the grounding electrode system and one or more of the following, as applicable:

(1) Grounded conductor in each service equipment disconnecting means enclosure
(2) Equipment grounding conductor installed with the feeder
(3) Supply-side bonding jumper

Each grounding electrode conductor shall be sized in accordance with 250.66 based on the service-entrance or feeder conductor(s) supplying the individual disconnecting means.

Grounding and Bonding 250.64

(3) Common Location. A grounding electrode conductor shall be connected in a wireway or other accessible enclosure on the supply side of the disconnecting means to one or more of the following, as applicable:

(1) Grounded service conductor(s)
(2) Equipment grounding conductor installed with the feeder
(3) Supply-side bonding jumper

The connection shall be made with exothermic welding or a connector listed as grounding and bonding equipment. The grounding electrode conductor shall be sized in accordance with 250.66 based on the service-entrance or feeder conductor(s) at the common location where the connection is made.

(E) Raceways and Enclosures for Grounding Electrode Conductors.

(1) General. Ferrous metal raceways, enclosures, and cable armor for grounding electrode conductors shall be electrically continuous from the point of attachment to cabinets or equipment to the grounding electrode and shall be securely fastened to the ground clamp or fitting. Ferrous metal raceways, enclosures, and cable armor shall be bonded at each end of the raceway or enclosure to the grounding electrode or grounding electrode conductor to create an electrically parallel path. Nonferrous metal raceways, enclosures, and cable armor shall not be required to be electrically continuous.

(2) Methods. Bonding shall be in compliance with 250.92(B) and ensured by one of the methods in 250.92(B)(2) through (B)(4).

(3) Size. The bonding jumper for a grounding electrode conductor(s), raceway(s), enclosure(s), or cable armor shall be the same size as, or larger than, the largest enclosed grounding electrode conductor.

(4) Wiring Methods. If a raceway is used as protection for a grounding electrode conductor, the installation shall comply with the requirements of the appropriate raceway article.

(F) Installation to Electrode(s). Grounding electrode conductor(s) and bonding jumpers interconnecting grounding electrodes shall be installed in accordance with (1), (2), or (3). The grounding electrode conductor shall be sized for the largest grounding electrode conductor required among all the electrodes connected to it.

(1) The grounding electrode conductor shall be permitted to be run to any convenient grounding electrode available in the grounding

electrode system where the other electrode(s), if any, is connected by bonding jumpers that are installed in accordance with 250.53(C).

(2) Grounding electrode conductor(s) shall be permitted to be run to one or more grounding electrode(s) individually.

(3) Bonding jumper(s) from grounding electrode(s) shall be permitted to be connected to an aluminum or copper busbar not less than 6 mm thick × 50 mm wide (¼ in. thick × 2 in wide.) and of sufficient length to accommodate the number of terminations necessary for the installation. The busbar shall be securely fastened and shall be installed in an accessible location. Connections shall be made by a listed connector or by the exothermic welding process. The grounding electrode conductor shall be permitted to be run to the busbar. Where aluminum busbars are used, the installation shall comply with 250.64(A).

250.66 Size of Alternating-Current Grounding Electrode Conductor. The size of the grounding electrode conductor at the service, at each building or structure where supplied by a feeder(s) or branch circuit(s), or at a separately derived system of a grounded or ungrounded ac system shall not be less than given in Table 250.66, except as permitted in 250.66(A) through (C).

(A) Connections to a Rod, Pipe, or Plate Electrode(s). If the grounding electrode conductor or bonding jumper connected to a single or multiple rod, pipe, or plate electrode(s), or any combination thereof, as described in 250.52(A)(5) or (A)(7), does not extend on to other types of electrodes that require a larger size conductor, the grounding electrode conductor shall not be required to be larger than 6 AWG copper wire or 4 AWG aluminum wire.

(B) Connections to Concrete-Encased Electrodes. If the grounding electrode conductor or bonding jumper connected to a single or multiple concrete-encased electrode(s), as described in 250.52(A)(3), does not extend on to other types of electrodes that require a larger size of conductor, the grounding electrode conductor shall not be required to be larger than 4 AWG copper wire.

(C) Connections to Ground Rings. If the grounding electrode conductor or bonding jumper connected to a ground ring, as described in 250.52(A)(4), does not extend on to other types of electrodes that require a larger size of conductor, the grounding electrode conductor shall not be required to be larger than the conductor used for the ground ring.

Grounding and Bonding 250.68

Table 250.66 Grounding Electrode Conductor for Alternating-Current Systems

Size of Largest Ungrounded Conductor or Equivalent Area for Parallel Conductors (AWG/kcmil)		Size of Grounding Electrode Conductor (AWG/kcmil)	
Copper	Aluminum or Copper-Clad Aluminum	Copper	Aluminum or Copper-Clad Aluminum
2 or smaller	1/0 or smaller	8	6
1 or 1/0	2/0 or 3/0	6	4
2/0 or 3/0	4/0 or 250	4	2
Over 3/0 through 350	Over 250 through 500	2	1/0
Over 350 through 600	Over 500 through 900	1/0	3/0
Over 600 through 1100	Over 900 through 1750	2/0	4/0
Over 1100	Over 1750	3/0	250

Notes:

1. If multiple sets of service-entrance conductors connect directly to a service drop, set of overhead service conductors, set of underground service conductors, or service lateral, the equivalent size of the largest service-entrance conductor shall be determined by the largest sum of the areas of the corresponding conductors of each set.

2. Where there are no service-entrance conductors, the grounding electrode conductor size shall be determined by the equivalent size of the largest service-entrance conductor required for the load to be served.

3. See installation restrictions in 250.64.

250.68 Grounding Electrode Conductor and Bonding Jumper Connection to Grounding Electrodes. The connection of a grounding electrode conductor at the service, at each building or structure where supplied by a feeder(s) or branch circuit(s), or at a separately derived system and associated bonding jumper(s) shall be made as specified 250.68(A) through (C).

(A) Accessibility. All mechanical elements used to terminate a grounding electrode conductor or bonding jumper to a grounding electrode shall be accessible.

Exception No. 1: An encased or buried connection to a concrete-encased, driven, or buried grounding electrode shall not be required to be accessible.

Exception No. 2: Exothermic or irreversible compression connections used at terminations, together with the mechanical means used to attach such terminations to fireproofed structural metal whether or not the mechanical means is reversible, shall not be required to be accessible.

(B) Effective Grounding Path. The connection of a grounding electrode conductor or bonding jumper to a grounding electrode shall be made in a manner that will ensure an effective grounding path. Where necessary to ensure the grounding path for a metal piping system used as a grounding electrode, bonding shall be provided around insulated joints and around any equipment likely to be disconnected for repairs or replacement. Bonding jumpers shall be of sufficient length to permit removal of such equipment while retaining the integrity of the grounding path.

(C) Grounding Electrode Conductor Connections. Grounding electrode conductors and bonding jumpers shall be permitted to be connected at the following locations and used to extend the connection to an electrode(s):

(1) Interior metal water piping that is electrically continuous with a metal underground water pipe electrode and is located not more than 1.52 m (5 ft) from the point of entrance to the building shall be permitted to extend the connection to an electrode(s). Interior metal water piping located more than 1.52 m (5 ft) from the point of entrance to the building shall not be used as a conductor to interconnect electrodes of the grounding electrode system.

(2) The metal structural frame of a building shall be permitted to be used as a conductor to interconnect electrodes that are part of the grounding electrode system, or as a grounding electrode conductor. Hold-down bolts securing the structural steel column that are connected to a concrete-encased electrode complying with 250.52(A)(3) and located in the support footing or foundation shall be permitted to connect the metal structural frame of a building or structure to the concrete encased grounding electrode. The hold-down bolts shall be connected to the concrete-encased electrode by welding, exothermic welding, the usual steel tie wires, or other approved means.

(3) A rebar-type concrete-encased electrode installed in accordance with 250.52(A)(3) with an additional rebar section extended from its

Grounding and Bonding 250.80

location within the concrete foundation or footing to an accessible location that is not subject to corrosion shall be permitted for connection of grounding electrode conductors and bonding jumpers in accordance with the following:

a. The additional rebar section shall be continuous with the grounding electrode rebar or shall be connected to the grounding electrode rebar and connected together by the usual steel tie wires, exothermic welding, welding, or other effective means.
b. The rebar extension shall not be exposed to contact with the earth without corrosion protection.
c. Rebar shall not be used as a conductor to interconnect the electrodes of grounding electrode systems.

250.70 Methods of Grounding and Bonding Conductor Connection to Electrodes. The grounding or bonding conductor shall be connected to the grounding electrode by exothermic welding, listed lugs, listed pressure connectors, listed clamps, or other listed means. Connections depending on solder shall not be used. Ground clamps shall be listed for the materials of the grounding electrode and the grounding electrode conductor and, where used on pipe, rod, or other buried electrodes, shall also be listed for direct soil burial or concrete encasement. Not more than one conductor shall be connected to the grounding electrode by a single clamp or fitting unless the clamp or fitting is listed for multiple conductors. One of the following methods shall be used:

(1) A pipe fitting, pipe plug, or other approved device screwed into a pipe or pipe fitting
(2) A listed bolted clamp of cast bronze or brass, or plain or malleable iron
(3) For indoor communications purposes only, a listed sheet metal strap-type ground clamp having a rigid metal base that seats on the electrode and having a strap of such material and dimensions that it is not likely to stretch during or after installation
(4) An equally substantial approved means

PART IV. ENCLOSURE, RACEWAY, AND SERVICE CABLE CONNECTIONS

250.80 Service Raceways and Enclosures. Metal enclosures and raceways for service conductors and equipment shall be connected to the grounded system conductor if the electrical system is grounded or to the grounding electrode conductor for electrical systems that are not grounded.

Exception: Metal components that are installed in a run of underground nonmetallic raceway(s) and are isolated from possible contact by a minimum cover of 450 mm (18 in.) to all parts of the metal components shall not be required to be connected to the grounded system conductor, supply side bonding jumper, or grounding electrode conductor.

PART V. BONDING

250.90 General. Bonding shall be provided where necessary to ensure electrical continuity and the capacity to conduct safely any fault current likely to be imposed.

250.92 Services.

(A) Bonding of Equipment for Services. The normally non–current-carrying metal parts of equipment indicated in 250.92(A)(1) and (A)(2) shall be bonded together.

(1) All raceways, cable trays, cablebus framework, auxiliary gutters, or service cable armor or sheath that enclose, contain, or support service conductors, except as permitted in 250.80
(2) All enclosures containing service conductors, including meter fittings, boxes, or the like, interposed in the service raceway or armor

(B) Method of Bonding at the Service. Bonding jumpers meeting the requirements of this article shall be used around impaired connections, such as reducing washers or oversized, concentric, or eccentric knockouts. Standard locknuts or bushings shall not be the only means for the bonding required by this section but shall be permitted to be installed to make a mechanical connection of the raceway(s).

Electrical continuity at service equipment, service raceways, and service conductor enclosures shall be ensured by one of the following methods:

(1) Bonding equipment to the grounded service conductor in a manner provided in 250.8
(2) Connections using threaded couplings or listed threaded hubs on enclosures if made up wrenchtight
(3) Threadless couplings and connectors if made up tight for metal raceways and metal-clad cables
(4) Other listed devices, such as bonding-type locknuts, bushings, or bushings with bonding jumpers

Grounding and Bonding 250.94

250.94 Bonding for Communication Systems. Communications system bonding terminations shall be connected in accordance with 250.94(A) or (B).

(A) The Intersystem Bonding Termination Device. An intersystem bonding termination (IBT) for connecting intersystem bonding conductors shall be provided external to enclosures at the service equipment or metering equipment enclosure and at the disconnecting means for any additional buildings or structures. If an IBT is used, it shall comply with the following:

(1) Be accessible for connection and inspection.
(2) Consist of a set of terminals with the capacity for connection of not less than three intersystem bonding conductors.
(3) Not interfere with opening the enclosure for a service, building or structure disconnecting means, or metering equipment.
(4) At the service equipment, be securely mounted and electrically connected to an enclosure for the service equipment, to the meter enclosure, or to an exposed nonflexible metallic service raceway, or be mounted at one of these enclosures and be connected to the enclosure or to the grounding electrode conductor with a minimum 6 AWG copper conductor.
(5) At the disconnecting means for a building or structure, be securely mounted and electrically connected to the metallic enclosure for the building or structure disconnecting means, or be mounted at the disconnecting means and be connected to the metallic enclosure or to the grounding electrode conductor with a minimum 6 AWG copper conductor.
(6) The terminals shall be listed as grounding and bonding equipment.

Exception: In existing buildings or structures where any of the intersystem bonding and grounding electrode conductors required by 770.100(B)(2), 800.100(B)(2), 810.21(F)(2), 820.100, and 830.100 exist, installation of the intersystem bonding termination is not required. An accessible means external to enclosures for connecting intersystem bonding and grounding electrode conductors shall be permitted at the service equipment and at the disconnecting means for any additional buildings or structures by at least one of the following means:

(1) Exposed nonflexible metallic raceways
(2) An exposed grounding electrode conductor
(3) Approved means for the external connection of a copper or other corrosion-resistant bonding or grounding electrode conductor to the grounded raceway or equipment

Informational Note No. 1: A 6 AWG copper conductor with one end bonded to the grounded nonflexible metallic raceway or equipment and with 150 mm (6 in.) or more of the other end made accessible on the outside wall is an example of the approved means covered in 250.94, Exception item (3).

Informational Note No. 2: See 770.100, 800.100, 810.21, 820.100, and 830.100 for intersystem bonding and grounding requirements for conductive optical fiber cables, communications circuits, radio and television equipment, CATV circuits and network-powered broadband communications systems, respectively.

(B) Other Means. Connections to an aluminum or copper busbar not less than 6 mm thick × 50 mm wide (¼ in. thick × 2 in. wide) and of sufficient length to accommodate at least three terminations for communication systems in addition to other connections. The busbar shall be securely fastened and shall be installed in an accessible location. Connections shall be made by a listed connector. If aluminum busbars are used, the installation shall also comply with 250.64(A).

Exception to (A) and (B): Means for connecting intersystem bonding conductors are not required where communications systems are not likely to be used.

Informational Note: The use of an IBT can reduce electrical noise on communication systems.

250.96 Bonding Other Enclosures.

(A) General. Metal raceways, cable trays, cable armor, cable sheath, enclosures, frames, fittings, and other metal non–current-carrying parts that are to serve as equipment grounding conductors, with or without the use of supplementary equipment grounding conductors, shall be bonded where necessary to ensure electrical continuity and the capacity to conduct safely any fault current likely to be imposed on them. Any nonconductive paint, enamel, or similar coating shall be removed at threads, contact points, and contact surfaces or shall be connected by means of fittings designed so as to make such removal unnecessary.

250.102 Grounded Conductor, Bonding Conductors, and Jumpers.

(A) Material. Bonding jumpers shall be of copper, aluminum, copper-clad aluminum, or other corrosion-resistant material. A bonding jumper shall be a wire, bus, screw, or similar suitable conductor.

Grounding and Bonding 250.102

(B) Attachment. Bonding jumpers shall be attached in the manner specified in 250.8 for circuits and equipment and in 250.70 for grounding electrodes.

(C) Size — Supply-Side Bonding Jumper.

(1) Size for Supply Conductors in a Single Raceway or Cable. The supply-side bonding jumper shall not be smaller than specified in Table 250.102(C)(1).

(2) Size for Parallel Conductor Installations in Two or More Raceways or Cables. Where the ungrounded supply conductors are paralleled in two or more raceways or cables, and an individual supply-side bonding jumper is used for bonding these raceways or cables, the size of the supply-side bonding jumper for each raceway or cable shall be selected from Table 250.102(C)(1) based on the size of the ungrounded supply conductors in each raceway or cable. A single supply-side bonding jumper installed for bonding two or more raceways or cables shall be sized in accordance with 250.102(C)(1).

> Informational Note No. 1: The term *supply conductors* includes ungrounded conductors that do not have overcurrent protection on their supply side and terminate at service equipment or the first disconnecting means of a separately derived system.

> Informational Note No. 2: See Chapter 9, Table 8, for the circular mil area of conductors 18 AWG through 4/0 AWG.

(D) Size — Equipment Bonding Jumper on Load Side of an Overcurrent Device. The equipment bonding jumper on the load side of an overcurrent device(s) shall be sized in accordance with 250.122.

A single common continuous equipment bonding jumper shall be permitted to connect two or more raceways or cables if the bonding jumper is sized in accordance with 250.122 for the largest overcurrent device supplying circuits therein.

(E) Installation. Bonding jumpers or conductors and equipment bonding jumpers shall be permitted to be installed inside or outside of a raceway or an enclosure.

(1) Inside a Raceway or an Enclosure. If installed inside a raceway, equipment bonding jumpers and bonding jumpers or conductors shall comply with the requirements of 250.119 and 250.148.

(2) Outside a Raceway or an Enclosure. If installed on the outside, the length of the bonding jumper or conductor or equipment bonding

Table 250.102(C)(1) Grounded Conductor, Main Bonding Jumper, System Bonding Jumper, and Supply-Side Bonding Jumper for Alternating-Current Systems

Size of Largest Ungrounded Conductor or Equivalent Area for Parallel Conductors (AWG/kcmil)		Size of Grounded Conductor or Bonding Jumper* (AWG/kcmil)	
Copper	Aluminum or Copper-Clad Aluminum	Copper	Aluminum or Copper-Clad Aluminum
2 or smaller	1/0 or smaller	8	6
1 or 1/0	2/0 or 3/0	6	4
2/0 or 3/0	4/0 or 250	4	2
Over 3/0 through 350	Over 250 through 500	2	1/0
Over 350 through 600	Over 500 through 900	1/0	3/0
Over 600 through 1100	Over 900 through 1750	2/0	4/0
Over 1100	Over 1750	See Notes 1 and 2.	

Notes:

1. If the ungrounded supply conductors are larger than 1100 kcmil copper or 1750 kcmil aluminum, the grounded conductor or bonding jumper shall have an area not less than 12½ percent of the area of the largest ungrounded supply conductor or equivalent area for parallel supply conductors. The grounded conductor or bonding jumper shall not be required to be larger than the largest ungrounded conductor or set of ungrounded conductors.

2. If the ungrounded supply conductors are larger than 1100 kcmil copper or 1750 kcmil aluminum and if the ungrounded supply conductors and the bonding jumper are of different materials (copper, aluminum, or copper-clad aluminum), the minimum size of the grounded conductor or bonding jumper shall be based on the assumed use of ungrounded supply conductors of the same material as the grounded conductor or bonding jumper and will have an ampacity equivalent to that of the installed ungrounded supply conductors.

3. If multiple sets of service-entrance conductors are used as permitted in 230.40, Exception No. 2, or if multiple sets of ungrounded supply conductors are installed for a separately derived system, the equivalent size of the largest ungrounded supply conductor(s) shall be determined by the largest sum of the areas of the corresponding conductors of each set.

4. If there are no service-entrance conductors, the supply conductor size shall be determined by the equivalent size of the largest service-entrance conductor required for the load to be served.

*For the purposes of applying this table and its notes, the term *bonding jumper* refers to main bonding jumpers, system bonding jumpers, and supply-side bonding jumpers.

Grounding and Bonding 250.104

jumper shall not exceed 1.8 m (6 ft) and shall be routed with the raceway or enclosure.

Exception: An equipment bonding jumper or supply-side bonding jumper longer than 1.8 m (6 ft) shall be permitted at outside pole locations for the purpose of bonding or grounding isolated sections of metal raceways or elbows installed in exposed risers of metal conduit or other metal raceway, and for bonding grounding electrodes, and shall not be required to be routed with a raceway or enclosure.

(3) Protection. Bonding jumpers or conductors and equipment bonding jumpers shall be installed in accordance with 250.64(A) and (B).

250.104 Bonding of Piping Systems and Exposed Structural Metal.

(A) Metal Water Piping. The metal water piping system shall be bonded as required in 250.104(A)(1), (A)(2), or (A)(3).

(1) General. Metal water piping system(s) installed in or attached to a building or structure shall be bonded to any of the following:

(1) Service equipment enclosure
(2) Grounded conductor at the service
(3) Grounding electrode conductor, if of sufficient size
(4) One or more grounding electrodes used, if the grounding electrode conductor or bonding jumper to the grounding electrode is of sufficient size

The bonding jumper(s) shall be installed in accordance with 250.64(A), 250.64(B), and 250.64(E). The points of attachment of the bonding jumper(s) shall be accessible. The bonding jumper(s) shall be sized in accordance with Table 250.102(C)(1) except that it shall not be required to be larger than 3/0 copper or 250 kcmil aluminum or copper-clad aluminum and except as permitted in 250.104(A)(2) and 250.104(A)(3).

(2) Buildings of Multiple Occupancy. In buildings of multiple occupancy where the metal water piping system(s) installed in or attached to a building or structure for the individual occupancies is metallically isolated from all other occupancies by use of nonmetallic water piping, the metal water piping system(s) for each occupancy shall be permitted to be bonded to the equipment grounding terminal of the switchgear, switchboard, or panelboard enclosure (other than service equipment) supplying that occupancy. The bonding jumper shall be sized in accordance with 250.102(D).

(3) Buildings or Structures Supplied by a Feeder(s) or Branch Circuit(s). The metal water piping system(s) installed in or attached to a building or structure shall be bonded to any of the following:

(1) Building or structure disconnecting means enclosure where located at the building or structure
(2) Equipment grounding conductor run with the supply conductors
(3) One or more grounding electrodes used

The bonding jumper(s) shall be sized in accordance with 250.102(D). The bonding jumper shall not be required to be larger than the largest ungrounded feeder or branch-circuit conductor supplying the building or structure.

(B) Other Metal Piping. If installed in or attached to a building or structure, a metal piping system(s), including gas piping, that is likely to become energized shall be bonded to any of the following:

(1) Equipment grounding conductor for the circuit that is likely to energize the piping system
(2) Service equipment enclosure
(3) Grounded conductor at the service
(4) Grounding electrode conductor, if of sufficient size
(5) One or more grounding electrodes used, if the grounding electrode conductor or bonding jumper to the grounding electrode is of sufficient size

The bonding conductor(s) or jumper(s) shall be sized in accordance with Table 250.122, and equipment grounding conductors shall be sized in accordance with Table 250.122 using the rating of the circuit that is likely to energize the piping system(s). The points of attachment of the bonding jumper(s) shall be accessible.

> Informational Note No. 1: Bonding all piping and metal air ducts within the premises will provide additional safety.
>
> Informational Note No. 2: Additional information for gas piping systems can be found in NFPA 54-2018, *National Fuel Gas Code*, and NFPA 780-2017, *Standard for the Installation of Lightning Protection Systems*.

(C) Structural Metal. Exposed structural metal that is interconnected to form a metal building frame, is not intentionally grounded or bonded, and is likely to become energized shall be bonded to any of the following:

Grounding and Bonding 250.112

(1) Service equipment enclosure
(2) Grounded conductor at the service
(3) Disconnecting means for buildings or structures supplied by a feeder or branch circuit
(4) Grounding electrode conductor, if of sufficient size
(5) One or more grounding electrodes used, if the grounding electrode conductor or bonding jumper to the grounding electrode is of sufficient size

The bonding conductor(s) or jumper(s) shall be sized in accordance with Table 250.102(C)(1), except that it shall not be required to be larger than 3/0 copper or 250 kcmil aluminum or copper-clad aluminum, and installed in accordance with 250.64(A), 250.64(B), and 250.64(E). The points of attachment of the bonding jumper(s) shall be accessible unless installed in compliance with 250.68(A) Exception No. 2.

PART VI. EQUIPMENT GROUNDING AND EQUIPMENT GROUNDING CONDUCTORS

250.110 Equipment Fastened in Place (Fixed) or Connected by Permanent Wiring Methods. Exposed, normally non–current-carrying metal parts of fixed equipment supplied by or enclosing conductors or components that are likely to become energized shall be connected to an equipment grounding conductor under any of the following conditions:

(1) Where within 2.5 m (8 ft) vertically or 1.5 m (5 ft) horizontally of ground or grounded metal objects and subject to contact by persons
(2) Where located in a wet or damp location and not isolated
(3) Where in electrical contact with metal
(4) Where in a hazardous (classified) location as covered by Articles 500 through 517
(5) Where supplied by a wiring method that provides an equipment grounding conductor, except as permitted by 250.86, Exception No. 2, for short sections of metal enclosures
(6) Where equipment operates with any terminal at over 150 volts to ground

250.112 Specific Equipment Fastened in Place (Fixed) or Connected by Permanent Wiring Methods. Except as permitted in 250.112(F) and (I), exposed, normally non–current-carrying metal parts of equipment described in 250.112(A) through (K), and normally non–current-carrying metal parts of equipment and enclosures described in 250.112(L) and

(M), shall be connected to an equipment grounding conductor, regardless of voltage.

(J) Luminaires. Luminaires as provided in Part V of Article 410.

(L) Motor-Operated Water Pumps. Motor-operated water pumps, including the submersible type.

(M) Metal Well Casings. Where a submersible pump is used in a metal well casing, the well casing shall be connected to the pump circuit equipment grounding conductor.

250.114 Equipment Connected by Cord and Plug. Exposed, normally non–current-carrying metal parts of cord-and-plug-connected equipment shall be connected to the equipment grounding conductor under any of the following conditions:

(3) In residential occupancies:

   a. Refrigerators, freezers, and air conditioners
   b. Clothes-washing, clothes-drying, and dish-washing machines; ranges; kitchen waste disposers; information technology equipment; sump pumps; and electrical aquarium equipment
   c. Hand-held motor-operated tools, stationary and fixed motor-operated tools, and light industrial motor-operated tools
   d. Motor-operated appliances of the following types: hedge clippers, lawn mowers, snow blowers, and wet scrubbers
   e. Portable handlamps and portable luminaires

250.116 Nonelectrical Equipment. The metal parts of the following nonelectrical equipment described in this section shall be connected to the equipment grounding conductor:

(1) Frames and tracks of electrically operated cranes and hoists
(2) Frames of nonelectrically driven elevator cars to which electrical conductors are attached
(3) Hand-operated metal shifting ropes or cables of electric elevators

   Informational Note: Where extensive metal in or on buildings or structures may become energized and is subject to personal contact, adequate bonding and grounding will provide additional safety.

250.118 Types of Equipment Grounding Conductors. The equipment grounding conductor run with or enclosing the circuit conductors shall be one or more or a combination of the following:

(1) A copper, aluminum, or copper-clad aluminum conductor. This conductor shall be solid or stranded; insulated, covered, or bare; and in the form of a wire or a busbar of any shape.
(2) Rigid metal conduit.
(3) Intermediate metal conduit.
(4) Electrical metallic tubing.
(5) Listed flexible metal conduit meeting all the following conditions:
   a. The conduit is terminated in listed fittings.
   b. The circuit conductors contained in the conduit are protected by overcurrent devices rated at 20 amperes or less.
   c. The size of the conduit does not exceed metric designator 35 (trade size 1¼).
   d. The combined length of flexible metal conduit, flexible metallic tubing, and liquidtight flexible metal conduit in the same effective ground-fault current path does not exceed 1.8 m (6 ft).
   e. If used to connect equipment where flexibility is necessary to minimize the transmission of vibration from equipment or to provide flexibility for equipment that requires movement after installation, a wire-type equipment grounding conductor shall be installed.
(6) Listed liquidtight flexible metal conduit meeting all the following conditions:
   a. The conduit is terminated in listed fittings.
   b. For metric designators 12 through 16 (trade sizes ⅜ through ½), the circuit conductors contained in the conduit are protected by overcurrent devices rated at 20 amperes or less.
   c. For metric designators 21 through 35 (trade sizes ¾ through 1¼), the circuit conductors contained in the conduit are protected by overcurrent devices rated not more than 60 amperes and there is no flexible metal conduit, flexible metallic tubing, or liquidtight flexible metal conduit in metric designators 12 through 16 (trade sizes ⅜ through ½) in the effective ground-fault current path.
   d. The combined length of flexible metal conduit, flexible metallic tubing, and liquidtight flexible metal conduit in the same effective ground-fault current path does not exceed 1.8 m (6 ft).
   e. If used to connect equipment where flexibility is necessary to minimize the transmission of vibration from equipment or to provide flexibility for equipment that requires movement after installation, a wire-type equipment grounding conductor shall be installed.

(7) Flexible metallic tubing where the tubing is terminated in listed fittings and meeting the following conditions:
   a. The circuit conductors contained in the tubing are protected by overcurrent devices rated at 20 amperes or less.
   b. The combined length of flexible metal conduit, flexible metallic tubing, and liquidtight flexible metal conduit in the same effective ground-fault current path does not exceed 1.8 m (6 ft).
(8) Armor of Type AC cable as provided in 320.108.
(9) The copper sheath of mineral-insulated, metal-sheathed cable Type MI.
(10) Type MC cable that provides an effective ground-fault current path in accordance with one or more of the following:
   a. It contains an insulated or uninsulated equipment grounding conductor in compliance with 250.118(1).
   b. The combined metallic sheath and uninsulated equipment grounding/bonding conductor of interlocked metal tape–type MC cable that is listed and identified as an equipment grounding conductor
   c. The metallic sheath or the combined metallic sheath and equipment grounding conductors of the smooth or corrugated tube-type MC cable that is listed and identified as an equipment grounding conductor

250.119 Identification of Equipment Grounding Conductors. Unless required elsewhere in this *Code*, equipment grounding conductors shall be permitted to be bare, covered, or insulated. Individually covered or insulated equipment grounding conductors shall have a continuous outer finish that is either green or green with one or more yellow stripes except as permitted in this section. Conductors with insulation or individual covering that is green, green with one or more yellow stripes, or otherwise identified as permitted by this section shall not be used for ungrounded or grounded circuit conductors.

Exception No. 1: Power-limited Class 2 or Class 3 cables, power-limited fire alarm cables, or communications cables containing only circuits operating at less than 50 volts ac or 60 volts dc where connected to equipment not required to be grounded shall be permitted to use a conductor with green insulation or green with one or more yellow stripes for other than equipment grounding purposes.

Exception No. 2: Flexible cords having an integral insulation and jacket without an equipment grounding conductor shall be permitted to have a continuous outer finish that is green.

Grounding and Bonding 250.120

(A) Conductors 4 AWG and Larger. Equipment grounding conductors 4 AWG and larger shall comply with 250.119(A)(1) and (A)(2).

(1) An insulated or covered conductor 4 AWG and larger shall be permitted, at the time of installation, to be permanently identified as an equipment grounding conductor at each end and at every point where the conductor is accessible.

Exception: Conductors 4 AWG and larger shall not be required to be marked in conduit bodies that contain no splices or unused hubs.

(2) Identification shall encircle the conductor and shall be accomplished by one of the following:

   a. Stripping the insulation or covering from the entire exposed length
   b. Coloring the insulation or covering green at the termination
   c. Marking the insulation or covering with green tape or green adhesive labels at the termination

250.120 Equipment Grounding Conductor Installation. An equipment grounding conductor shall be installed in accordance with 250.120(A), (B), and (C).

(A) Raceway, Cable Trays, Cable Armor, Cablebus, or Cable Sheaths. Where it consists of a raceway, cable tray, cable armor, cablebus framework, or cable sheath or where it is a wire within a raceway or cable, it shall be installed in accordance with the applicable provisions in this *Code* using fittings for joints and terminations approved for use with the type raceway or cable used. All connections, joints, and fittings shall be made tight using suitable tools.

(B) Aluminum and Copper-Clad Aluminum Conductors. Equipment grounding conductors of bare, covered, or insulated aluminum or copper-clad aluminum shall comply with the following:

(1) Unless part of a suitable Chapter 3 cable wiring method, bare or covered conductors shall not be installed where subject to corrosive conditions or be installed in direct contact with concrete, masonry, or the earth.
(2) Terminations made within outdoor enclosures that are listed and identified for the environment shall be permitted within 450 mm (18 in.) of the bottom of the enclosure.
(3) Aluminum or copper-clad aluminum conductors external to buildings or enclosures shall not be terminated within 450 mm (18 in.) of the earth, unless terminated within a listed wire connector system.

(C) Equipment Grounding Conductors Smaller Than 6 AWG. Where not routed with circuit conductors as permitted in 250.130(C) and 250.134(A) Exception No. 2, equipment grounding conductors smaller than 6 AWG shall be protected from physical damage by an identified raceway or cable armor unless installed within hollow spaces of the framing members of buildings or structures and where not subject to physical damage.

250.121 Restricted Use of Equipment Grounding Conductors.

(A) Grounding Electrode Conductor. An equipment grounding conductor shall not be used as a grounding electrode conductor.

Exception: A wire-type equipment grounding conductor installed in compliance with 250.6(A) and the applicable requirements for both the equipment grounding conductor and the grounding electrode conductor in Parts II, III, and VI of this article shall be permitted to serve as both an equipment grounding conductor and a grounding electrode conductor.

250.122 Size of Equipment Grounding Conductors.

(A) General. Copper, aluminum, or copper-clad aluminum equipment grounding conductors of the wire type shall not be smaller than shown in Table 250.122. The equipment grounding conductor shall not be required to be larger than the circuit conductors supplying the equipment. If a cable tray, a raceway, or a cable armor or sheath is used as the equipment grounding conductor, as provided in 250.118 and 250.134(A), it shall comply with 250.4(A)(5) or (B)(4).

Equipment grounding conductors shall be permitted to be sectioned within a multiconductor cable, provided the combined circular mil area complies with Table 250.122.

(B) Increased in Size. If ungrounded conductors are increased in size for any reason other than as required in 310.15(B) or 310.15(C), wire-type equipment grounding conductors, if installed, shall be increased in size proportionately to the increase in circular mil area of the ungrounded conductors.

(C) Multiple Circuits. A single equipment grounding conductor shall be permitted to be installed for multiple circuits that are installed in the same raceway, cable, trench, or cable tray. It shall be sized from Table 250.122 for the largest overcurrent device protecting circuit conductors in the raceway, cable, trench, or cable tray. Equipment grounding conductors installed in cable trays shall meet the minimum requirements of 392.10(B)(1)(c).

Grounding and Bonding 250.122

(F) Conductors in Parallel. For circuits of parallel conductors as permitted in 310.10(G), the equipment grounding conductor shall be installed in accordance with 250.122(F)(1) or (F)(2).

(1) Conductor Installations in Raceways, Auxiliary Gutters, or Cable Trays.

(a) *Single Raceway or Cable Tray, Auxiliary Gutter, or Cable Tray.* If circuit conductors are connected in parallel in the same raceway, auxiliary gutter, or cable tray, a single wire-type conductor shall be permitted as the equipment grounding conductor. The wire-type equipment grounding conductor shall be sized in accordance with 250.122, based on the overcurrent protective device for the feeder or branch circuit.

(b) *Multiple Raceways.* If conductors are installed in multiple raceways and are connected in parallel, a wire-type equipment grounding conductor, if used, shall be installed in each raceway and shall be connected in parallel. The equipment grounding conductor installed in each raceway shall be sized in accordance with 250.122 based on the rating of the overcurrent protective device for the feeder or branch circuit.

(2) Multiconductor Cables.

(a) Except as provided in 250.122(F)(2)(c) for raceway or cable tray installations, the equipment grounding conductor in each multiconductor cable shall be sized in accordance with 250.122 based on the overcurrent protective device for the feeder or branch circuit.

(b) If circuit conductors of multiconductor cables are connected in parallel, the equipment grounding conductor(s) in each cable shall be connected in parallel.

(c) If multiconductor cables are paralleled in the same raceway, auxiliary gutter, or cable tray, a single equipment grounding conductor that is sized in accordance with 250.122 shall be permitted in combination with the equipment grounding conductors provided within the multiconductor cables and shall all be connected together.

(d) Equipment grounding conductors installed in cable trays shall meet the minimum requirements of 392.10(B)(1)(c). Cable trays complying with 392.60(B), metal raceways in accordance with 250.118, or auxiliary gutters shall be permitted as the equipment grounding conductor.

(G) Feeder Taps. Equipment grounding conductors installed with feeder taps shall not be smaller than shown in Table 250.122 based on the rating of the overcurrent device ahead of the feeder on the supply side ahead of the tap but shall not be required to be larger than the tap conductors.

Table 250.122 Minimum Size Equipment Grounding Conductors for Grounding Raceway and Equipment

Rating or Setting of Automatic Overcurrent Device in Circuit Ahead of Equipment, Conduit, etc., Not Exceeding (Amperes)	Size (AWG or kcmil)	
	Copper	Aluminum or Copper-Clad Aluminum*
15	14	12
20	12	10
60	10	8
100	8	6
200	6	4
300	4	2
400	3	1
500	2	1/0
600	1	2/0
800	1/0	3/0
1000	2/0	4/0
1200	3/0	250
1600	4/0	350
2000	250	400
2500	350	600
3000	400	600
4000	500	750
5000	700	1250
6000	800	1250

Note: Where necessary to comply with 250.4(A)(5) or (B)(4), the equipment grounding conductor shall be sized larger than given in this table.

*See installation restrictions in 250.120.

250.126 Identification of Wiring Device Terminals. The terminal for the connection of the equipment grounding conductor shall be identified by one of the following:

(1) A green, not readily removable terminal screw with a hexagonal head.
(2) A green, hexagonal, not readily removable terminal nut.

Grounding and Bonding 250.130

(3) A green pressure wire connector. If the terminal for the equipment grounding conductor is not visible, the conductor entrance hole shall be marked with the word *green* or *ground*, the letters *G* or *GR*, a grounding symbol, or otherwise identified by a distinctive green color. If the terminal for the equipment grounding conductor is readily removable, the area adjacent to the terminal shall be similarly marked.

PART VII. METHODS OF EQUIPMENT GROUNDING CONDUCTOR CONNECTIONS

250.130 Equipment Grounding Conductor Connections. Equipment grounding conductor connections at the source of separately derived systems shall be made in accordance with 250.30(A)(1). Equipment grounding conductor connections at service equipment shall be made as indicated in 250.130(A) or (B). For replacement of non–grounding-type receptacles with grounding-type receptacles and for branch-circuit extensions only in existing installations that do not have an equipment grounding conductor in the branch circuit, connections shall be permitted as indicated in 250.130(C).

(A) For Grounded Systems. The connection shall be made by bonding the equipment grounding conductor to the grounded service conductor and the grounding electrode conductor.

(B) For Ungrounded Systems. The connection shall be made by bonding the equipment grounding conductor to the grounding electrode conductor.

(C) Nongrounding Receptacle Replacement or Branch Circuit Extensions. The equipment grounding conductor of a grounding-type receptacle or a branch-circuit extension shall be permitted to be connected to any of the following:

(1) Any accessible point on the grounding electrode system as described in 250.50
(2) Any accessible point on the grounding electrode conductor
(3) The equipment grounding terminal bar within the enclosure where the branch circuit for the receptacle or branch circuit originates
(4) An equipment grounding conductor that is part of another branch circuit that originates from the enclosure where the branch circuit for the receptacle or branch circuit originates

(5) For grounded systems, the grounded service conductor within the service equipment enclosure
(6) For ungrounded systems, the grounding terminal bar within the service equipment enclosure

250.132 Short Sections of Raceway. Isolated sections of metal raceway or cable armor, if required to be connected to an equipment grounding conductor, shall be connected in accordance with 250.134.

250.134 Equipment Fastened in Place or Connected by Permanent Wiring Methods (Fixed). Unless connected to the grounded circuit conductor as permitted by 250.32, 250.140, and 250.142, non–current-carrying metal parts of equipment, raceways, and other enclosures, if grounded, shall be connected to an equipment grounding conductor by one of the following methods:

(1) By connecting to any of the equipment grounding conductors permitted by 250.118(2) through (14)
(2) By connecting to an equipment grounding conductor of the wire type that is contained within the same raceway, contained within the same cable, or otherwise run with the circuit conductors

250.138 Cord-and-Plug-Connected Equipment. Non–current-carrying metal parts of cord-and-plug-connected equipment, if required to be connected to an equipment grounding conductor, shall be connected by one of the methods in 250.138(A) or (B).

(A) By Means of an Equipment Grounding Conductor. By means of an equipment grounding conductor run with the power supply conductors in a cable assembly or flexible cord properly terminated in a grounding-type attachment plug with one fixed grounding contact.

Exception: The grounding contacting pole of grounding-type plug-in ground-fault circuit interrupters shall be permitted to be of the movable, self-restoring type on circuits operating at not over 150 volts between any two conductors or over 150 volts between any conductor and ground.

(B) By Means of a Separate Flexible Wire or Strap. By means of a separate flexible wire or strap, insulated or bare, connected to an equipment grounding conductor, and protected as well as practicable against physical damage, where part of equipment.

Grounding and Bonding 250.142

250.140 Frames of Ranges and Clothes Dryers. Frames of electric ranges, wall-mounted ovens, counter-mounted cooking units, clothes dryers, and outlet or junction boxes that are part of the circuit for these appliances shall be connected to the equipment grounding conductor in the manner specified by 250.134 or 250.138.

Exception: For existing branch-circuit installations only where an equipment grounding conductor is not present in the outlet or junction box, the frames of electric ranges, wall-mounted ovens, counter-mounted cooking units, clothes dryers, and outlet or junction boxes that are part of the circuit for these appliances shall be permitted to be connected to the grounded circuit conductor if all the following conditions are met.

(1) The supply circuit is 120/240-volt, single-phase, 3-wire; or 208Y/120-volt derived from a 3-phase, 4-wire, wye-connected system.

(2) The grounded conductor is not smaller than 10 AWG copper or 8 AWG aluminum.

(3) The grounded conductor is insulated, or the grounded conductor is uninsulated and part of a Type SE service-entrance cable and the branch circuit originates at the service equipment.

(4) Grounding contacts of receptacles furnished as part of the equipment are bonded to the equipment.

250.142 Use of Grounded Circuit Conductor for Grounding Equipment.

(A) Supply-Side Equipment. A grounded circuit conductor shall be permitted to be connected to non–current-carrying metal parts of equipment, raceways, and other enclosures at any of the following locations:

(1) On the supply side or within the enclosure of the ac service disconnecting means
(2) On the supply side or within the enclosure of the main disconnecting means for separate buildings as provided in 250.32(B)(1) Exception No. 1

(B) Load-Side Equipment. Except as permitted in 250.30(A)(1), 250.32(B)(1), Exception No. 1, and Part X of Article 250, a grounded circuit conductor shall not be connected to non–current-carrying metal parts of equipment on the load side of the service disconnecting means or on the load side of a separately derived system disconnecting means or the overcurrent devices for a separately derived system not having a main disconnecting means.

Exception No. 1: The frames of ranges, wall-mounted ovens, countermounted cooking units, and clothes dryers under the conditions permitted for existing installations by 250.140 shall be permitted to be connected to the grounded circuit conductor.

Exception No. 2: It shall be permissible to connect meter enclosures to the grounded circuit conductor on the load side of the service disconnect if all of the following conditions apply:

(1) Ground-fault protection of equipment is not installed.
(2) All meter enclosures are located immediately adjacent to the service disconnecting means.
(3) The size of the grounded circuit conductor is not smaller than the size specified in Table 250.122 for equipment grounding conductors.

Exception No. 3: Electrode-type boilers operating at over 1000 volts shall be grounded as required in 490.72(E)(1) and 490.74.

250.146 Connecting Receptacle Grounding Terminal to an Equipment Grounding Conductor. An equipment bonding jumper shall be used to connect the grounding terminal of a grounding-type receptacle to a metal box that is connected to an equipment grounding conductor, except as permitted in 250.146(A) through (D). The equipment bonding jumper shall be sized in accordance with Table 250.122.

(A) Surface-Mounted Box. If a metal box is mounted on the surface, the direct metal-to-metal contact between the device yoke or strap to the box shall be permitted to provide the required effective ground fault current path. At least one of the insulating washers shall be removed from receptacles that do not have a contact yoke or device to ensure direct metal-to-metal contact. Direct metal-to-metal contact for providing continuity applies to cover-mounted receptacles if the box and cover combination are listed as providing satisfactory continuity between the box and the receptacle. A listed exposed work cover shall be permitted to be the grounding and bonding means under both of the following conditions:

(1) The device is attached to the cover with at least two fasteners that are permanent (such as a rivet) or have a thread locking or screw or nut locking means
(2) The cover mounting holes are located on a flat non-raised portion of the cover

(B) Contact Devices or Yokes. Contact devices or yokes designed and listed as self-grounding shall be permitted in conjunction with the

Grounding and Bonding 250.148

supporting screws to establish equipment bonding between the device yoke and flush-type boxes.

(C) Floor Boxes. Floor boxes designed for and listed as providing satisfactory continuity between the box and the device shall be permitted.

(D) Isolated Ground Receptacles. Where installed for the reduction of electromagnetic interference on the equipment grounding conductor, a receptacle in which the grounding terminal is purposely insulated from the receptacle mounting means shall be permitted. The receptacle grounding terminal shall be connected to an insulated equipment grounding conductor run with the circuit conductors. This equipment grounding conductor shall be permitted to pass through one or more panelboards without a connection to the panelboard grounding terminal bar as permitted in 408.40, Exception, so as to terminate within the same building or structure directly at an equipment grounding conductor terminal of the applicable derived system or service. Where installed in accordance with this section, this equipment grounding conductor shall also be permitted to pass through boxes, wireways, or other enclosures without being connected to such enclosures.

> Informational Note: Use of an isolated equipment grounding conductor does not relieve the requirement for connecting the raceway system and outlet box to an equipment grounding conductor.

250.148 Continuity of Equipment Grounding Conductors and Attachment in Boxes. If circuit conductors are spliced within a box or terminated on equipment within or supported by a box, all wire-type equipment grounding conductor(s) associated with any of those circuit conductors shall be connected within the box or to the box in accordance with 250.8 and 250.148(A) through (D).

Exception: The equipment grounding conductor permitted in 250.146(D) shall not be required to be connected to the other equipment grounding conductors or to the box.

(A) Connections and Splices. Connections and splices shall be made in accordance with 110.14(B) except that insulation shall not be required.

(B) Equipment Grounding Conductor Continuity. The arrangement of grounding connections shall be such that the disconnection or the removal of a luminaire, receptacle, or other device fed from the box does not interrupt the electrical continuity of the equipment grounding conductor(s) providing an effective ground-fault current path.

(C) Metal Boxes. A connection used for no other purpose shall be made between the metal box and the equipment grounding conductor(s) in accordance with 250.8.

(D) Nonmetallic Boxes. One or more equipment grounding conductors brought into a nonmetallic outlet box shall be arranged such that a connection can be made to any fitting or device in that box requiring connection to an equipment grounding conductor.

CHAPTER 12

BOXES AND ENCLOSURES

INTRODUCTION

This chapter contains requirements from Article 314 — Outlet, Device, Pull, and Junction Boxes; Conduit Bodies; Fittings; and Handhole Enclosures. This chapter also includes requirements from Article 312 — Cabinets, Cutout Boxes, and Meter Socket Enclosures. Article 314 requirements cover the installation of all boxes, conduit bodies, and handhole enclosures used as outlet, device, pull, or junction points in a wiring system. This article also includes installation requirements for fittings used to join raceways and to connect raceways (and cables) to boxes and conduit bodies. Article 312 covers the installation of cabinets, cutout boxes, and meter sockets.

Article 314's requirements range from selecting the size and type of box to installing the cover or canopy. Provisions for sizing boxes enclosing 18 through 6 AWG conductors are covered by 314.16. Enclosures (outlet, device, and junction boxes) covered by 314.16 are calculated based on the sizes and numbers of conductors, plus other devices and hardware that occupy space in the box. Requirements on sizing boxes containing large conductors (4 AWG and larger) are located in 314.28. Pull and junction boxes covered by 314.28 are calculated using the size and number of raceways entering an enclosure. Where cables are used, the determination is based on transposing conductor size into the minimum size raceway required. The objective is to provide sufficient space to bend and route conductors within an enclosure without damaging conductor insulation.

ARTICLE 314
Outlet, Device, Pull, and Junction Boxes; Conduit Bodies; Fittings; and Handhole Enclosures

PART I. SCOPE AND GENERAL

314.2 Round Boxes. Round boxes shall not be used where conduits or connectors requiring the use of locknuts or bushings are to be connected to the side of the box.

314.3 Nonmetallic Boxes. Nonmetallic boxes shall be permitted only with open wiring on insulators, concealed knob-and-tube wiring, cabled wiring methods with entirely nonmetallic sheaths, flexible cords, and nonmetallic raceways.

Exception No. 1: Where internal bonding means are provided between all entries, nonmetallic boxes shall be permitted to be used with metal raceways or metal-armored cables.

Exception No. 2: Where integral bonding means with a provision for attaching an equipment bonding jumper inside the box are provided between all threaded entries in nonmetallic boxes listed for the purpose, nonmetallic boxes shall be permitted to be used with metal raceways or metal-armored cables.

314.4 Metal Boxes. Metal boxes shall be grounded and bonded in accordance with Parts I, IV, V, VI, VII, and X of Article 250 as applicable, except as permitted in 250.112(I).

PART II. INSTALLATION

314.15 Damp or Wet Locations. In damp or wet locations, boxes, conduit bodies, outlet box hoods, and fittings shall be placed or equipped so as to prevent moisture from entering or accumulating within the box, conduit body, or fitting. Boxes, conduit bodies, outlet box hoods, and fittings installed in wet locations shall be listed for use in wet locations. Approved drainage openings not smaller than 3 mm (⅛ in.) and not larger than 6 mm (¼ in.) in diameter shall be permitted to be installed in the field in boxes or conduit bodies listed for use in damp or wet locations. For installation of listed drain fittings, larger openings are permitted to be installed in the field in accordance with manufacturer's instructions.

Boxes and Enclosures 314.16

314.16 Number of Conductors in Outlet, Device, and Junction Boxes, and Conduit Bodies. Boxes and conduit bodies shall be of an approved size to provide free space for all enclosed conductors. In no case shall the volume of the box, as calculated in 314.16(A), be less than the fill calculation as calculated in 314.16(B). The minimum volume for conduit bodies shall be as calculated in 314.16(C).

The provisions of this section shall not apply to terminal housings supplied with motors or generators.

(A) Box Volume Calculations. The volume of a wiring enclosure (box) shall be the total volume of the assembled sections and, where used, the space provided by plaster rings, domed covers, extension rings, and so forth, that are marked with their volume or are made from boxes the dimensions of which are listed in Table 314.16(A). Where a box is provided with one or more securely installed barriers, the volume shall be apportioned to each of the resulting spaces. Each barrier, if not marked with its volume, shall be considered to take up 8.2 cm^3 (½ in.^{3}) if metal, and 16.4 cm^3 (1.0 in.^{3}) if nonmetallic.

(1) Standard Boxes. The volumes of standard boxes that are not marked with their volume shall be as given in Table 314.16(A).

(2) Other Boxes. Boxes 1650 cm^3 (100 in.^{3}) or less, other than those described in Table 314.16(A), and nonmetallic boxes shall be durably and legibly marked by the manufacturer with their volume(s). Boxes described in Table 314.16(A) that have a volume larger than is designated in the table shall be permitted to have their volume marked as required by this section.

(B) Box Fill Calculations. The volumes in paragraphs 314.16(B)(1) through (B)(5), as applicable, shall be added together. No allowance shall be required for small fittings such as locknuts and bushings. Each space within a box installed with a barrier shall be calculated separately.

(1) Conductor Fill. Each conductor that originates outside the box and terminates or is spliced within the box shall be counted once, and each conductor that passes through the box without splice or termination shall be counted once. Each loop or coil of unbroken conductor not less than twice the minimum length required for free conductors in 300.14 shall be counted twice. The conductor fill shall be calculated using Table 314.16(B). A conductor, no part of which leaves the box, shall not be counted.

Table 314.16(A) Metal Boxes

Box Trade Size			Minimum Volume		Maximum Number of Conductors* (arranged by AWG size)						
mm	in.		cm³	in.³	18	16	14	12	10	8	6
100 × 32	(4 × 1¼)	round/octagonal	205	12.5	8	7	6	5	5	5	2
100 × 38	(4 × 1½)	round/octagonal	254	15.5	10	8	7	6	6	5	3
100 × 54	(4 × 2⅛)	round/octagonal	353	21.5	14	12	10	9	8	7	4
100 × 32	(4 × 1¼)	square	295	18.0	12	10	9	8	7	6	3
100 × 38	(4 × 1½)	square	344	21.0	14	12	10	9	8	7	4
100 × 54	(4 × 2⅛)	square	497	30.3	20	17	15	13	12	10	6
120 × 32	(4¹¹⁄₁₆ × 1¼)	square	418	25.5	17	14	12	11	10	8	5
120 × 38	(4¹¹⁄₁₆ × 1½)	square	484	29.5	19	16	14	13	11	9	5
120 × 54	(4¹¹⁄₁₆ × 2⅛)	square	689	42.0	28	24	21	18	16	14	8
75 × 50 × 38	(3 × 2 × 1½)	device	123	7.5	5	4	3	3	3	2	1
75 × 50 × 50	(3 × 2 × 2)	device	164	10.0	6	5	5	4	4	3	2
75 × 50 × 57	(3 × 2 × 2¼)	device	172	10.5	7	6	5	4	4	3	2

Box Trade Size			Minimum Volume		Maximum Number of Conductors* (arranged by AWG size)						
mm	in.		cm³	in.³	18	16	14	12	10	8	6
75 × 50 × 65	(3 × 2 × 2½)	device	205	12.5	8	7	6	5	5	4	2
75 × 50 × 70	(3 × 2 × 2¾)	device	230	14.0	9	8	7	6	5	4	2
75 × 50 × 90	(3 × 2 × 3½)	device	295	18.0	12	10	9	8	7	6	3
100 × 54 × 38	(4 × 2⅛ × 1½)	device	169	10.3	6	5	5	4	4	3	2
100 × 54 × 48	(4 × 2⅛ × 1⅞)	device	213	13.0	8	7	6	5	5	4	2
100 × 54 × 54	(4 × 2⅛ × 2⅛)	device	238	14.5	9	8	7	6	5	4	2
95 × 50 × 65	(3¾ × 2 × 2½)	masonry box/gang	230	14.0	9	8	7	6	5	4	2
95 × 50 × 90	(3¾ × 2 × 3½)	masonry box/gang	344	21.0	14	12	10	9	8	7	4
min. 44.5 depth	FS — single cover/gang (1¾)		221	13.5	9	7	6	6	5	4	2
min. 60.3 depth	FD — single cover/gang (2⅜)		295	18.0	12	10	9	8	7	6	3
min. 44.5 depth	FS — multiple cover/gang (1¾)		295	18.0	12	10	9	8	7	6	3
min. 60.3 depth	FD — multiple cover/gang (2⅜)		395	24.0	16	13	12	10	9	8	4

*Where no volume allowances are required by 314.16(B)(2) through (B)(5).

Exception: An equipment grounding conductor or conductors or not over four fixture wires smaller than 14 AWG, or both, shall be permitted to be omitted from the calculations where they enter a box from a domed luminaire or similar canopy and terminate within that box.

(2) Clamp Fill. Where one or more internal cable clamps, whether factory or field supplied, are present in the box, a single volume allowance in accordance with Table 314.16(B) shall be made based on the largest conductor present in the box. No allowance shall be required for a cable connector with its clamping mechanism outside the box.

A clamp assembly that incorporates a cable termination for the cable conductors shall be listed and marked for use with specific nonmetallic boxes. Conductors that originate within the clamp assembly shall be included in conductor fill calculations covered in 314.16(B)(1) as though they entered from outside the box. The clamp assembly shall not require a fill allowance, but the volume of the portion of the assembly that remains within the box after installation shall be excluded from the box volume as marked in 314.16(A)(2).

(3) Support Fittings Fill. Where one or more luminaire studs or hickeys are present in the box, a single volume allowance in accordance with Table 314.16(B) shall be made for each type of fitting based on the largest conductor present in the box.

(4) Device or Equipment Fill. For each yoke or strap containing one or more devices or equipment, a double volume allowance in accordance with Table 314.16(B) shall be made for each yoke or strap based on the largest conductor connected to a device(s) or equipment supported by that yoke or strap. A device or utilization equipment wider than a single

Table 314.16(B) Volume Allowance Required per Conductor

Size of Conductor (AWG)	Free Space Within Box for Each Conductor	
	cm^3	in.^{3}
18	24.6	1.50
16	28.7	1.75
14	32.8	2.00
12	36.9	2.25
10	41.0	2.50
8	49.2	3.00
6	81.9	5.00

Boxes and Enclosures 314.17

50 mm (2 in.) device box as described in Table 314.16(A) shall have double volume allowances provided for each gang required for mounting.

(5) Equipment Grounding Conductor Fill. Where up to four equipment grounding conductors or equipment bonding jumpers enter a box, a single volume allowance in accordance with Table 314.16(B) shall be made based on the largest equipment grounding conductor or equipment bonding jumper entering the box. A ¼ volume allowance shall be made for each additional equipment grounding conductor or equipment bonding jumper that enters the box, based on the largest equipment grounding conductor or equipment bonding conductor.

(C) Conduit Bodies.

(1) General. Conduit bodies enclosing 6 AWG conductors or smaller, other than short-radius conduit bodies as described in 314.16(C)(3), shall have a cross-sectional area not less than twice the cross-sectional area of the largest conduit or tubing to which they can be attached. The maximum number of conductors permitted shall be the maximum number permitted by Table 1 of Chapter 9 for the conduit or tubing to which it is attached.

(2) With Splices, Taps, or Devices. Only those conduit bodies that are durably and legibly marked by the manufacturer with their volume shall be permitted to contain splices, taps, or devices. The maximum number of conductors shall be calculated in accordance with 314.16(B). Conduit bodies shall be supported in a rigid and secure manner.

(3) Short Radius Conduit Bodies. Conduit bodies such as capped elbows and service-entrance elbows that enclose conductors 6 AWG or smaller, and are only intended to enable the installation of the raceway and the contained conductors, shall not contain splices, taps, or devices and shall be of an approved size to provide free space for all conductors enclosed in the conduit body.

314.17 Conductors Entering Boxes, Conduit Bodies, or Fittings. Conductors entering boxes, conduit bodies, or fittings shall be protected from abrasion and shall comply with 314.17(A) through (D).

(A) Openings to Be Closed. Openings through which conductors enter shall be closed in a manner identified for the application.

(B) Boxes and Conduit Bodies. The installation of the conductors in boxes and conduit bodies shall comply with 314.17(B)(1) through (B)(4).

(1) Conductors Entering Through Individual Holes or Through Flexible Tubing. For messenger-supported wiring, open wiring on insulators, or concealed knob-and-tube wiring, the conductors shall enter the box through individual holes. In installations where metal boxes or conduit bodies are used with conductors unprotected by flexible tubing, the individual openings shall be provided with insulating bushings. Where flexible tubing is used to enclose the conductors, the tubing shall extend from the last insulating support to not less than 6 mm (¼ in.) inside the box or conduit body and beyond any cable clamp. The wiring method shall be secured to the box or conduit body.

(2) Conductors Entering Through Cable Clamps. Where cable assemblies with nonmetallic sheaths are used, the sheath shall extend not less than 6 mm (¼ in.) inside the box and beyond any cable clamp. Except as provided in 300.15(C), the wiring method shall be secured to the box or conduit body.

Exception: Where nonmetallic-sheathed cable is used with single gang nonmetallic boxes not larger than a nominal size 57 mm × 100 mm (2¼ in. × 4 in.) mounted in walls or ceilings, and where the cable is fastened within 200 mm (8 in.) of the box measured along the sheath and where the sheath extends through a cable knockout not less than 6 mm (¼ in.), securing the cable to the box shall not be required. Multiple cable entries shall be permitted in a single cable knockout opening.

(3) Conductors Entering Through Raceways. Where the raceway is complete between boxes, conduit bodies, or both and encloses individual conductors or nonmetallic cable assemblies or both, the conductors or cable assemblies shall not be required to be additionally secured. Where raceways enclose cable assemblies as provided in 300.15(C), the cable assembly shall not be required to be additionally secured within the box or conduit body.

(4) Temperature Limitation. Nonmetallic boxes and conduit bodies shall be suitable for the lowest temperature-rated conductor entering the box or conduit body.

(C) Conductors 4 AWG or Larger. Installation shall comply with 300.4(G).

> Informational Note: See 110.12(A) for requirements on closing unused cable and raceway knockout openings.

314.19 Boxes Enclosing Flush Devices or Flush Equipment. Boxes used to enclose flush devices or flush equipment shall be of such design

Boxes and Enclosures

that the devices or equipment will be completely enclosed on the back and sides, and substantial support for the devices or equipment will be provided. Screws for supporting the box shall not also be used to attach a device or equipment.

314.20 Flush-Mounted Installations. Installations within or behind a surface of concrete, tile, gypsum, plaster, or other noncombustible material, including boxes employing a flush-type cover or faceplate, shall be made so that the front edge of the box, plaster ring, extension ring, or listed extender will not be set back of the finished surface more than 6 mm (¼ in.).

Installations within a surface of wood or other combustible surface material, boxes, plaster rings, extension rings, or listed extenders shall extend to the finished surface or project therefrom.

314.21 Repairing Noncombustible Surfaces. Noncombustible surfaces that are broken or incomplete around boxes employing a flush-type cover or faceplate shall be repaired so there will be no gaps or open spaces greater than 3 mm (⅛ in.) at the edge of the box.

314.22 Surface Extensions. Surface extensions shall be made by mounting and mechanically securing an extension ring over the box. Equipment grounding shall be in accordance with Part VI of Article 250.

Exception: A surface extension shall be permitted to be made from the cover of a box where the cover is designed so it is unlikely to fall off or be removed if its securing means becomes loose. The wiring method shall be flexible for an approved length that permits removal of the cover and provides access to the box interior and shall be arranged so that any grounding continuity is independent of the connection between the box and cover.

314.23 Supports. Enclosures within the scope of this article shall be supported in accordance with one or more of the provisions in 314.23(A) through (H).

(A) Surface Mounting. An enclosure mounted on a building or other surface shall be rigidly and securely fastened in place. If the surface does not provide rigid and secure support, additional support in accordance with other provisions of this section shall be provided.

(B) Structural Mounting. An enclosure supported from a structural member or from grade shall be rigidly supported either directly or by using a metal, polymeric, or wood brace.

(1) Nails and Screws. Nails and screws, where used as a fastening means, shall secure boxes by using brackets on the outside of the enclosure, or by using mounting holes in the back or in one or more sides of the enclosure, or they shall pass through the interior within 6 mm (¼ in.) of the back or ends of the enclosure. Screws shall not be permitted to pass through the box unless exposed threads in the box are protected using approved means to avoid abrasion of conductor insulation. Mounting holes made in the field shall be approved.

(2) Braces. Metal braces shall be protected against corrosion and formed from metal that is not less than 0.51 mm (0.020 in.) thick uncoated. Wood braces shall have a cross section not less than nominal 25 mm × 50 mm (1 in. × 2 in.). Wood braces in wet locations shall be treated for the conditions. Polymeric braces shall be identified as being suitable for the use.

(C) Mounting in Finished Surfaces. An enclosure mounted in a finished surface shall be rigidly secured thereto by clamps, anchors, or fittings identified for the application.

(D) Suspended Ceilings. An enclosure mounted to structural or supporting elements of a suspended ceiling shall be not more than 1650 cm^3 (100 in.^{3}) in size and shall be securely fastened in place in accordance with either 314.23(D)(1) or (D)(2).

(1) Framing Members. An enclosure shall be fastened to the framing members by mechanical means such as bolts, screws, or rivets, or by the use of clips or other securing means identified for use with the type of ceiling framing member(s) and enclosure(s) employed. The framing members shall be supported in an approved manner and securely fastened to each other and to the building structure.

(2) Support Wires. The installation shall comply with 300.11(A). The enclosure shall be secured, using identified methods, to ceiling support wire(s), including any additional support wire(s) installed for ceiling support. Support wire(s) used for enclosure support shall be fastened at each end so as to be taut within the ceiling cavity.

(E) Raceway-Supported Enclosure, Without Devices, Luminaires, or Lampholders. An enclosure that does not contain a device(s), other than splicing devices, or supports a luminaire(s), a lampholder, or other equipment and is supported by entering raceways shall not exceed 1650 cm^3 (100 in.^{3}) in size. It shall have threaded entries or identified hubs. It shall be supported by two or more conduits threaded wrenchtight

Boxes and Enclosures 314.23

into the enclosure or hubs. Each conduit shall be secured within 900 mm (3 ft) of the enclosure, or within 450 mm (18 in.) of the enclosure if all conduit entries are on the same side.

Exception: The following wiring methods shall be permitted to support a conduit body of any size, including a conduit body constructed with only one conduit entry, provided that the trade size of the conduit body is not larger than the largest trade size of the conduit or tubing:

(1) Intermediate metal conduit, Type IMC
(2) Rigid metal conduit, Type RMC
(3) Rigid polyvinyl chloride conduit, Type PVC
(4) Reinforced thermosetting resin conduit, Type RTRC
(5) Electrical metallic tubing, Type EMT

(F) Raceway-Supported Enclosures, with Devices, Luminaires, or Lampholders. An enclosure that contains a device(s), other than splicing devices, or supports a luminaire(s), a lampholder, or other equipment and is supported by entering raceways shall not exceed 1650 cm^3 (100 in.^{3}) in size. It shall have threaded entries or identified hubs. It shall be supported by two or more conduits threaded wrenchtight into the enclosure or hubs. Each conduit shall be secured within 450 mm (18 in.) of the enclosure.

Exception No. 1: Rigid metal or intermediate metal conduit shall be permitted to support a conduit body of any size, including a conduit body constructed with only one conduit entry, provided the trade size of the conduit body is not larger than the largest trade size of the conduit.

(G) Enclosures in Concrete or Masonry. An enclosure supported by embedment shall be identified as suitably protected from corrosion and securely embedded in concrete or masonry.

(H) Pendant Boxes. An enclosure supported by a pendant shall comply with 314.23(H)(1) or (H)(2).

(1) Flexible Cord. A box shall be supported from a multiconductor cord or cable in an approved manner that protects the conductors against strain, such as a strain-relief connector threaded into a box with a hub.

(2) Conduit. A box supporting lampholders or luminaires, or wiring enclosures within luminaires used in lieu of boxes in accordance with 300.15(B), shall be supported by rigid or intermediate metal conduit stems. For stems longer than 450 mm (18 in.), the stems shall be connected to the wiring system with listed swivel hangers suitable for the location. At the luminaire end, the conduit(s) shall be threaded wrenchtight into the box, wiring enclosure, or identified hubs.

Where supported by only a single conduit, the threaded joints shall be prevented from loosening by the use of set-screws or other effective means, or the luminaire, at any point, shall be at least 2.5 m (8 ft) above grade or standing area and at least 900 mm (3 ft) measured horizontally to the 2.5 m (8 ft) elevation from windows, doors, porches, fire escapes, or similar locations. A luminaire supported by a single conduit shall not exceed 300 mm (12 in.) in any horizontal direction from the point of conduit entry.

314.24 Depth of Boxes. Outlet and device boxes shall have an approved depth to allow equipment installed within them to be mounted properly and without likelihood of damage to conductors within the box.

(A) Outlet Boxes Without Enclosed Devices or Utilization Equipment. Outlet boxes that do not enclose devices or utilization equipment shall have a minimum internal depth of 12.7 mm (½ in.).

(B) Outlet and Device Boxes with Enclosed Devices or Utilization Equipment. Outlet and device boxes that enclose devices or utilization equipment shall have a minimum internal depth that accommodates the rearward projection of the equipment and the size of the conductors that supply the equipment. The internal depth shall include, where used, that of any extension boxes, plaster rings, or raised covers. The internal depth shall comply with all applicable provisions of 314.24(B)(1) through (B)(5).

(1) Large Equipment. Boxes that enclose devices or utilization equipment that projects more than 48 mm (1⅞ in.) rearward from the mounting plane of the box shall have a depth that is not less than the depth of the equipment plus 6 mm (¼ in.).

(2) Conductors Larger Than 4 AWG. Boxes that enclose devices or utilization equipment supplied by conductors larger than 4 AWG shall be identified for their specific function.

Exception to (2): Devices or utilization equipment supplied by conductors larger than 4 AWG shall be permitted to be mounted on or in junction and pull boxes larger than 1650 cm³ (100 in.³) if the spacing at the terminals meets the requirements of 312.6.

(3) Conductors 8, 6, or 4 AWG. Boxes that enclose devices or utilization equipment supplied by 8, 6, or 4 AWG conductors shall have an internal depth that is not less than 52.4 mm (2 1/16 in.).

(4) Conductors 12 or 10 AWG. Boxes that enclose devices or utilization equipment supplied by 12 or 10 AWG conductors shall have an internal depth that is not less than 30.2 mm ($1^{3}/_{16}$ in.). Where the equipment projects rearward from the mounting plane of the box by more than 25 mm (1 in.), the box shall have a depth not less than that of the equipment plus 6 mm ($1/_{4}$ in.).

(5) Conductors 14 AWG and Smaller. Boxes that enclose devices or utilization equipment supplied by 14 AWG or smaller conductors shall have a depth that is not less than 23.8 mm ($15/_{16}$ in.).

Exception to (1) through (5): Devices or utilization equipment that is listed to be installed with specified boxes shall be permitted.

314.25 Covers and Canopies. In completed installations, each box shall have a cover, faceplate, lampholder, or luminaire canopy, except where the installation complies with 410.24(B). Screws used for the purpose of attaching covers, or other equipment, to the box shall be either machine screws matching the thread gauge and size that is integral to the box or shall be in accordance with the manufacturer's instructions.

(A) Nonmetallic or Metal Covers and Plates. Nonmetallic or metal covers and plates shall be permitted. Where metal covers or plates are used, they shall comply with the grounding requirements of 250.110.

Informational Note: For additional grounding requirements, see 410.42 for metal luminaire canopies, and 404.12 and 406.6(B) for metal faceplates.

(B) Exposed Combustible Wall or Ceiling Finish. Where a luminaire canopy or pan is used, any combustible wall or ceiling finish exposed between the edge of the canopy or pan and the outlet box shall be covered with noncombustible material if required by 410.23.

(C) Flexible Cord Pendants. Covers of outlet boxes and conduit bodies having holes through which flexible cord pendants pass shall be provided with identified bushings or shall have smooth, well-rounded surfaces on which the cords may bear. So-called hard rubber or composition bushings shall not be used.

314.27 Outlet Boxes.

(A) Boxes at Luminaire or Lampholder Outlets. Outlet boxes or fittings designed for the support of luminaires and lampholders, and installed as required by 314.23, shall be permitted to support a luminaire or lampholder.

(1) Vertical Surface Outlets. Boxes used at luminaire or lampholder outlets in or on a vertical surface shall be identified and marked on the interior of the box to indicate the maximum weight of the luminaire that is permitted to be supported by the box if other than 23 kg (50 lb).

Exception: A vertically mounted luminaire or lampholder weighing not more than 3 kg (6 lb) shall be permitted to be supported on other boxes or plaster rings that are secured to other boxes, provided that the luminaire or its supporting yoke, or the lampholder, is secured to the box with no fewer than two No. 6 or larger screws.

(2) Ceiling Outlets. At every outlet used exclusively for lighting, the box shall be designed or installed so that a luminaire or lampholder may be attached. Boxes shall be required to support a luminaire weighing a minimum of 23 kg (50 lb). A luminaire that weighs more than 23 kg (50 lb) shall be supported independently of the outlet box, unless the outlet box is listed for not less than the weight to be supported. The interior of the box shall be marked by the manufacturer to indicate the maximum weight the box shall be permitted to support.

(B) Floor Boxes. Boxes listed specifically for this application shall be used for receptacles located in the floor.

Exception: Where the authority having jurisdiction judges them free from likely exposure to physical damage, moisture, and dirt, boxes located in elevated floors of show windows and similar locations shall be permitted to be other than those listed for floor applications. Receptacles and covers shall be listed as an assembly for this type of location.

(C) Boxes at Ceiling-Suspended (Paddle) Fan Outlets. Outlet boxes or outlet box systems used as the sole support of a ceiling-suspended (paddle) fan shall be listed, shall be marked by their manufacturer as suitable for this purpose, and shall not support ceiling-suspended (paddle) fans that weigh more than 32 kg (70 lb). For outlet boxes or outlet box systems designed to support ceiling-suspended (paddle) fans that weigh more than 16 kg (35 lb), the required marking shall include the maximum weight to be supported.

Outlet boxes mounted in the ceilings of habitable rooms of dwelling occupancies in a location acceptable for the installation of a ceiling-suspended (paddle) fan shall comply with one of the following:

(1) Listed for the sole support of ceiling-suspended (paddle) fans
(2) An outlet box complying with the applicable requirements of 314.27 and providing access to structural framing capable of supporting of a ceiling-suspended (paddle) fan bracket or equivalent

Boxes and Enclosures   314.28

314.28 Pull and Junction Boxes and Conduit Bodies. Boxes and conduit bodies used as pull or junction boxes shall comply with 314.28(A) through (E).

(A) Minimum Size. For raceways containing conductors of 4 AWG or larger that are required to be insulated, and for cables containing conductors of 4 AWG or larger, the minimum dimensions of pull or junction boxes installed in a raceway or cable run shall comply with 314.28(A)(1) through (A)(3). Where an enclosure dimension is to be calculated based on the diameter of entering raceways, the diameter shall be the metric designator (trade size) expressed in the units of measurement employed.

(1) Straight Pulls. In straight pulls, the length of the box or conduit body shall not be less than eight times the metric designator (trade size) of the largest raceway.

(2) Angle or U Pulls, or Splices. Where splices or where angle or U pulls are made, the distance between each raceway entry inside the box or conduit body and the opposite wall of the box or conduit body shall not be less than six times the metric designator (trade size) of the largest raceway in a row. This distance shall be increased for additional entries by the amount of the sum of the diameters of all other raceway entries in the same row on the same wall of the box. Each row shall be calculated individually, and the single row that provides the maximum distance shall be used.

Exception: Where a raceway or cable entry is in the wall of a box or conduit body opposite a removable cover, the distance from that wall to the cover shall be permitted to comply with the distance required for one wire per terminal in Table 312.6(A).

The distance between raceway entries enclosing the same conductor shall not be less than six times the metric designator (trade size) of the larger raceway.

When transposing cable size into raceway size in 314.28(A)(1) and (A)(2), the minimum metric designator (trade size) raceway required for the number and size of conductors in the cable shall be used.

(3) Smaller Dimensions. Listed boxes or listed conduit bodies of dimensions less than those required in 314.28(A)(1) and (A)(2) shall be permitted for installations of combinations of conductors that are less than the maximum conduit or tubing fill (of conduits or tubing being used) permitted by Table 1 of Chapter 9.

Listed conduit bodies of dimensions less than those required in 314.28(A)(2), and having a radius of the curve to the centerline not less than that indicated in Table 2 of Chapter 9 for one-shot and full-shoe benders, shall be permitted for installations of combinations of conductors permitted by Table 1 of Chapter 9. These conduit bodies shall be marked to show they have been specifically evaluated in accordance with this provision.

Where the permitted combinations of conductors for which the box or conduit body has been listed are less than the maximum conduit or tubing fill permitted by Table 1 of Chapter 9, the box or conduit body shall be permanently marked with the maximum number and maximum size of conductors permitted. For other conductor sizes and combinations, the total cross-sectional area of the fill shall not exceed the cross-sectional area of the conductors specified in the marking, based on the type of conductor identified as part of the product listing.

> Informational Note: Unless otherwise specified, the applicable product standards evaluate the fill markings covered here based on conductors with Type XHHW insulation.

314.29 Boxes, Conduit Bodies, and Handhole Enclosures to Be Accessible. Boxes, conduit bodies, and handhole enclosures shall be installed so that wiring contained in them can be rendered accessible in accordance with 314.29(A) and (B).

(A) In Buildings and Other Structures. Boxes and conduit bodies shall be installed so the contained wiring can be accessed without removing any part of the building or structure.

(B) Underground. Underground boxes and handhole enclosures shall be installed so they are accessible without excavating sidewalks, paving, earth, or other substance that is to be used to establish the finished grade.

Exception: Listed boxes and handhole enclosures shall be permitted where covered by gravel, light aggregate, or noncohesive granulated soil if their location is effectively identified and accessible for excavation.

314.30 Handhole Enclosures. Handhole enclosures shall be designed and installed to withstand all loads likely to be imposed on them. They shall be identified for use in underground systems.

> Informational Note: See ANSI/SCTE 77-2013, *Specification for Underground Enclosure Integrity*, for additional information on deliberate and nondeliberate traffic loading that can be expected to bear on underground enclosures.

Boxes and Enclosures 312.3

(A) Size. Handhole enclosures shall be sized in accordance with 314.28(A) for conductors operating at 1000 volts or below, and in accordance with 314.71 for conductors operating at over 1000 volts. For handhole enclosures without bottoms where the provisions of 314.28(A)(2), Exception, or 314.71(B)(1), Exception No. 1, apply, the measurement to the removable cover shall be taken from the end of the conduit or cable assembly.

(B) Wiring Entries. Underground raceways and cable assemblies entering a handhole enclosure shall extend into the enclosure, but they shall not be required to be mechanically connected to the enclosure.

(C) Enclosed Wiring. All enclosed conductors and any splices or terminations, if present, shall be listed as suitable for wet locations.

(D) Covers. Handhole enclosure covers shall have an identifying mark or logo that prominently identifies the function of the enclosure, such as "electric." Handhole enclosure covers shall require the use of tools to open, or they shall weigh over 45 kg (100 lb). Metal covers and other exposed conductive surfaces shall be bonded in accordance with 250.92 if the conductors in the handhole are service conductors, or in accordance with 250.96(A) if the conductors in the handhole are feeder or branch-circuit conductors.

ARTICLE 312
Cabinets, Cutout Boxes, and Meter Socket Enclosures

PART I. SCOPE AND INSTALLATION

312.2 Damp and Wet Locations. In damp or wet locations, surface-type enclosures within the scope of this article shall be placed or equipped so as to prevent moisture or water from entering and accumulating within the cabinet or cutout box, and shall be mounted so there is at least 6-mm (¼-in.) airspace between the enclosure and the wall or other supporting surface. Enclosures installed in wet locations shall be weatherproof. For enclosures in wet locations, raceways or cables entering above the level of uninsulated live parts shall use fittings listed for wet locations.

312.3 Position in Wall. In walls of concrete, tile, or other noncombustible material, cabinets shall be installed so that the front edge of the cabinet is not set back of the finished surface more than 6 mm (¼ in.). In walls constructed of wood or other combustible material, cabinets shall be flush with the finished surface or project therefrom.

312.4 Repairing Noncombustible Surfaces. Noncombustible surfaces that are broken or incomplete shall be repaired so there will be no gaps or open spaces greater than 3 mm (⅛ in.) at the edge of the cabinet or cutout box employing a flush-type cover.

312.5 Cabinets, Cutout Boxes, and Meter Socket Enclosures. Conductors entering enclosures within the scope of this article shall be protected from abrasion and shall comply with 312.5(A) through (C).

(A) Openings to Be Closed. Openings through which conductors enter shall be closed in an approved manner.

(B) Metal Cabinets, Cutout Boxes, and Meter Socket Enclosures. Where metal enclosures within the scope of this article are installed with messenger-supported wiring, open wiring on insulators, or concealed knob-and-tube wiring, conductors shall enter through insulating bushings or, in dry locations, through flexible tubing extending from the last insulating support and firmly secured to the enclosure.

(C) Cables. Where cable is used, each cable shall be secured to the cabinet, cutout box, or meter socket enclosure.

Exception No. 1: Cables with entirely nonmetallic sheaths shall be permitted to enter the top of a surface-mounted enclosure through one or more nonflexible raceways not less than 450 mm (18 in.) and not more than 3.0 m (10 ft) in length, provided all of the following conditions are met:

(1) Each cable is fastened within 300 mm (12 in.), measured along the sheath, of the outer end of the raceway.

(2) The raceway extends directly above the enclosure and does not penetrate a structural ceiling.

(3) A fitting is provided on each end of the raceway to protect the cable(s) from abrasion and the fittings remain accessible after installation.

(4) The raceway is sealed or plugged at the outer end using approved means so as to prevent access to the enclosure through the raceway.

(5) The cable sheath is continuous through the raceway and extends into the enclosure beyond the fitting not less than 6 mm (¼ in.).

(6) The raceway is fastened at its outer end and at other points in accordance with the applicable article.

Boxes and Enclosures 312.5

(7) Where installed as conduit or tubing, the cable fill does not exceed the amount that would be permitted for complete conduit or tubing systems by Table 1 of Chapter 9 of this Code and all applicable notes thereto. Note 2 to the tables in Chapter 9 does not apply to this condition.

CHAPTER 13

CABLES

INTRODUCTION

This chapter contains requirements from Article 334 — Nonmetallic-Sheathed Cable: Types NM, NMC, and NMS; Article 330 — Metal Clad Cable: Type MC; Article 336 — Power and Control Tray Cable: Type TC; Article 338 — Service Entrance Cable: Types SE and USE; and Article 340 — Underground Feeder and Branch-Circuit Cable: Type UF. Contained in each article are requirements pertaining to the use and installation for the cable listed. Part I of the respective cable articles contains the definition for the respective cable types plus the requirement for the wiring method to be *listed* (see definition of *listed* in Article 100), and Part II provides the installation requirements. Some of the topics covered in Part II include uses permitted, uses not permitted, bending radius, securing and supporting, fittings, and ampacity. Each of these articles also contains construction requirements, but such requirements are outside the purview of this *Pocket Guide*.

While service-entrance cable used as service-entrance conductors must be installed as required by Article 230, where used for interior wiring it must comply with many of the installation requirements in Article 334. Underground feeder and branch-circuit cable installed as nonmetallic-sheathed cable must also comply with the provisions of Article 334.

Although the specifications in each article may be different, the arrangement and numbering in each article follow the same format. For example, the uses permitted for each cable are in 3__.10, and the uses not permitted are in 3__.12.

ARTICLE 334
Nonmetallic-Sheathed Cable: Types NM and NMC

PART I. GENERAL

334.2 Definitions. The definitions in this section shall apply within this article and throughout the *Code*.

Nonmetallic-Sheathed Cable. A factory assembly of two or more insulated conductors enclosed within an overall nonmetallic jacket.

Type NM. Insulated conductors enclosed within an overall nonmetallic jacket.

Type NMC. Insulated conductors enclosed within an overall, corrosion resistant, nonmetallic jacket.

PART II. INSTALLATION

334.10 Uses Permitted. Type NM and Type NMC cables shall be permitted to be used in the following, except as prohibited in 334.12:

(1) One- and two-family dwellings and their attached or detached garages, and their storage buildings.
(2) Multi-family dwellings permitted to be of Types III, IV, and V construction.

(A) Type NM. Type NM cable shall be permitted as follows:

(1) For both exposed and concealed work in normally dry locations except as prohibited in 334.10(3)
(2) To be installed or fished in air voids in masonry block or tile walls

(B) Type NMC. Type NMC cable shall be permitted as follows:

(1) For both exposed and concealed work in dry, moist, damp, or corrosive locations, except as prohibited by 334.10(3)
(2) In outside and inside walls of masonry block or tile
(3) In a shallow chase in masonry, concrete, or adobe protected against nails or screws by a steel plate at least 1.59 mm (1/16 in.) thick and covered with plaster, adobe, or similar finish

334.12 Uses Not Permitted.

(A) Types NM and NMC. Types NM and NMC cables shall not be permitted as follows:

Cables 334.15

(1) In any dwelling or structure not specifically permitted in 334.10(1), (2), (3), and (5)
(2) Exposed within a dropped or suspended ceiling cavity in other than one- and two-family and multifamily dwellings
(3) As service-entrance cable
(4) In commercial garages having hazardous (classified) locations as defined in 511.3
(5) In theaters and similar locations, except where permitted in 518.4(B)
(6) In motion picture studios
(7) In storage battery rooms
(8) In hoistways or on elevators or escalators
(9) Embedded in poured cement, concrete, or aggregate
(10) In hazardous (classified) locations, except where specifically permitted by other articles in this *Code*

(B) Type NM. Type NM cables shall not be used under the following conditions or in the following locations:

(1) Where exposed to corrosive fumes or vapors
(2) Where embedded in masonry, concrete, adobe, fill, or plaster
(3) In a shallow chase in masonry, concrete, or adobe and covered with plaster, adobe, or similar finish
(4) In wet or damp locations

334.15 Exposed Work. In exposed work, except as provided in 300.11(B), cable shall be installed as specified in 334.15(A) through (C).

(A) To Follow Surface. Cable shall closely follow the surface of the building finish or of running boards.

(B) Protection from Physical Damage. Cable shall be protected from physical damage where necessary by rigid metal conduit, intermediate metal conduit, electrical metallic tubing, Schedule 80 PVC conduit, Type RTRC marked with the suffix -XW, or other approved means. Where passing through a floor, the cable shall be enclosed in rigid metal conduit, intermediate metal conduit, electrical metallic tubing, Schedule 80 PVC conduit, Type RTRC marked with the suffix -XW, or other approved means extending at least 150 mm (6 in.) above the floor.

Type NMC cable installed in shallow chases or grooves in masonry, concrete, or adobe shall be protected in accordance with the requirements in 300.4(F) and covered with plaster, adobe, or similar finish.

(C) In Unfinished Basements and Crawl Spaces. Where cable is run at angles with joists in unfinished basements and crawl spaces, it shall be permissible to secure cables not smaller than two 6 AWG or three 8 AWG conductors directly to the lower edges of the joists. Smaller cables shall be run either through bored holes in joists or on running boards. Nonmetallic-sheathed cable installed on the wall of an unfinished basement shall be permitted to be installed in a listed conduit or tubing or shall be protected in accordance with 300.4. Conduit or tubing shall be provided with a suitable insulating bushing or adapter at the point the cable enters the raceway. The sheath of the nonmetallic-sheathed cable shall extend through the conduit or tubing and into the outlet or device box not less than 6 mm (1/4 in.). The cable shall be secured within 300 mm (12 in.) of the point where the cable enters the conduit or tubing. Metal conduit, tubing, and metal outlet boxes shall be connected to an equipment grounding conductor complying with the provisions of 250.86 and 250.148.

334.17 Through or Parallel to Framing Members. Types NM and NMC cable shall be protected in accordance with 300.4 where installed through or parallel to framing members. Grommets used as required in 300.4(B)(1) shall remain in place and be listed for the purpose of cable protection.

334.23 In Accessible Attics. The installation of cable in accessible attics or roof spaces shall also comply with 320.23.

334.24 Bending Radius. Bends in Types NM and NMC cable shall be so made that the cable will not be damaged. The radius of the curve of the inner edge of any bend during or after installation shall not be less than five times the diameter of the cable.

334.30 Securing and Supporting. Nonmetallic-sheathed cable shall be supported and secured by staples, cable ties listed and identified for securement and support, or straps, hangers, or similar fittings designed and installed so as not to damage the cable, at intervals not exceeding 1.4 m (4 1/2 ft) and within 300 mm (12 in.) of every cable entry into enclosures such as outlet boxes, junction boxes, cabinets, or fittings. The cable length between the cable entry and the closest cable support shall not exceed 450 mm (18 in.). Flat cables shall not be stapled on edge.

Sections of cable protected from physical damage by raceway shall not be required to be secured within the raceway.

(A) Horizontal Runs Through Holes and Notches. In other than vertical runs, cables installed in accordance with 300.4 shall be considered to be supported and secured where such support does not exceed 1.4-m (4½-ft) intervals and the nonmetallic-sheathed cable is securely fastened in place by an approved means within 300 mm (12 in.) of each box, cabinet, conduit body, or other nonmetallic-sheathed cable termination.

Informational Note: See 314.17(B)(1) for support where nonmetallic boxes are used.

(B) Unsupported Cables. Nonmetallic-sheathed cable shall be permitted to be unsupported where the cable:

(1) Is fished between access points through concealed spaces in finished buildings or structures and supporting is impracticable.
(2) Is not more than 1.4 m (4½ ft) from the last point of cable support to the point of connection to a luminaire or other piece of electrical equipment and the cable and point of connection are within an accessible ceiling in one-, two-, or multifamily dwellings.

(C) Wiring Device Without a Separate Outlet Box. A wiring device identified for the use, without a separate outlet box, and incorporating an integral cable clamp shall be permitted where the cable is secured in place at intervals not exceeding 1.4 m (4½ ft) and within 300 mm (12 in.) from the wiring device wall opening, and there shall be at least a 300 mm (12 in.) loop of unbroken cable or 150 mm (6 in.) of a cable end available on the interior side of the finished wall to permit replacement.

334.40 Boxes and Fittings.

(A) Boxes of Insulating Material. Nonmetallic outlet boxes shall be permitted as provided by 314.3.

(B) Devices of Insulating Material. Self-contained switches, self-contained receptacles, and nonmetallic-sheathed cable interconnector devices of insulating material that are listed shall be permitted to be used without boxes in exposed cable wiring and for repair wiring in existing buildings where the cable is concealed. Openings in such devices shall form a close fit around the outer covering of the cable, and the device shall fully enclose the part of the cable from which any part of the covering has been removed. Where connections to conductors are by binding-screw terminals, there shall be available as many terminals as conductors.

(C) Devices with Integral Enclosures. Wiring devices with integral enclosures identified for such use shall be permitted as provided by 300.15(E).

334.80 Ampacity.
The ampacity of Types NM and NMC cable shall be determined in accordance with 310.14. The ampacity shall not exceed

that of a 60°C (140°F) rated conductor. The 90°C (194°F) rating shall be permitted to be used for ampacity adjustment and correction calculations, provided the final calculated ampacity does not exceed that of a 60°C (140°F) rated conductor. The ampacity of Types NM and NMC cable installed in cable trays shall be determined in accordance with 392.80(A).

Where more than two NM cables containing two or more current-carrying conductors are installed, without maintaining spacing between the cables, through the same opening in wood framing that is to be sealed with thermal insulation, caulk, or sealing foam, the ampacity of each conductor shall be adjusted in accordance with Table 310.15(C)(1) and the provisions of 310.14(A)(2), Exception, shall not apply.

Where more than two NM cables containing two or more current-carrying conductors are installed in contact with thermal insulation without maintaining spacing between cables, the ampacity of each conductor shall be adjusted in accordance with Table 310.15(C)(1).

ARTICLE 330
Metal-Clad Cable: Type MC

PART I. GENERAL

330.2 Definition. The definition in this section shall apply within this article and throughout the *Code*.

Metal Clad Cable, Type MC. A factory assembly of one or more insulated circuit conductors with or without optical fiber members enclosed in an armor of interlocking metal tape, or a smooth or corrugated metallic sheath.

PART II. INSTALLATION

330.10 Uses Permitted.

(A) General Uses. Type MC cable shall be permitted as follows:

(1) For services, feeders, and branch circuits.
(2) For power, lighting, control, and signal circuits.
(3) Indoors or outdoors.
(4) Exposed or concealed.
(5) To be direct buried where identified for such use.
(7) In any raceway.
(8) As aerial cable on a messenger.

Cables 330.15

(10) In dry locations and embedded in plaster finish on brick or other masonry except in damp or wet locations.
(11) In wet locations where a corrosion-resistant jacket is provided over the metallic covering and any of the following conditions are met:

   a. The metallic covering is impervious to moisture.
   b. A jacket resistant to moisture is provided under the metal covering.
   c. The insulated conductors under the metallic covering are listed for use in wet locations.

(12) Where single-conductor cables are used, all phase conductors and, where used, the grounded conductor shall be grouped together to minimize induced voltage on the sheath.

(B) Specific Uses. Type MC cable shall be permitted to be installed in compliance with Parts II and III of Article 725 and 770.133 as applicable and in accordance with 330.10(B)(1) through (B)(4).

Informational Note: The "Uses Permitted" is not an all-inclusive list.

(2) Direct Buried. Direct-buried cable shall comply with 300.5 or 300.50, as appropriate.

(3) Installed as Service-Entrance Cable. Type MC cable installed as service-entrance cable shall be permitted in accordance with 230.43.

(4) Installed Outside of Buildings or Structures or as Aerial Cable. Type MC cable installed outside of buildings or structures or as aerial cable shall comply with 225.10, 396.10, and 396.12.

330.12 Uses Not Permitted. Type MC cable shall not be used under either of the following conditions:

(1) Where subject to physical damage
(2) Where exposed to any of the destructive corrosive conditions in (a) or (b), unless the metallic sheath or armor is resistant to the conditions or is protected by material resistant to the conditions:

   a. Direct buried in the earth or embedded in concrete unless identified for direct burial
   b. Exposed to cinder fills, strong chlorides, caustic alkalis, or vapors of chlorine or of hydrochloric acids

330.15 Exposed Work. Exposed runs of cable, except as provided in 300.11(B), shall closely follow the surface of the building finish or of running boards. Exposed runs shall also be permitted to be installed on the underside of joists where supported at each joist and located so as not to be subject to physical damage.

330.17 Through or Parallel to Framing Members. Type MC cable shall be protected in accordance with 300.4(A), (C), and (D) where installed through or parallel to framing members.

330.23 In Accessible Attics. The installation of Type MC cable in accessible attics or roof spaces shall also comply with 320.23.

330.24 Bending Radius. Bends in Type MC cable shall be so made that the cable will not be damaged. The radius of the curve of the inner edge of any bend shall not be less than required in 330.24(A) through (C).

(A) Smooth Sheath.

(1) Ten times the external diameter of the metallic sheath for cable not more than 19 mm (¾ in.) in external diameter
(2) Twelve times the external diameter of the metallic sheath for cable more than 19 mm (¾ in.) but not more than 38 mm (1½ in.) in external diameter
(3) Fifteen times the external diameter of the metallic sheath for cable more than 38 mm (1½ in.) in external diameter

(B) Interlocked-Type Armor or Corrugated Sheath. Seven times the external diameter of the metallic sheath.

(C) Shielded Conductors. Twelve times the overall diameter of one of the individual conductors or seven times the overall diameter of the multiconductor cable, whichever is greater.

330.30 Securing and Supporting.

(A) General. Type MC cable shall be supported and secured by staples; cable ties listed and identified for securement and support; straps, hangers, or similar fittings; or other approved means designed and installed so as not to damage the cable.

(B) Securing. Unless otherwise provided, cables shall be secured at intervals not exceeding 1.8 m (6 ft). Cables containing four or fewer conductors sized no larger than 10 AWG shall be secured within 300 mm (12 in.) of every box, cabinet, fitting, or other cable termination. In vertical installations, listed cables with ungrounded conductors 250 kcmil and larger shall be permitted to be secured at intervals not exceeding 3 m (10 ft).

Cables 336.2

(C) Supporting. Unless otherwise provided, cables shall be supported at intervals not exceeding 1.8 m (6 ft).

Horizontal runs of Type MC cable installed in wooden or metal framing members or similar supporting means shall be considered supported and secured where such support does not exceed 1.8-m (6-ft) intervals.

(D) Unsupported Cables. Type MC cable shall be permitted to be unsupported and unsecured where the cable complies with any of the following:

(1) Is fished between access points through concealed spaces in finished buildings or structures and supporting is impractical.
(2) Is not more than 1.8 m (6 ft) in length from the last point of cable support to the point of connection to luminaires or other electrical equipment and the cable and point of connection are within an accessible ceiling.
(3) Is Type MC of the interlocked armor type in lengths not exceeding 900 mm (3 ft) from the last point where it is securely fastened and is used to connect equipment where flexibility is necessary to minimize the transmission of vibration from equipment or to provide flexibility for equipment that requires movement after installation.

330.80 Ampacity. The ampacity of Type MC cable shall be determined in accordance with 310.14 or 311.60 for 14 AWG and larger conductors and in accordance with Table 402.5 for 18 AWG and 16 AWG conductors. The installation shall not exceed the temperature ratings of terminations and equipment.

ARTICLE 336
Power and Control Tray Cable: Type TC

PART I. GENERAL

336.2 Definition. The definition in this section shall apply within this article and throughout this *Code*.

Power and Control Tray Cable, Type TC. A factory assembly of two or more insulated conductors, with or without associated bare or covered equipment grounding conductors, under a nonmetallic jacket.

PART II. INSTALLATION

336.10 Uses Permitted. Type TC cable shall be permitted to be used as follows:

(8) Type TC cable shall be resistant to moisture and corrosive agents where installed in wet locations.
(9) In one- and two-family dwelling units, Type TC-ER-JP cable containing both power and control conductors shall be permitted for branch circuits and feeders. Type TC-ER-JP cable used as interior wiring shall be installed per the requirements of Part II of Article 334 and where installed as exterior wiring shall be installed per the requirements of Part II of Article 340.

Exception: Where used to connect a generator and associated equipment having terminals rated 75°C (140°F) or higher, the cable shall not be limited in ampacity by 334.80 or 340.80.

Informational Note: See 725.136 for limitations on Class 2 or 3 circuits contained within the same cable with conductors of electric light, power, or Class 1 circuits.

336.12 Uses Not Permitted. Type TC tray cable shall not be installed or used as follows:

(1) Installed where it will be exposed to physical damage
(3) Used where exposed to direct rays of the sun, unless identified as sunlight resistant

336.24 Bending Radius. Bends in Type TC cable shall be made so as not to damage the cable. For Type TC cable without metal shielding, the minimum bending radius shall be as follows:

(1) Four times the overall diameter for cables 25 mm (1 in.) or less in diameter
(2) Five times the overall diameter for cables larger than 25 mm (1 in.) but not more than 50 mm (2 in.) in diameter
(3) Six times the overall diameter for cables larger than 50 mm (2 in.) in diameter

Type TC cables with metallic shielding shall have a minimum bending radius of not less than 12 times the cable overall diameter.

336.80 Ampacity. The ampacity of Type TC tray cable shall be determined in accordance with 392.80(A) for 14 AWG and larger conductors, in accordance with 402.5 for 18 AWG through 16 AWG conductors where installed in cable trays, and in accordance with 310.14 where installed outside of cable trays, where permitted.

ARTICLE 338
Service-Entrance Cable: Types SE and USE

PART I. GENERAL

338.2 Definitions. The definition in this section shall apply within this article and throughout the *Code*.

Service-Entrance Cable. A single conductor or multiconductor cable provided with an overall covering, primarily used for services, and of the following types:

Type SE. Service-entrance cable having a flame-retardant, moisture-resistant covering.

Type USE. Service-entrance cable, identified for underground use, having a moisture-resistant covering, but not required to have a flame-retardant covering.

Service-Entrance Conductor Assembly. Multiple single-insulated conductors twisted together without an overall covering, other than an optional binder intended only to keep the conductors together.

338.6 Listing Requirements. Type SE and USE cables and associated fittings shall be listed.

PART II. INSTALLATION

338.10 Uses Permitted.

(A) Service-Entrance Conductors. Service-entrance cable shall be permitted to be used as service-entrance conductors and shall be installed in accordance with 230.6, 230.7, and Parts II, III, and IV of Article 230.

(B) Branch Circuits or Feeders.

(1) Grounded Conductor Insulated. Type SE service-entrance cables shall be permitted in wiring systems where all of the circuit conductors of the cable are of the thermoset or thermoplastic type.

(2) Use of Uninsulated Conductor. Type SE service-entrance cable shall be permitted for use where the insulated conductors are used for circuit wiring and the uninsulated conductor is used only for equipment grounding purposes.

Exception: In existing installations, uninsulated conductors shall be permitted as a grounded conductor in accordance with 250.32 and 250.140, where the uninsulated grounded conductor of the cable originates in service equipment, and with 225.30 through 225.40.

(3) Temperature Limitations. Type SE service-entrance cable used to supply appliances shall not be subject to conductor temperatures in excess of the temperature specified for the type of insulation involved.

(4) Installation Methods for Branch Circuits and Feeders.

   (a) *Interior Installations.*

(1) In addition to the provisions of this article, Type SE service-entrance cable used for interior wiring shall comply with the installation requirements of Part II of Article 334, excluding 334.80.
(2) Where more than two Type SE cables containing two or more current-carrying conductors in each cable are installed in contact with thermal insulation, caulk, or sealing foam without maintaining spacing between cables, the ampacity of each conductor shall be adjusted in accordance with Table 310.15(C)(1).
(3) For Type SE cable with ungrounded conductor sizes 10 AWG and smaller, where installed in contact with thermal insulation, the ampacity shall be in accordance with 60°C (140°F) conductor temperature rating. The maximum conductor temperature rating shall be permitted to be used for ampacity adjustment and correction purposes, if the final ampacity does not exceed that for a 60°C (140°F) rated conductor.

   (b) *Exterior Installations.*

(1) In addition to the provisions of this article, service-entrance cable used for feeders or branch circuits, where installed as exterior wiring, shall be installed in accordance with Part I of Article 225. The cable shall be supported in accordance with 334.30.
(2) Type USE cable installed as underground feeder and branch circuit cable shall comply with Part II of Article 340.

Exception: Single-conductor Type USE and multi-rated USE conductors shall not be subject to the ampacity limitations of Part II of Article 340.

338.12 Uses Not Permitted.

(A) Service-Entrance Cable. Service-entrance cable (SE) shall not be used under the following conditions or in the following locations:

(1) Where subject to physical damage unless protected in accordance with 230.50(B)
(2) Underground with or without a raceway
(3) For exterior branch circuits and feeder wiring unless the installation complies with the provisions of Part I of Article 225 and is supported in accordance with 334.30 or is used as messenger-supported wiring as permitted in Part II of Article 396

(B) Underground Service-Entrance Cable. Underground service-entrance cable (USE) shall not be used under the following conditions or in the following locations:

(1) For interior wiring
(2) For aboveground installations except where USE cable emerges from the ground and is terminated in an enclosure at an outdoor location and the cable is protected in accordance with 300.5(D)
(3) As aerial cable unless it is a multiconductor cable identified for use aboveground and installed as messenger-supported wiring in accordance with 225.10 and Part II of Article 396

338.24 Bending Radius. Bends in Types USE and SE cable shall be so made that the cable will not be damaged. The radius of the curve of the inner edge of any bend, during or after installation, shall not be less than five times the diameter of the cable.

ARTICLE 340
Underground Feeder and Branch-Circuit Cable: Type UF

PART I. GENERAL

340.2 Definition. The definition in this section shall apply within this article and throughout the *Code*.

Underground Feeder and Branch-Circuit Cable, Type UF. A factory assembly of one or more insulated conductors with an integral or an overall covering of nonmetallic material suitable for direct burial in the earth.

PART II. INSTALLATION

340.10 Uses Permitted. Type UF cable shall be permitted as follows:

(1) For use underground, including direct burial in the earth.
(2) As single-conductor cables. Where installed as single-conductor cables, all conductors of the feeder or branch circuit, including the grounded conductor and equipment grounding conductor, if any, shall be installed in accordance with 300.3.
(3) For wiring in wet, dry, or corrosive locations.
(4) Installed as nonmetallic-sheathed cable. Where so installed, the installation and conductor requirements shall comply with Parts II and III of Article 334 and shall be of the multiconductor type.
(5) As single-conductor cables as the nonheating leads for heating cables as provided in 424.43
(6) Supported by cable trays. Type UF cable supported by cable trays shall be of the multiconductor type.

340.12 Uses Not Permitted. Type UF cable shall not be used as follows:

(1) As service-entrance cable
(8) Embedded in poured cement, concrete, or aggregate, except where embedded in plaster as nonheating leads where permitted in 424.43
(9) Where exposed to direct rays of the sun, unless identified as sunlight resistant
(10) Where subject to physical damage
(11) As overhead cable, except where installed as messenger-supported wiring in accordance with Part II of Article 396

340.24 Bending Radius. Bends in Type UF cable shall be so made that the cable is not damaged. The radius of the curve of the inner edge of any bend shall not be less than five times the diameter of the cable.

340.80 Ampacity. The ampacity of Type UF cable shall be that of 60°C (140°F) conductors in accordance with 310.14.

CHAPTER 14

RACEWAYS

INTRODUCTION

This chapter contains requirements from Article 358 — Electrical Metallic Tubing: Type EMT; Article 344 — Rigid Metal Conduit: Type RMC; Article 352 — Rigid Polyvinyl Chloride Conduit: Type PVC; Article 362 — Electrical Nonmetallic Tubing: Type ENT; Article 348 — Flexible Metal Conduit: Type FMC; Article 350 — Liquidtight Flexible Metal Conduit: Type LFMC; and Article 356 — Liquidtight Flexible Nonmetallic Conduit: Type LFNC. Contained in each article are requirements pertaining to the use and installation for the conduit or tubing listed. Part I of each article covers the raceway's scope and definition. Other topics covered in most of these articles include: uses (permitted and not permitted), size (minimum and maximum), number of conductors, bends (how made and number in one run), securing and supporting, and grounding.

Although the requirements in each article may be different, the arrangement and numbering in each article follow the same format. For example, in each article, 3__.10 covers uses permitted, 3__.12 covers uses not permitted, 3__.20 covers size, 3__.22 covers number of conductors, 3__.30 covers securing and supporting, and 3__.60 covers grounding.

ARTICLE 358
Electrical Metallic Tubing: Type EMT

PART I. GENERAL

358.2 Definition. The definition in this section shall apply within this article and throughout the *Code*.

Electrical Metallic Tubing (EMT). An unthreaded thinwall raceway of circular cross section designed for the physical protection and routing of conductors and cables and for use as an equipment grounding conductor when installed utilizing appropriate fittings.

PART II. INSTALLATION

358.10 Uses Permitted.

(A) Exposed and Concealed. The use of EMT shall be permitted for both exposed and concealed work for the following:

(1) In concrete, in direct contact with the earth or in areas subject to severe corrosive influences where installed in accordance with 358.10(B)
(2) In dry, damp, and wet locations
(3) In any hazardous (classified) location as permitted by other articles in this *Code*

(B) Corrosive Environments.

(1) Galvanized Steel and Stainless Steel EMT, Elbows, and Fittings. Galvanized steel and stainless steel EMT, elbows, and fittings shall be permitted to be installed in concrete, in direct contact with the earth, or in areas subject to severe corrosive influences where protected by corrosion protection and approved as suitable for the condition.

(2) Supplementary Protection of Aluminum EMT. Aluminum EMT shall be provided with approved supplementary corrosion protection where encased in concrete or in direct contact with the earth.

(C) Cinder Fill. Galvanized steel and stainless steel EMT shall be permitted to be installed in cinder concrete or cinder fill where subject to permanent moisture when protected on all sides by a layer of noncinder concrete at least 50 mm (2 in.) thick or when the tubing is installed at least 450 mm (18 in.) under the fill.

(D) Wet Locations. All supports, bolts, straps, screws, and so forth shall be of corrosion-resistant materials or protected against corrosion by corrosion-resistant materials.

358.12 Uses Not Permitted. EMT shall not be used under the following conditions:

(1) Where subject to severe physical damage
(2) For the support of luminaires or other equipment except conduit bodies no larger than the largest trade size of the tubing

358.24 Bends — How Made. Bends shall be made so that the tubing is not damaged and the internal diameter of the tubing is not effectively reduced. The radius of the curve of any field bend to the centerline of the tubing shall not be less than shown in Table 2, Chapter 9 for one-shot and full shoe benders.

358.26 Bends — Number In One Run. There shall not be more than the equivalent of four quarter bends (360 degrees total) between pull points, for example, conduit bodies and boxes.

358.28 Reaming and Threading.

(A) Reaming. All cut ends of EMT shall be reamed or otherwise finished to remove rough edges.

358.30 Securing and Supporting. EMT shall be installed as a complete system in accordance with 300.18 and shall be securely fastened in place and supported in accordance with 358.30(A) and (B).

(A) Securely Fastened. EMT shall be securely fastened in place at intervals not to exceed 3 m (10 ft). In addition, each EMT run between termination points shall be securely fastened within 900 mm (3 ft) of each outlet box, junction box, device box, cabinet, conduit body, or other tubing termination.

Exception No. 1: Fastening of unbroken lengths shall be permitted to be increased to a distance of 1.5 m (5 ft) where structural members do not readily permit fastening within 900 mm (3 ft).

Exception No. 2: For concealed work in finished buildings or prefinished wall panels where such securing is impracticable, unbroken lengths (without coupling) of EMT shall be permitted to be fished.

(B) Supports. Horizontal runs of EMT supported by openings through framing members at intervals not greater than 3 m (10 ft) and securely fastened within 900 mm (3 ft) of termination points shall be permitted.

358.42 Couplings and Connectors. Couplings and connectors used with EMT shall be made up tight. Where buried in masonry or concrete, they shall be concretetight type. Where installed in wet locations, they shall comply with 314.15.

358.60 Grounding. EMT shall be permitted as an equipment grounding conductor.

ARTICLE 344
Rigid Metal Conduit: Type RMC

PART I. GENERAL

344.2 Definition. The definition in this section shall apply within this article and throughout the *Code*.

Rigid Metal Conduit (RMC). A threadable raceway of circular cross section designed for the physical protection and routing of conductors and cables and for use as an equipment grounding conductor when installed with its integral or associated coupling and appropriate fittings.

PART II. INSTALLATION

344.10 Uses Permitted.

(A) Atmospheric Conditions and Occupancies.

(1) Galvanized Steel, Stainless Steel, and Red Brass RMC. Galvanized steel, stainless steel, and red brass RMC shall be permitted under all atmospheric conditions and occupancies.

(2) Aluminum RMC. Aluminum RMC shall be permitted to be installed where approved for the environment. Rigid aluminum conduit encased in concrete or in direct contact with the earth shall be provided with approved supplementary corrosion protection.

(3) Ferrous Raceways and Fittings. Ferrous raceways and fittings protected from corrosion solely by enamel shall be permitted only indoors and in occupancies not subject to severe corrosive influences.

(B) Corrosive Environments.

(1) Galvanized Steel, Stainless Steel, and Red Brass RMC, Elbows, Couplings, and Fittings. Galvanized steel, stainless steel, and red

Raceways 344.30

brass RMC elbows, couplings, and fittings shall be permitted to be installed in concrete, in direct contact with the earth, or in areas subject to severe corrosive influences where protected by corrosion protection approved for the condition.

(2) Supplementary Protection of Aluminum RMC. Aluminum RMC shall be provided with approved supplementary corrosion protection where encased in concrete or in direct contact with the earth.

(C) Cinder Fill. Galvanized steel, stainless steel, and red brass RMC shall be permitted to be installed in or under cinder fill where subject to permanent moisture where protected on all sides by a layer of noncinder concrete not less than 50 mm (2 in.) thick; where the conduit is not less than 450 mm (18 in.) under the fill; or where protected by corrosion protection approved for the condition.

(D) Wet Locations. All supports, bolts, straps, screws, and so forth, shall be of corrosion-resistant materials or protected against corrosion by corrosion-resistant materials.

Informational Note: See 300.6 for protection against corrosion.

(E) Severe Physical Damage. RMC shall be permitted to be installed where subject to severe physical damage.

344.24 Bends — How Made. Bends of RMC shall be so made that the conduit will not be damaged and so that the internal diameter of the conduit will not be effectively reduced. The radius of the curve of any field bend to the centerline of the conduit shall not be less than indicated in Table 2, Chapter 9.

344.26 Bends — Number in One Run. There shall be not more than the equivalent of four quarter bends (360 degrees total) between pull points, for example, conduit bodies and boxes.

344.28 Reaming and Threading. All cut ends shall be reamed or otherwise finished to remove rough edges. Where conduit is threaded in the field, a standard cutting die with a 1 in 16 taper (¾ in. taper per foot) shall be used.

344.30 Securing and Supporting. RMC shall be installed as a complete system in accordance with 300.18 and shall be securely fastened in place and supported in accordance with 344.30(A) and (B).

(A) Securely Fastened. RMC shall be secured in accordance with one of the following:

(1) RMC shall be securely fastened within 900 mm (3 ft) of each outlet box, junction box, device box, cabinet, conduit body, or other conduit termination.

(2) Fastening shall be permitted to be increased to a distance of 1.5 m (5 ft) where structural members do not readily permit fastening within 900 mm (3 ft).

(3) Where approved, conduit shall not be required to be securely fastened within 900 mm (3 ft) of the service head for above-the-roof termination of a mast.

(B) Supports. RMC shall be supported in accordance with one of the following:

(1) Conduit shall be supported at intervals not exceeding 3 m (10 ft).
(2) The distance between supports for straight runs of conduit shall be permitted in accordance with Table 344.30(B)(2), provided the conduit is made up with threaded couplings and supports that prevent transmission of stresses to termination where conduit is deflected between supports.
(3) Exposed vertical risers from industrial machinery or fixed equipment shall be permitted to be supported at intervals not exceeding 6 m (20 ft) if the conduit is made up with threaded couplings, the conduit is supported and securely fastened at the top and bottom of the riser, and no other means of intermediate support is readily available.
(4) Horizontal runs of RMC supported by openings through framing members at intervals not exceeding 3 m (10 ft) and securely fastened within 900 mm (3 ft) of termination points shall be permitted.

Table 344.30(B)(2) Supports for Rigid Metal Conduit

Conduit Size		Maximum Distance Between Rigid Metal Conduit Supports	
Metric Designator	Trade Size	m	ft
16–21	½–¾	3.0	10
27	1	3.7	12
35–41	1¼–1½	4.3	14
53–63	2–2½	4.9	16
78 and larger	3 and larger	6.1	20

Raceways

344.42 Couplings and Connectors.

(A) Threadless. Threadless couplings and connectors used with conduit shall be made tight. Where buried in masonry or concrete, they shall be the concrete tight type. Where installed in wet locations, they shall comply with 314.15. Threadless couplings and connectors shall not be used on threaded conduit ends unless listed for the purpose.

(B) Running Threads. Running threads shall not be used on conduit for connection at couplings.

344.46 Bushings. Where a conduit enters a box, fitting, or other enclosure, a bushing shall be provided to protect the wires from abrasion unless the box, fitting, or enclosure is designed to provide such protection.

Informational Note: See 300.4(G) for the protection of conductors sizes 4 AWG and larger at bushings.

344.56 Splices and Taps. Splices and taps shall be made in accordance with 300.15.

344.60 Grounding. RMC shall be permitted as an equipment grounding conductor.

ARTICLE 352
Rigid Polyvinyl Chloride Conduit: Type PVC

PART I. GENERAL

352.2 Definition. The definition in this section shall apply within this article and throughout the *Code*.

Rigid Polyvinyl Chloride Conduit (PVC). A rigid nonmetallic raceway of circular cross section, with integral or associated couplings, connectors, and fittings for the installation of electrical conductors and cables.

PART II. INSTALLATION

352.10 Uses Permitted. The use of PVC conduit shall be permitted in accordance with 352.10(A) through (I).

(A) Concealed. PVC conduit shall be permitted in walls, floors, and ceilings.

(B) Corrosive Influences. PVC conduit shall be permitted in locations subject to severe corrosive influences as covered in 300.6 and where subject to chemicals for which the materials are specifically approved.

(C) Cinders. PVC conduit shall be permitted in cinder fill.

(D) Wet Locations. PVC conduit shall be permitted in portions of dairies, laundries, canneries, or other wet locations, and in locations where walls are frequently washed, the entire conduit system, including boxes and fittings used therewith, shall be installed and equipped so as to prevent water from entering the conduit. All supports, bolts, straps, screws, and so forth, shall be of corrosion-resistant materials or be protected against corrosion by approved corrosion-resistant materials.

(E) Dry and Damp Locations. PVC conduit shall be permitted for use in dry and damp locations not prohibited by 352.12.

(F) Exposed. PVC conduit shall be permitted for exposed work. PVC conduit used exposed in areas of physical damage shall be identified for the use.

> Informational Note: PVC Conduit, Type Schedule 80, is identified for areas of physical damage.

(G) Underground Installations. For underground installations, PVC shall be permitted for direct burial and underground encased in concrete. See 300.5 and 300.50.

(H) Support of Conduit Bodies. PVC conduit shall be permitted to support nonmetallic conduit bodies not larger than the largest trade size of an entering raceway. These conduit bodies shall not support luminaires or other equipment and shall not contain devices other than splicing devices as permitted by 110.14(B) and 314.16(C)(2).

(I) Insulation Temperature Limitations. Conductors or cables rated at a temperature higher than the listed temperature rating of PVC conduit shall be permitted to be installed in PVC conduit, provided the conductors or cables are not operated at a temperature higher than the listed temperature rating of the PVC conduit.

352.12 Uses Not Permitted. PVC conduit shall not be used under the conditions specified in 352.12(A) through (E).

(B) Support of Luminaires. For the support of luminaires or other equipment not described in 352.10(H).

(C) Physical Damage. Where subject to physical damage unless identified for such use.

(D) Ambient Temperatures. Where subject to ambient temperatures in excess of 50°C (122°F) unless listed otherwise.

352.24 Bends — How Made. Bends shall be so made that the conduit will not be damaged and the internal diameter of the conduit will not be effectively reduced. Field bends shall be made only with identified bending equipment. The radius of the curve to the centerline of such bends shall not be less than shown in Table 2, Chapter 9.

352.26 Bends — Number in One Run. There shall not be more than the equivalent of four quarter bends (360 degrees total) between pull points, for example, conduit bodies and boxes.

352.28 Trimming. All cut ends shall be trimmed inside and outside to remove rough edges.

352.30 Securing and Supporting. PVC conduit shall be installed as a complete system as provided in 300.18 and shall be fastened so that movement from thermal expansion or contraction is permitted. PVC conduit shall be securely fastened and supported in accordance with 352.30(A) and (B).

(A) Securely Fastened. PVC conduit shall be securely fastened within 900 mm (3 ft) of each outlet box, junction box, device box, conduit body, or other conduit termination. Conduit listed for securing at other than 900 mm (3 ft) shall be permitted to be installed in accordance with the listing.

(B) Supports. PVC conduit shall be supported as required in Table 352.30. Conduit listed for support at spacings other than as shown in Table 352.30 shall be permitted to be installed in accordance with the listing. Horizontal runs of PVC conduit supported by openings through framing members at intervals not exceeding those in Table 352.30 and securely fastened within 900 mm (3 ft) of termination points shall be permitted.

352.44 Expansion Fittings. Expansion fittings for PVC conduit shall be provided to compensate for thermal expansion and contraction where the length change, in accordance with Table 352.44, is expected to be 6 mm (¼ in.) or greater in a straight run between securely mounted items such as boxes, cabinets, elbows, or other conduit terminations.

Table 352.30 Support of Rigid Polyvinyl Chloride Conduit (PVC)

Conduit Size		Maximum Spacing Between Supports	
Metric Designator	Trade Size	mm or m	ft
16–27	½–1	900 mm	3
35–53	1¼–2	1.5 m	5
63–78	2½–3	1.8 m	6
91–129	3½–5	2.1 m	7
155	6	2.5 m	8

352.46 Bushings. Where a conduit enters a box, fitting, or other enclosure, a bushing or adapter shall be provided to protect the wire from abrasion unless the box, fitting, or enclosure design provides equivalent protection.

Informational Note: See 300.4(G) for the protection of conductors 4 AWG and larger at bushings.

352.48 Joints. All joints between lengths of conduit, and between conduit and couplings, fittings, and boxes, shall be made by an approved method.

352.60 Grounding. Where equipment grounding is required, a separate equipment grounding conductor shall be installed in the conduit.

ARTICLE 362
Electrical Nonmetallic Tubing: Type ENT

PART I. GENERAL

362.2 Definition. The definition in this section shall apply within this article and throughout the *Code*.

Electrical Nonmetallic Tubing (ENT). A nonmetallic, pliable, corrugated raceway of circular cross section with integral or associated couplings, connectors, and fittings for the installation of electrical conductors. ENT is composed of a material that is resistant to moisture and chemical atmospheres and is flame retardant.

A pliable raceway is a raceway that can be bent by hand with a reasonable force but without other assistance.

Table 352.44 Expansion Characteristics of PVC Rigid Nonmetallic Conduit Coefficient of Thermal Expansion = 6.084×10^{-5} mm/mm/°C (3.38×10^{-5} in./in./°F)

Temperature Change (°C)	Length Change of PVC Conduit (mm/m)	Temperature Change (°F)	Length Change of PVC Conduit (in./100 ft)	Temperature Change (°F)	Length Change of PVC Conduit (in./100 ft)
5	0.30	5	0.20	105	4.26
10	0.61	10	0.41	110	4.46
15	0.91	15	0.61	115	4.66
20	1.22	20	0.81	120	4.87
25	1.52	25	1.01	125	5.07
30	1.83	30	1.22	130	5.27
35	2.13	35	1.42	135	5.48
40	2.43	40	1.62	140	5.68
45	2.74	45	1.83	145	5.88
50	3.04	50	2.03	150	6.08
55	3.35	55	2.23	155	6.29
60	3.65	60	2.43	160	6.49
65	3.95	65	2.64	165	6.69
70	4.26	70	2.84	170	6.90
75	4.56	75	3.04	175	7.10
80	4.87	80	3.24	180	7.30
85	5.17	85	3.45	185	7.50
90	5.48	90	3.65	190	7.71
95	5.78	95	3.85	195	7.91
100	6.08	100	4.06	200	8.11

PART II. INSTALLATION

362.10 Uses Permitted. For the purpose of this article, the first floor of a building shall be that floor that has 50 percent or more of the exterior wall surface area level with or above finished grade. One additional level that is the first level and not designed for human habitation and used only for vehicle parking, storage, or similar use shall be permitted. The use of ENT and fittings shall be permitted in the following:

(1) In any building not exceeding three floors above grade as follows:

   a. For exposed work, where not prohibited by 362.12
   b. Concealed within walls, floors, and ceilings

(2) In any building exceeding three floors above grade, ENT shall be concealed within walls, floors, and ceilings where the walls, floors, and ceilings provide a thermal barrier of material that has at least a 15-minute finish rating as identified in listings of fire-rated assemblies. The 15-minute-finish-rated thermal barrier shall be permitted to be used for combustible or noncombustible walls, floors, and ceilings.

   Exception to (2): Where a fire sprinkler system(s) is installed in accordance with NFPA 13-2016, Standard for the Installation of Sprinkler Systems, on all floors, ENT shall be permitted to be used within walls, floors, and ceilings, exposed or concealed, in buildings exceeding three floors abovegrade.

(3) In locations subject to severe corrosive influences as covered in 300.6 and where subject to chemicals for which the materials are specifically approved.
(4) In concealed, dry, and damp locations not prohibited by 362.12.
(5) Above suspended ceilings where the suspended ceilings provide a thermal barrier of material that has at least a 15-minute finish rating as identified in listings of fire-rated assemblies, except as permitted in 362.10(1)a.

   Exception to (5): ENT shall be permitted to be used above suspended ceilings in buildings exceeding three floors above grade where the building is protected throughout by a fire sprinkler system installed in accordance with NFPA 13-2016, Standard for the Installation of Sprinkler Systems.

Raceways 362.28

(6) Encased in poured concrete, or embedded in a concrete slab on grade where ENT is placed on sand or approved screenings, provided fittings identified for this purpose are used for connections.

(7) For wet locations indoors as permitted in this section or in a concrete slab on or belowgrade, with fittings listed for the purpose.

(8) Metric designator 16 through 27 (trade size ½ through 1) as listed manufactured prewired assembly.

(9) Conductors or cables rated at a temperature higher than the listed temperature rating of ENT shall be permitted to be installed in ENT, if the conductors or cables are not operated at a temperature higher than the listed temperature rating of the ENT.

362.12 Uses Not Permitted. ENT shall not be used in the following:

(2) For the support of luminaires and other equipment
(3) Where subject to ambient temperatures in excess of 50°C (122°F) unless listed otherwise
(4) For direct earth burial
(5) In exposed locations, except as permitted by 362.10(1), 362.10(5), and 362.10(7)
(7) Where exposed to the direct rays of the sun, unless identified as sunlight resistant
(8) Where subject to physical damage

Informational Note: Extreme cold may cause some types of nonmetallic conduits to become brittle and therefore more susceptible to damage from physical contact.

362.24 Bends — How Made. Bends shall be so made that the tubing will not be damaged and the internal diameter of the tubing will not be effectively reduced. Bends shall be permitted to be made manually without auxiliary equipment, and the radius of the curve to the centerline of such bends shall not be less than shown in Table 2, Chapter 9 using the column "Other Bends."

362.26 Bends — Number in One Run. There shall not be more than the equivalent of four quarter bends (360 degrees total) between pull points, for example, conduit bodies and boxes.

362.28 Trimming. All cut ends shall be trimmed inside and outside to remove rough edges.

362.30 Securing and Supporting. ENT shall be installed as a complete system in accordance with 300.18 and shall be securely fastened in place by an approved means and supported in accordance with 362.30(A) and (B).

(A) Securely Fastened. ENT shall be securely fastened at intervals not exceeding 900 mm (3 ft). In addition, ENT shall be securely fastened in place within 900 mm (3 ft) of each outlet box, device box, junction box, cabinet, or fitting where it terminates. Where used, cable ties shall be listed for the application and for securing and supporting.

Exception No. 1: Lengths not exceeding a distance of 1.8 m (6 ft) from a luminaire terminal connection for tap connections to lighting luminaires shall be permitted without being secured.

Exception No. 2: Lengths not exceeding 1.8 m (6 ft) from the last point where the raceway is securely fastened for connections within an accessible ceiling to luminaire(s) or other equipment.

Exception No. 3: For concealed work in finished buildings or prefinished wall panels where such securing is impracticable, unbroken lengths (without coupling) of ENT shall be permitted to be fished.

(B) Supports. Horizontal runs of ENT supported by openings in framing members at intervals not exceeding 900 mm (3 ft) and securely fastened within 900 mm (3 ft) of termination points shall be permitted.

362.46 Bushings. Where a tubing enters a box, fitting, or other enclosure, a bushing or adapter shall be provided to protect the wire from abrasion unless the box, fitting, or enclosure design provides equivalent protection.

> Informational Note: See 300.4(G) for the protection of conductors size 4 AWG or larger.

362.48 Joints. All joints between lengths of tubing and between tubing and couplings, fittings, and boxes shall be by an approved method.

362.60 Grounding. Where equipment grounding is required, a separate equipment grounding conductor shall be installed in the raceway in compliance with Article 250, Part VI.

ARTICLE 348
Flexible Metal Conduit: Type FMC

PART I. GENERAL

348.2 Definition. The definition in this section shall apply within this article and throughout the *Code*.

Flexible Metal Conduit (FMC). A raceway of circular cross section made of helically wound, formed, interlocked metal strip.

PART II. INSTALLATION

348.10 Uses Permitted. FMC shall be permitted to be used in exposed and concealed locations.

348.12 Uses Not Permitted. FMC shall not be used in the following:

(1) In wet locations
(5) Where exposed to materials having a deteriorating effect on the installed conductors, such as oil or gasoline
(6) Underground or embedded in poured concrete or aggregate
(7) Where subject to physical damage

348.22 Number of Conductors. The number of conductors shall not exceed that permitted by the percentage fill specified in Table 1, Chapter 9, or as permitted in Table 348.22, or for metric designator 12 (trade size ⅜).

Cables shall be permitted to be installed where such use is not prohibited by the respective cable articles. The number of cables shall not exceed the allowable percentage fill specified in Table 1, Chapter 9.

348.26 Bends — Number In One Run. There shall not be more than the equivalent of four quarter bends (360 degrees total) between pull points, for example, conduit bodies and boxes.

348.28 Trimming. All cut ends shall be trimmed or otherwise finished to remove rough edges, except where fittings that thread into the convolutions are used.

348.30 Securing and Supporting. FMC shall be securely fastened in place and supported in accordance with 348.30(A) and (B).

Table 348.22 Maximum Number of Insulated Conductors in Metric Designator 12 (Trade Size ½) Flexible Metal Conduit (FMC)*

Size (AWG)	Types RFH-2, SF-2		Types TF, XHHW, TW		Types TFN, THHN, THWN		Types FEP, FEBB, PF, PGF	
	Fittings Inside Conduit	Fittings Outside Conduit	Fittings Inside Conduit	Fittings Outside Conduit	Fittings Inside Conduit	Fittings Outside Conduit	Fittings Inside Conduit	Fittings Outside Conduit
18	2	3	3	5	5	8	5	8
16	1	2	3	4	4	6	4	6
14	1	2	2	3	3	4	3	4
12	—	1	1	2	2	3	2	3
10	—	1	1	1	1	2	1	2

*In addition, one insulated, covered, or bare equipment grounding conductor of the same size shall be permitted.

(A) Securely Fastened. FMC shall be securely fastened in place by an approved means within 300 mm (12 in.) of each box, cabinet, conduit body, or other conduit termination and shall be supported and secured at intervals not to exceed 1.4 m (4½ ft). Where used, cable ties shall be listed and be identified for securement and support.

Exception No. 1: Where FMC is fished between access points through concealed spaces in finished buildings or structures and supporting is impracticable.

Exception No. 2: Where flexibility is necessary after installation, lengths from the last point where the raceway is securely fastened shall not exceed the following:

(1) 900 mm (3 ft) for metric designators 16 through 35 (trade sizes ½ through 1¼)
(2) 1200 mm (4 ft) for metric designators 41 through 53 (trade sizes 1½ through 2)
(3) 1500 mm (5 ft) for metric designators 63 (trade size 2½) and larger

Exception No. 3: Lengths not exceeding 1.8 m (6 ft) from a luminaire terminal connection for tap connections to luminaires as permitted in 410.117(C).

Exception No. 4: Lengths not exceeding 1.8 m (6 ft) from the last point where the raceway is securely fastened for connections within an accessible ceiling to a luminaire(s) or other equipment. For the purposes of this exception, listed flexible metal conduit fittings shall be permitted as a means of securement and support.

(B) Supports. Horizontal runs of FMC supported by openings through framing members at intervals not greater than 1.4 m (4½ ft) and securely fastened within 300 mm (12 in.) of termination points shall be permitted.

348.42 Couplings and Connectors. Angle connectors shall not be concealed.

348.60 Grounding and Bonding. If used to connect equipment where flexibility is necessary to minimize the transmission of vibration from equipment or to provide flexibility for equipment that requires movement after installation, an equipment grounding conductor shall be installed.

Where flexibility is not required after installation, FMC shall be permitted to be used as an equipment grounding conductor when installed in accordance with 250.118(5).

Where required or installed, equipment grounding conductors shall be installed in accordance with 250.134.

Where required or installed, equipment bonding jumpers shall be installed in accordance with 250.102.

ARTICLE 350
Liquidtight Flexible Metal Conduit: Type LFMC

PART I. GENERAL

350.2 Definition. The definition in this section shall apply within this article and throughout the *Code*.

Liquidtight Flexible Metal Conduit (LFMC). A raceway of circular cross section having an outer liquidtight, nonmetallic, sunlight-resistant jacket over an inner flexible metal core with associated couplings, connectors, and fittings for the installation of electric conductors.

PART II. INSTALLATION

350.10 Uses Permitted. LFMC shall be permitted to be used in exposed or concealed locations as follows:

(1) Where conditions of installation, operation, or maintenance require flexibility or protection from machine oils, liquids, vapors, or solids.
(2) In hazardous (classified) locations where specifically permitted by Chapter 5.
(3) For direct burial where listed and marked for the purpose.

350.12 Uses Not Permitted. LFMC shall not be used where subject to physical damage.

350.26 Bends — Number In One Run. There shall not be more than the equivalent of four quarter bends (360 degrees total) between pull points, for example, conduit bodies and boxes.

350.28 Trimming. All cut ends of conduit shall be trimmed inside and outside to remove rough edges.

350.30 Securing and Supporting. LFMC shall be securely fastened in place and supported in accordance with 350.30(A) and (B).

(A) Securely Fastened. LFMC shall be securely fastened in place by an approved means within 300 mm (12 in.) of each box, cabinet, conduit

Raceways 350.60

body, or other conduit termination and shall be supported and secured at intervals not to exceed 1.4 m (4½ ft). Where used, cable ties shall be listed and be identified for securement and support.

Exception No. 1: Where LFMC is fished between access points through concealed spaces in finished buildings or structures and supporting is impractical.

Exception No. 2: Where flexibility is necessary after installation, lengths from the last point where the raceway is securely fastened shall not exceed the following:

(1) 900 mm (3 ft) for metric designators 16 through 35 (trade sizes ½ through 1¼)
(2) 1200 mm (4 ft) for metric designators 41 through 53 (trade sizes 1½ through 2)
(3) 1500 mm (5 ft) for metric designators 63 (trade size 2½) and larger

Exception No. 3: Lengths not exceeding 1.8 m (6 ft) from a luminaire terminal connection for tap conductors to luminaires, as permitted in 410.117(C).

Exception No. 4: Lengths not exceeding 1.8 m (6 ft) from the last point where the raceway is securely fastened for connections within an accessible ceiling to luminaire(s) or other equipment.

(B) Supports. Horizontal runs of LFMC supported by openings through framing members at intervals not greater than 1.4 m (4½ ft) and securely fastened within 300 mm (12 in.) of termination points shall be permitted.

350.42 Couplings and Connectors. Only fittings listed for use with LFMC shall be used. Angle connectors shall not be concealed. Straight LFMC fittings shall be permitted for direct burial where marked.

350.60 Grounding and Bonding. If used to connect equipment where flexibility is necessary to minimize the transmission of vibration from equipment or to provide flexibility for equipment that requires movement after installation, an equipment grounding conductor shall be installed.

Where flexibility is not required after installation, LFMC shall be permitted to be used as an equipment grounding conductor when installed in accordance with 250.118(6).

Where required or installed, equipment grounding conductors shall be installed in accordance with 250.134.

Where required or installed, equipment bonding jumpers shall be installed in accordance with 250.102.

ARTICLE 356
Liquidtight Flexible Nonmetallic Conduit: Type LFNC

PART I. GENERAL

356.2 Definition. The definition in this section shall apply within this article and throughout the *Code*.

Liquidtight Flexible Nonmetallic Conduit (LFNC). A raceway of circular cross section of various types as follows:

(1) A smooth seamless inner core and cover bonded together and having one or more reinforcement layers between the core and covers, designated as Type LFNC-A
(2) A smooth inner surface with integral reinforcement within the raceway wall, designated as Type LFNC-B
(3) A corrugated internal and external surface without integral reinforcement within the raceway wall, designated as Type LFNC-C

PART II. INSTALLATION

356.10 Uses Permitted. LFNC shall be permitted to be used in exposed or concealed locations for the following purposes:

(1) Where flexibility is required for installation, operation, or maintenance.
(2) Where protection of the contained conductors is required from vapors, machine oils, liquids, or solids.
(3) For outdoor locations where listed and marked as suitable for the purpose.
(4) For direct burial where listed and marked for the purpose.
(5) Type LFNC shall be permitted to be installed in lengths longer than 1.8 m (6 ft) where secured in accordance with 356.30.
(6) Type LFNC-B as a listed manufactured prewired assembly, metric designator 16 through 27 (trade size ½ through 1) conduit.
(7) For encasement in concrete where listed for direct burial and installed in compliance with 356.42.

356.12 Uses Not Permitted. LFNC shall not be used as follows:

(1) Where subject to physical damage
(2) Where any combination of ambient and conductor temperatures is in excess of that for which it is listed

(3) In lengths longer than 1.8 m (6 ft), except as permitted by 356.10(5) or where a longer length is approved as essential for a required degree of flexibility

356.26 Bends — Number in One Run. There shall not be more than the equivalent of four quarter bends (360 degrees total) between pull points, for example, conduit bodies and boxes.

356.28 Trimming. All cut ends of conduit shall be trimmed inside and outside to remove rough edges.

356.30 Securing and Supporting. Type LFNC shall be securely fastened and supported in accordance with one of the following:

(1) Where installed in lengths exceeding 1.8 m (6 ft), the conduit shall be securely fastened at intervals not exceeding 900 mm (3 ft) and within 300 mm (12 in.) on each side of every outlet box, junction box, cabinet, or fitting. Where used, cable ties shall be listed for the application and for securing and supporting.
(2) Securing or supporting of the conduit shall not be required where it is fished, installed in lengths not exceeding 900 mm (3 ft) at terminals where flexibility is required, or installed in lengths not exceeding 1.8 m (6 ft) from a luminaire terminal connection for tap conductors to luminaires permitted in 410.117(C).
(3) Horizontal runs of LFNC supported by openings through framing members at intervals not exceeding 900 mm (3 ft) and securely fastened within 300 mm (12 in.) of termination points shall be permitted.
(4) Securing or supporting of LFNC shall not be required where installed in lengths not exceeding 1.8 m (6 ft) from the last point where the raceway is securely fastened for connections within an accessible ceiling to a luminaire(s) or other equipment. For the purpose of 356.30, listed liquidtight flexible nonmetallic conduit fittings shall be permitted as a means of support.

356.42 Couplings and Connectors. Only fittings listed for use with LFNC shall be used. Angle connectors shall not be used for concealed raceway installations. Straight LFNC fittings are permitted for direct burial or encasement in concrete.

356.60 Grounding. Where equipment grounding is required, a separate equipment grounding conductor shall be installed in the conduit.

CHAPTER 15

WIRING METHODS AND CONDUCTORS

INTRODUCTION

This chapter contains requirements from Articles 300 and 310. Article 300 covers wiring methods for all wiring installations unless modified by other articles. This article covers some general installation requirements for both cables and raceways. Topics include protection against physical damage; underground installations; securing and supporting; and boxes, conduit bodies, or fittings.

Article 310 covers general requirements for conductors, including their ampacity ratings. For dwelling unit applications, conductor ampacities specified in Table 310.16 are typically used. Under specified conditions in dwelling units, Section 310.12 includes a provision allowing a .83 reduction in ampacity for dwelling services and feeder conductors that supply the entire load. Where no adjustment or correction factors are required, Table 310.12 is permitted to be applied. Service conductors sized in accordance with this rule must be supplied by a three-wire, 120/240 volt, single-phase supply system. Three-wire feeders consisting of two ungrounded (hot) conductors and an ungrounded (neutral) conductor that supply an individual dwelling unit can be supplied from either a 120/240 volt single phase, three-wire, or a 208Y/120 volt, three-phase, four-wire supply system. These conductors can be installed in a raceway or cable, with or without an equipment grounding conductor.

ARTICLE 300
General Requirements for Wiring Methods and Materials

PART I. GENERAL REQUIREMENTS

300.3 Conductors.

(A) Single Conductors. Single conductors specified in Table 310.4(A) shall only be installed where part of a recognized wiring method of Chapter 3.

(B) Conductors of the Same Circuit. All conductors of the same circuit and, where used, the grounded conductor and all equipment grounding conductors and bonding conductors shall be contained within the same raceway, auxiliary gutter, cable tray, cablebus assembly, trench, cable, or cord, unless otherwise permitted in accordance with 300.3(B)(1) through (B)(4).

300.4 Protection Against Physical Damage. Where subject to physical damage, conductors, raceways, and cables shall be protected.

(A) Cables and Raceways Through Wood Members.

(1) Bored Holes. In both exposed and concealed locations, where a cable- or raceway-type wiring method is installed through bored holes in joists, rafters, or wood members, holes shall be bored so that the edge of the hole is not less than 32 mm (1¼ in.) from the nearest edge of the wood member. Where this distance cannot be maintained, the cable or raceway shall be protected from penetration by screws or nails by a steel plate(s) or bushing(s), at least 1.6 mm (¹⁄₁₆ in.) thick, and of appropriate length and width installed to cover the area of the wiring.

Exception No. 1: Steel plates shall not be required to protect rigid metal conduit, intermediate metal conduit, rigid nonmetallic conduit, or electrical metallic tubing.

Exception No. 2: A listed and marked steel plate less than 1.6 mm (¹⁄₁₆ in.) thick that provides equal or better protection against nail or screw penetration shall be permitted.

(2) Notches in Wood. Where there is no objection because of weakening the building structure, in both exposed and concealed locations, cables or raceways shall be permitted to be laid in notches in wood studs, joists, rafters, or other wood members where the cable or raceway at those points is protected against nails or screws by a steel plate at least

Wiring Methods and Conductors 300.4

1.6 mm (¹⁄₁₆ in.) thick, and of appropriate length and width, installed to cover the area of the wiring. The steel plate shall be installed before the building finish is applied.

Exception No. 1: Steel plates shall not be required to protect rigid metal conduit, intermediate metal conduit, rigid nonmetallic conduit, or electrical metallic tubing.

Exception No. 2: A listed and marked steel plate less than 1.6 mm (¹⁄₁₆ in.) thick that provides equal or better protection against nail or screw penetration shall be permitted.

(B) Nonmetallic-Sheathed Cables and Electrical Nonmetallic Tubing Through Metal Framing Members.

(1) Nonmetallic-Sheathed Cable. In both exposed and concealed locations where nonmetallic-sheathed cables pass through either factory- or field-punched, cut, or drilled slots or holes in metal members, the cable shall be protected by listed bushings or listed grommets covering all metal edges that are securely fastened in the opening prior to installation of the cable.

(2) Nonmetallic-Sheathed Cable and Electrical Nonmetallic Tubing. Where nails or screws are likely to penetrate nonmetallic-sheathed cable or electrical nonmetallic tubing, a steel sleeve, steel plate, or steel clip not less than 1.6 mm (¹⁄₁₆ in.) in thickness shall be used to protect the cable or tubing.

Exception: A listed and marked steel plate less than 1.6 mm (¹⁄₁₆ in.) thick that provides equal or better protection against nail or screw penetration shall be permitted.

(C) Cables Through Spaces Behind Panels Designed to Allow Access. Cables or raceway-type wiring methods, installed behind panels designed to allow access, shall be supported according to their applicable articles.

(D) Cables and Raceways Parallel to Framing Members and Furring Strips. In both exposed and concealed locations, where a cable- or raceway-type wiring method is installed parallel to framing members, such as joists, rafters, or studs, or is installed parallel to furring strips, the cable or raceway shall be installed and supported so that the nearest outside surface of the cable or raceway is not less than 32 mm (1¼ in.) from the nearest edge of the framing member or furring strips where nails or screws are likely to penetrate. Where this distance cannot be maintained, the cable or raceway shall be protected from penetration by nails or screws by a steel plate, sleeve, or equivalent at least 1.6 mm (¹⁄₁₆ in.) thick.

Exception No. 1: Steel plates, sleeves, or the equivalent shall not be required to protect rigid metal conduit, intermediate metal conduit, rigid nonmetallic conduit, or electrical metallic tubing.

Exception No. 2: For concealed work in finished buildings, or finished panels for prefabricated buildings where such supporting is impracticable, it shall be permissible to fish the cables between access points.

Exception No. 3: A listed and marked steel plate less than 1.6 mm (¹⁄₁₆ in.) thick that provides equal or better protection against nail or screw penetration shall be permitted.

(E) Cables, Raceways, or Boxes Installed In or Under Roof Decking. A cable, raceway, or box, installed in exposed or concealed locations under metal-corrugated sheet roof decking, shall be installed and supported so there is not less than 38 mm (1½ in.) measured from the lowest surface of the roof decking to the top of the cable, raceway, or box. A cable, raceway, or box shall not be installed in concealed locations in metal-corrugated, sheet decking–type roof.

Exception: Rigid metal conduit and intermediate metal conduit shall not be required to comply with 300.4(E).

(F) Cables and Raceways Installed in Shallow Grooves. Cable- or raceway-type wiring methods installed in a groove, to be covered by wallboard, siding, paneling, carpeting, or similar finish, shall be protected by 1.6 mm (¹⁄₁₆ in.) thick steel plate, sleeve, or equivalent or by not less than 32-mm (1¼-in.) free space for the full length of the groove in which the cable or raceway is installed.

Exception No. 1: Steel plates, sleeves, or the equivalent shall not be required to protect rigid metal conduit, intermediate metal conduit, rigid nonmetallic conduit, or electrical metallic tubing.

Exception No. 2: A listed and marked steel plate less than 1.6 mm (¹⁄₁₆ in.) thick that provides equal or better protection against nail or screw penetration shall be permitted.

(G) Fittings. Where raceways contain 4 AWG or larger insulated circuit conductors, and these conductors enter a cabinet, a box, an enclosure, or a raceway, the conductors shall be protected in accordance with any of the following:

Wiring Methods and Conductors 300.5

(1) An identified fitting providing a smoothly rounded insulating surface
(2) A listed metal fitting that has smoothly rounded edges
(3) Separation from the fitting or raceway using an identified insulating material that is securely fastened in place
(4) Threaded hubs or bosses that are an integral part of a cabinet, box, enclosure, or raceway providing a smoothly rounded or flared entry for conductors

Conduit bushings constructed wholly of insulating material shall not be used to secure a fitting or raceway. The insulating fitting or insulating material shall have a temperature rating not less than the insulation temperature rating of the installed conductors.

(H) Structural Joints. A listed expansion/deflection fitting or other approved means shall be used where a raceway crosses a structural joint intended for expansion, contraction or deflection, used in buildings, bridges, parking garages, or other structures.

300.5 Underground Installations.

(A) Minimum Cover Requirements. Direct-buried cable, conduit, or other raceways shall be installed to meet the minimum cover requirements of Table 300.5.

(B) Wet Locations. The interior of enclosures or raceways installed underground shall be considered to be a wet location. Insulated conductors and cables installed in these enclosures or raceways in underground installations shall comply with 310.10(C).

(C) Underground Cables and Conductors Under Buildings. Underground cable and conductors installed under a building shall be in a raceway.

(D) Protection from Damage. Direct-buried conductors and cables shall be protected from damage in accordance with 300.5(D)(1) through (D)(4).

(1) Emerging from Grade. Direct-buried conductors and cables emerging from grade and specified in columns 1 and 4 of Table 300.5 shall be protected by enclosures or raceways extending from the minimum cover distance below grade required by 300.5(A) to a point at least 2.5 m (8 ft) above finished grade. In no case shall the protection be required to exceed 450 mm (18 in.) below finished grade.

(2) Conductors Entering Buildings. Conductors entering a building shall be protected to the point of entrance.

Table 300.5 Minimum Cover Requirements, 0 to 1000 Volts, Nominal, Burial in Millimeters (Inches)

Location of Wiring Method or Circuit	Column 1 Direct Burial Cables or Conductors		Column 2 Rigid Metal Conduit or Intermediate Metal Conduit		Column 3 Nonmetallic Raceways Listed for Direct Burial Without Concrete Encasement or Other Approved Raceways		Column 4 Residential Branch Circuits Rated 120 Volts or Less with GFCI Protection and Maximum Overcurrent Protection of 20 Amperes		Column 5 Circuits for Control of Irrigation and Landscape Lighting Limited to Not More Than 30 Volts and Installed with Type UF or in Other Identified Cable or Raceway	
	mm	in.	mm	in.	mm	in.	mm	in.	mm	in.
All locations not specified below	600	24	150	6	450	18	300	12	150[a, b]	6[a, b]
In trench below 50 mm (2 in.) thick concrete or equivalent	450	18	150	6	300	12	150	6	150	6
Under a building	0 (in raceway or Type MC or Type MI cable identified for direct burial)	0	0	0	0	0	0 (in raceway or Type MC or Type MI cable identified for direct burial)	0	0 (in raceway or Type MC or Type MI cable identified for direct burial)	0
Under minimum of 102 mm (4 in.) thick concrete exterior slab with no vehicular traffic and the slab extending not less than 152 mm (6 in.)	450	18	100	4	100	4	150 (direct burial) 100 (in raceway)	6 (direct burial) 4 (in raceway)	150 (direct burial) 100 (in raceway)	6 (direct burial) 4 (in raceway)

Wiring Methods and Conductors — 300.5

Location of Wiring Method or Circuit	Column 1 mm	Column 1 in.	Column 2 mm	Column 2 in.	Column 3 mm	Column 3 in.	Column 4 mm	Column 4 in.	Column 5 mm	Column 5 in.
Under streets, highways, roads, alleys, driveways, and parking lots	600	24	600	24	600	24	600	24	600	24
One- and two-family dwelling driveways and outdoor parking areas, and used only for dwelling-related purposes	450	18	450	18	450	18	300	12	450	18
In or under airport runways, including adjacent areas where trespassing prohibited	450	18	450	18	450	18	450	18	450	18

[a] A lesser depth shall be permitted where specified in the installation instructions of a listed low-voltage lighting system.

[b] A depth of 150 mm (6 in.) shall be permitted for pool, spa, and fountain lighting, installed in a nonmetallic raceway, limited to not more than 30 volts where part of a listed low-voltage lighting system.

Notes:

1. Cover is defined as the shortest distance in mm (in.) measured between a point on the top surface of any direct-buried conductor, cable, conduit, or other raceway and the top surface of finished grade, concrete, or similar cover.

2. Raceways approved for burial only where concrete encased shall require concrete envelope not less than 50 mm (2 in.) thick.

3. Lesser depths shall be permitted where cables and conductors rise for terminations or splices or where access is otherwise required.

4. Where one of the wiring method types listed in Columns 1 through 3 is used for one of the circuit types in Columns 4 and 5, the shallowest depth of burial shall be permitted.

5. Where solid rock prevents compliance with the cover depths specified in this table, the wiring shall be installed in a metal raceway, or a nonmetallic raceway permitted for direct burial. The raceways shall be covered by a minimum of 50 mm (2 in.) of concrete extending down to rock.

(3) Service Conductors. Underground service conductors that are not encased in concrete and that are buried 450 mm (18 in.) or more below grade shall have their location identified by a warning ribbon that is placed in the trench at least 300 mm (12 in.) above the underground installation.

(4) Enclosure or Raceway Damage. Where the enclosure or raceway is subject to physical damage, the conductors shall be installed in electrical metallic tubing, rigid metal conduit, intermediate metal conduit, RTRC-XW, Schedule 80 PVC conduit, or equivalent.

(E) Splices and Taps. Direct-buried conductors or cables shall be permitted to be spliced or tapped without the use of splice boxes. The splices or taps shall be made in accordance with 110.14(B).

(F) Backfill. Backfill that contains large rocks, paving materials, cinders, large or sharply angular substances, or corrosive material shall not be placed in an excavation where materials may damage raceways, cables, conductors, or other substructures or prevent adequate compaction of fill or contribute to corrosion of raceways, cables, or other substructures.

Where necessary to prevent physical damage to the raceway, cable, or conductor, protection shall be provided in the form of granular or selected material, suitable running boards, suitable sleeves, or other approved means.

(G) Raceway Seals. Conduits or raceways through which moisture may contact live parts shall be sealed or plugged at either or both ends. Spare or unused raceways shall also be sealed. Sealants shall be identified for use with the cable insulation, conductor insulation, bare conductor, shield, or other components.

(H) Bushing. A bushing, or terminal fitting, with an integral bushed opening shall be used at the end of a conduit or other raceway that terminates underground where the conductors or cables emerge as a direct burial wiring method. A seal incorporating the physical protection characteristics of a bushing shall be permitted to be used in lieu of a bushing.

(I) Conductors of the Same Circuit. All conductors of the same circuit and, where used, the grounded conductor and all equipment grounding conductors shall be installed in the same raceway or cable or shall be installed in close proximity in the same trench.

Exception No. 1: Conductors shall be permitted to be installed in parallel in raceways, multiconductor cables, or direct-buried single conductor cables. Each raceway or multiconductor cable shall contain all

Wiring Methods and Conductors 300.11

conductors of the same circuit, including equipment grounding conductors. Each direct-buried single conductor cable shall be located in close proximity in the trench to the other single conductor cables in the same parallel set of conductors in the circuit, including equipment grounding conductors.

(J) Earth Movement. Where direct-buried conductors, raceways, or cables are subject to movement by settlement or frost, direct-buried conductors, raceways, or cables shall be arranged so as to prevent damage to the enclosed conductors or to equipment connected to the raceways.

> Informational Note: This section recognizes "S" loops in underground direct burial cables and conductors to raceway transitions, expansion fittings in raceway risers to fixed equipment, and, generally, the provision of flexible connections to equipment subject to settlement or frost heaves.

(K) Directional Boring. Cables or raceways installed using directional boring equipment shall be approved for the purpose.

300.9 Raceways in Wet Locations Abovegrade. Where raceways are installed in wet locations abovegrade, the interior of these raceways shall be considered to be a wet location. Insulated conductors and cables installed in raceways in wet locations abovegrade shall comply with 310.10(C).

300.10 Electrical Continuity of Metal Raceways and Enclosures. Metal raceways, cable armor, and other metal enclosures for conductors shall be metallically joined together into a continuous electrical conductor and shall be connected to all boxes, fittings, and cabinets so as to provide effective electrical continuity. Unless specifically permitted elsewhere in this *Code*, raceways and cable assemblies shall be mechanically secured to boxes, fittings, cabinets, and other enclosures.

Exception No. 1: Short sections of raceways used to provide support or protection of cable assemblies from physical damage shall not be required to be made electrically continuous.

300.11 Securing and Supporting.

(A) Secured in Place. Raceways, cable assemblies, boxes, cabinets, and fittings shall be securely fastened in place.

(B) Wiring Systems Installed Above Suspended Ceilings. Support wires that do not provide secure support shall not be permitted as the sole support. Support wires and associated fittings that provide secure support and that are installed in addition to the ceiling grid support wires shall be permitted as the sole support. Where independent support wires are used, they shall be secured at both ends. Cables and raceways shall not be supported by ceiling grids.

(1) Fire-Rated Assemblies. Wiring located within the cavity of a fire-rated floor–ceiling or roof–ceiling assembly shall not be secured to, or supported by, the ceiling assembly, including the ceiling support wires. An independent means of secure support shall be provided and shall be permitted to be attached to the assembly. Where independent support wires are used, they shall be distinguishable by color, tagging, or other effective means from those that are part of the fire-rated design.

Exception: The ceiling support system shall be permitted to support wiring and equipment that have been tested as part of the fire-rated assembly.

(2) Non–Fire-Rated Assemblies. Wiring located within the cavity of a non–fire-rated floor–ceiling or roof–ceiling assembly shall not be secured to, or supported by, the ceiling assembly, including the ceiling support wires. An independent means of secure support shall be provided and shall be permitted to be attached to the assembly. Where independent support wires are used, they shall be distinguishable by color, tagging, or other effective means.

Exception: The ceiling support system shall be permitted to support branch-circuit wiring and associated equipment where installed in accordance with the ceiling system manufacturer's instructions.

(C) Raceways Used as Means of Support. Raceways shall be used only as a means of support for other raceways, cables, or nonelectrical equipment under any of the following conditions:

(1) Where the raceway or means of support is identified as a means of support
(2) Where the raceway contains power supply conductors for electrically controlled equipment and is used to support Class 2 circuit conductors or cables that are solely for the purpose of connection to the equipment control circuits
(3) Where the raceway is used to support boxes or conduit bodies in accordance with 314.23 or to support luminaires in accordance with 410.36(E)

300.12 Mechanical Continuity — Raceways and Cables. Raceways, cable armors, and cable sheaths shall be continuous between cabinets, boxes, fittings, or other enclosures or outlets.

Wiring Methods and Conductors 300.15

Exception No. 1: Short sections of raceways used to provide support or protection of cable assemblies from physical damage shall not be required to be mechanically continuous.

300.13 Mechanical and Electrical Continuity — Conductors.

(A) General. Conductors in raceways shall be continuous between outlets, boxes, devices, and so forth. There shall be no splice or tap within a raceway unless permitted by 300.15, 368.56(A), 376.56, 378.56, 384.56, 386.56, 388.56, or 390.56.

(B) Device Removal. In multiwire branch circuits, the continuity of a grounded conductor shall not depend on device connections such as lampholders, receptacles, and so forth, where the removal of such devices would interrupt the continuity.

300.14 Length of Free Conductors at Outlets, Junctions, and Switch Points.
At least 150 mm (6 in.) of free conductor, measured from the point in the box where it emerges from its raceway or cable sheath, shall be left at each outlet, junction, and switch point for splices or the connection of luminaires or devices. Where the opening to an outlet, junction, or switch point is less than 200 mm (8 in.) in any dimension, each conductor shall be long enough to extend at least 75 mm (3 in.) outside the opening.

Exception: Conductors that are not spliced or terminated at the outlet, junction, or switch point shall not be required to comply with 300.14.

300.15 Boxes, Conduit Bodies, or Fittings — Where Required.
A box shall be installed at each outlet and switch point for concealed knob-and-tube wiring.

Fittings and connectors shall be used only with the specific wiring methods for which they are designed and listed.

Where the wiring method is conduit, tubing, Type AC cable, Type MC cable, Type MI cable, nonmetallic-sheathed cable, or other cables, a box or conduit body shall be installed at each conductor splice point, outlet point, switch point, junction point, termination point, or pull point, unless otherwise permitted in 300.15(A) through (L).

(A) Wiring Methods with Interior Access. A box or conduit body shall not be required for each splice, junction, switch, pull, termination, or outlet points in wiring methods with removable covers, such as wireways, multioutlet assemblies, auxiliary gutters, and surface raceways. The covers shall be accessible after installation.

(B) Equipment. An integral junction box or wiring compartment as part of approved equipment shall be permitted in lieu of a box.

(C) Protection. A box or conduit body shall not be required where cables enter or exit from conduit or tubing that is used to provide cable support or protection against physical damage. A fitting shall be provided on the end(s) of the conduit or tubing to protect the cable from abrasion.

(E) Integral Enclosure. A wiring device with integral enclosure identified for the use, having brackets that securely fasten the device to walls or ceilings of conventional on-site frame construction, for use with nonmetallic-sheathed cable, shall be permitted in lieu of a box or conduit body.

(F) Fitting. A fitting identified for the use shall be permitted in lieu of a box or conduit body where conductors are not spliced or terminated within the fitting. The fitting shall be accessible after installation, unless listed for concealed installation.

(G) Direct-Buried Conductors. As permitted in 300.5(E), a box or conduit body shall not be required for splices and taps in direct-buried conductors and cables.

(H) Insulated Devices. As permitted in 334.40(B), a box or conduit body shall not be required for insulated devices supplied by nonmetallic-sheathed cable.

(J) Luminaires. A box or conduit body shall not be required where a luminaire is used as a raceway as permitted in 410.64.

300.16 Raceway or Cable to Open or Concealed Wiring.

(A) Box, Conduit Body, or Fitting. A box, conduit body, or terminal fitting having a separately bushed hole for each conductor shall be used wherever a change is made from conduit, electrical metallic tubing, electrical nonmetallic tubing, nonmetallic-sheathed cable, Type AC cable, Type MC cable, or mineral-insulated, metal-sheathed cable and surface raceway wiring to open wiring or to concealed knob-and-tube wiring. A fitting used for this purpose shall contain no taps or splices and shall not be used at luminaire outlets. A conduit body used for this purpose shall contain no taps or splices, unless it complies with 314.16(C)(2).

(B) Bushing. A bushing shall be permitted in lieu of a box or terminal where the conductors emerge from a raceway and enter or terminate at equipment, such as open switchboards, unenclosed control equipment, or similar equipment. The bushing shall be of the insulating type for other than lead-sheathed conductors.

Wiring Methods and Conductors 300.22

300.18 Raceway Installations.

(A) Complete Runs. Raceways, other than busways or exposed raceways having hinged or removable covers, shall be installed complete between outlet, junction, or splicing points prior to the installation of conductors. Where required to facilitate the installation of utilization equipment, the raceway shall be permitted to be initially installed without a terminating connection at the equipment. Prewired raceway assemblies shall be permitted only where specifically permitted in this *Code* for the applicable wiring method.

Exception: Short sections of raceways used to contain conductors or cable assemblies for protection from physical damage shall not be required to be installed complete between outlet, junction, or splicing points.

300.20 Induced Currents in Ferrous Metal Enclosures or Ferrous Metal Raceways.

(A) Conductors Grouped Together. Where conductors carrying alternating current are installed in ferrous metal enclosures or ferrous metal raceways, they shall be arranged so as to avoid heating the surrounding ferrous metal by induction. To accomplish this, all phase conductors and, where used, the grounded conductor and all equipment grounding conductors shall be grouped together.

300.21 Spread of Fire or Products of Combustion.
Electrical installations in hollow spaces, vertical shafts, and ventilation or air-handling ducts shall be made so that the possible spread of fire or products of combustion will not be substantially increased. Openings around electrical penetrations into or through fire-resistant-rated walls, partitions, floors, or ceilings shall be firestopped using approved methods to maintain the fire resistance rating.

300.22 Wiring in Ducts Not Used for Air Handling, Fabricated Ducts for Environmental Air, and Other Spaces for Environmental Air (Plenums).
The provisions of this section shall apply to the installation and uses of electrical wiring and equipment in ducts used for dust, loose stock, or vapor removal; ducts specifically fabricated for environmental air; and other spaces used for environmental air (plenums).

(C) Other Spaces Used for Environmental Air (Plenums). This section shall apply to spaces not specifically fabricated for environmental air-handling purposes but used for air-handling purposes as a plenum. This section shall not apply to habitable rooms or areas of buildings, the prime purpose of which is not air handling.

ARTICLE 310

Conductors for General Wiring

PART I. GENERAL

310.1 Scope. This article covers general requirements for conductors rated up to and including 2000 volts and their type designations, insulations, markings, mechanical strengths, ampacity ratings, and uses. These requirements do not apply to conductors that form an integral part of equipment, such as motors, motor controllers, and similar equipment, or to conductors specifically provided for elsewhere in this *Code*.

PART III. INSTALLATION

310.10 Uses Permitted. The conductors described in 310.4 shall be permitted for use in any of the wiring methods covered in Chapter 3 and as specified in their respective tables or as permitted elsewhere in this *Code*.

(A) Dry Locations. Insulated conductors and cables used in dry locations shall be any of the types identified in this *Code*.

(B) Dry and Damp Locations. Insulated conductors and cables used in dry and damp locations shall be Types FEP, FEPB, MTW, PFA, RHH, RHW, RHW-2, SA, THHN, THW, THW-2, THHW, THWN, THWN-2, TW, XHH, XHHW, XHHW-2, XHHN, XHWN, XHWN-2, Z, or ZW.

(C) Wet Locations. Insulated conductors and cables used in wet locations shall comply with one of the following:

(1) Be moisture-impervious metal-sheathed
(2) Be types MTW, RHW, RHW-2, TW, THW, THW-2, THHW, THWN, THWN-2, XHHW, XHHW-2, XHWN, XHWN-2 or ZW
(3) Be of a type listed for use in wet locations

(D) Locations Exposed to Direct Sunlight. Insulated conductors or cables used where exposed to direct rays of the sun shall comply with (D)(1) or (D)(2):

(1) Conductors and cables shall be listed, or listed and marked, as being sunlight resistant
(2) Conductors and cables shall be covered with insulating material, such as tape or sleeving, that is listed, or listed and marked, as being sunlight resistant

Wiring Methods and Conductors 310.10

(F) Corrosive Conditions. Conductors exposed to oils, greases, vapors, gases, fumes, liquids, or other substances having a deleterious effect on the conductor or insulation shall be of a type suitable for the application.

(G) Conductors in Parallel.

(1) General. Aluminum, copper-clad aluminum, or copper conductors for each phase, polarity, neutral, or grounded circuit shall be permitted to be connected in parallel (electrically joined at both ends) only in sizes 1/0 AWG and larger where installed in accordance with 310.10(G)(2) through (G)(6).

(2) Conductor and Installation Characteristics. The paralleled conductors in each phase, polarity, neutral, grounded circuit conductor, equipment grounding conductor, or equipment bonding jumper shall comply with all of the following:

(1) Be the same length
(2) Consist of the same conductor material
(3) Be the same size in circular mil area
(4) Have the same insulation type
(5) Be terminated in the same manner

(3) Separate Cables or Raceways. Where run in separate cables or raceways, the cables or raceways with conductors shall have the same number of conductors and shall have the same electrical characteristics. Conductors of one phase, polarity, neutral, grounded circuit conductor, or equipment grounding conductor shall not be required to have the same physical characteristics as those of another phase, polarity, neutral, grounded circuit conductor, or equipment grounding conductor.

(4) Ampacity Adjustment. Conductors installed in parallel shall comply with the provisions of 310.15(C)(1).

(5) Equipment Grounding Conductors. Where parallel equipment grounding conductors are used, they shall be sized in accordance with 250.122. Sectioned equipment grounding conductors smaller than 1/0 AWG shall be permitted in multiconductor cables, if the combined circular mil area of the sectioned equipment grounding conductors in each cable complies with 250.122.

(6) Bonding Jumpers. Where parallel equipment bonding jumpers or supply-side bonding jumpers are installed in raceways, they shall be sized and installed in accordance with 250.102.

310.12 Single-Phase Dwelling Services and Feeders. For one-family dwellings and the individual dwelling units of two-family and multifamily dwellings, service and feeder conductors supplied by a single-phase, 120/240-volt system shall be permitted to be sized in accordance with 310.12(A) through (D).

For one-family dwellings and the individual dwelling units of two-family and multifamily dwellings, single-phase feeder conductors consisting of two ungrounded conductors and the neutral conductor from a 208Y/120 volt system shall be permitted to be sized in accordance with 310.12(A) through (C).

(A) Services. For a service rated 100 amperes through 400 amperes, the service conductors supplying the entire load associated with a one-family dwelling, or the service conductors supplying the entire load associated with an individual dwelling unit in a two-family or multifamily dwelling, shall be permitted to have an ampacity not less than 83 percent of the service rating. If no adjustment or correction factors are required, Table 310.12 shall be permitted to be applied.

Table 310.12 Single-Phase Dwelling Services and Feeders

	Conductor (AWG or kcmil)	
Service or Feeder Rating (Amperes)	Copper	Aluminum or Copper-Clad Aluminum
100	4	2
110	3	1
125	2	1/0
150	1	2/0
175	1/0	3/0
200	2/0	4/0
225	3/0	250
250	4/0	300
300	250	350
350	350	500
400	400	600

Note: If no adjustment or correction factors are required, this table shall be permitted to be applied.

Wiring Methods and Conductors 310.15

(B) Feeders. For a feeder rated 100 amperes through 400 amperes, the feeder conductors supplying the entire load associated with a one-family dwelling, or the feeder conductors supplying the entire load associated with an individual dwelling unit in a two-family or multifamily dwelling, shall be permitted to have an ampacity not less than 83 percent of the feeder rating. If no adjustment or correction factors are required, Table 310.12 shall be permitted to be applied.

(C) Feeder Ampacities. In no case shall a feeder for an individual dwelling unit be required to have an ampacity greater than that specified in 310.12(A) or (B).

(D) Grounded Conductors. Grounded conductors shall be permitted to be sized smaller than the ungrounded conductors, if the requirements of 220.61 and 230.42 for service conductors or the requirements of 215.2 and 220.61 for feeder conductors are met.

Where correction or adjustment factors are required by 310.15(B) or (C), they shall be permitted to be applied to the ampacity associated with the temperature rating of the conductor.

310.15 Ampacity Tables.

(A) General. Ampacities for conductors rated 0 volts to 2000 volts shall be as specified in the Ampacity Table 310.16 through Table 310.21, as modified by 310.15(A) through (F) and 310.12. Under engineering supervision, ampacities of sizes not shown in ampacity tables for conductors meeting the general wiring requirements shall be permitted to be determined by interpolation of the adjacent conductors based on the conductor's area.

The temperature correction and adjustment factors shall be permitted to be applied to the ampacity for the temperature rating of the conductor, if the corrected and adjusted ampacity does not exceed the ampacity for the temperature rating of the termination in accordance with the provisions of 110.14(C).

Informational Note No. 1: Table 310.16 through Table 310.19 are application tables for use in determining conductor sizes on loads calculated in accordance with Part II, Part III, Part IV, or Part V of Article 220. Ampacities result from consideration of one or more of the following:

(1) Temperature compatibility with connected equipment, especially the connection points.
(2) Coordination with circuit and system overcurrent protection.

(3) Compliance with the requirements of product listings or certifications. See 110.3(B).

(4) Preservation of the safety benefits of established industry practices and standardized procedures.

Informational Note No. 2: For conductor area see Chapter 9, Table 8, Conductor Properties. Interpolation is based on the conductor area and not the conductor overall area.

Informational Note No. 3: For the ampacities of flexible cords and cables, see 400.5. For the ampacities of fixture wires, see 402.5.

Informational Note No. 4: For explanation of type letters used in tables and for recognized sizes of conductors for the various conductor insulations, see Table 310.4(A) and Table 310.4(B). For installation requirements, see 310.1 through 310.14 and the various articles of this *Code*. For flexible cords, see Table 400.4, Table 400.5(A)(a), and Table 400.5(A)(b).

(B) Ambient Temperature Correction Factors.

(1) General. Ampacities for ambient temperatures other than those shown in the ampacity tables shall be corrected in accordance with Table 310.15(B)(1).

(C) Adjustment Factors.

(1) More than Three Current-Carrying Conductors. The ampacity of each conductor shall be reduced as shown in Table 310.15(C)(1) where the number of current-carrying conductors in a raceway or cable exceeds three, or where single conductors or multiconductor cables not installed in raceways are installed without maintaining spacing for a continuous length longer than 600 mm (24 in.). Each current-carrying conductor of a paralleled set of conductors shall be counted as a current-carrying conductor.

Where conductors of different systems, as provided in 300.3, are installed in a common raceway or cable, the adjustment factors shown in Table 310.15(C)(1) shall apply only to the number of power and lighting conductors (Articles 210, 215, 220, and 230).

(b) Adjustment factors shall not apply to conductors in raceways having a length not exceeding 600 mm (24 in.).

(c) Adjustment factors shall not apply to underground conductors entering or leaving an outdoor trench if those conductors have physical protection in the form of rigid metal conduit, intermediate metal conduit,

Wiring Methods and Conductors 310.15

Table 310.15(B)(1) Ambient Temperature Correction Factors Based on 30°C (86°F)

For ambient temperatures other than 30°C (86°F), multiply the ampacities specified in the ampacity tables by the appropriate correction factor shown below.

Ambient Temperature (°C)	Temperature Rating of Conductor			Ambient Temperature (°F)
	60°C	75°C	90°C	
10 or less	1.29	1.20	1.15	50 or less
11–15	1.22	1.15	1.12	51–59
16–20	1.15	1.11	1.08	60–68
21–25	1.08	1.05	1.04	69–77
26–30	1.00	1.00	1.00	78–86
31–35	0.91	0.94	0.96	87–95
36–40	0.82	0.88	0.91	96–104
41–45	0.71	0.82	0.87	105–113
46–50	0.58	0.75	0.82	114–122
51–55	0.41	0.67	0.76	123–131
56–60	—	0.58	0.71	132–140
61–65	—	0.47	0.65	141–149
66–70	—	0.33	0.58	150–158
71–75	—	—	0.50	159–167
76–80	—	—	0.41	168–176
81–85	—	—	0.29	177–185

rigid polyvinyl chloride conduit (PVC), or reinforced thermosetting resin conduit (RTRC) having a length not exceeding 3.05 m (10 ft), and if the number of conductors does not exceed four.

(D) Bare or Covered Conductors. Where bare or covered conductors are installed with insulated conductors, the temperature rating of the bare or covered conductor shall be equal to the lowest temperature rating of the insulated conductors for the purpose of determining ampacity.

(E) Neutral Conductor. Neutral conductors shall be considered current carrying in accordance with any of the following:

(1) A neutral conductor that carries only the unbalanced current from other conductors of the same circuit shall not be required to be counted when applying the provisions of 310.15(C)(1).

(F) Grounding or Bonding Conductor. A grounding or bonding conductor shall not be counted when applying the provisions of 310.15(C)(1).

Table 310.15(C)(1) Adjustment Factors for More Than Three Current-Carrying Conductors

Number of Conductors*	Percent of Values in Table 310.16 Through Table 310.19 as Adjusted for Ambient Temperature if Necessary
4–6	80
7–9	70
10–20	50
21–30	45
31–40	40
41 and above	35

*Number of conductors is the total number of conductors in the raceway or cable, including spare conductors. The count shall be adjusted in accordance with 310.15(E) and (F). The count shall not include conductors that are connected to electrical components that cannot be simultaneously energized.

310.16 Ampacities of Insulated Conductors in Raceway, Cable, or Earth (Directly Buried). The ampacities shall be as specified in Table 310.16 where all of the following conditions apply:

(1) Conductors are rated 0 volts through 2000 volts.
(2) Conductors are rated 60°C (140°F), 75°C (167°F), or 90°C (194°F).
(3) Wiring is installed in a 30°C (86°F) ambient temperature.
(4) There are not more than three current-carrying conductors.

PART II. CONSTRUCTION SPECIFICATIONS

310.3 Conductors.

(B) Conductor Material. Conductors in this article shall be of aluminum, copper-clad aluminum, or copper unless otherwise specified.

(C) Stranded Conductors. Where installed in raceways, conductors 8 AWG and larger shall be stranded, unless specifically permitted or required elsewhere in this *Code* to be solid.

(D) Insulated. Conductors not specifically permitted elsewhere in this *Code* to be covered or bare shall be insulated.

Table 310.16 Ampacities of Insulated Conductors with Not More Than Three Current-Carrying Conductors in Raceway, Cable, or Earth (Directly Buried)

[See Table 310.4(A).]

Size AWG or kcmil	Temperature Rating of Conductor						Size AWG or kcmil
	60°C (140°F)	75°C (167°F)	90°C (194°F)	60°C (140°F)	75°C (167°F)	90°C (194°F)	
	Types TW, UF	Types RHW, THHW, THW, THWN, XHHW, USE, ZW	Types TBS, SA, SIS, FEP, FEPB, MI, PFA, RHH, RHW-2, THHN, THHW, THW-2, THWN-2, USE-2, XHH, XHHW, XHHW-2, XHWN, XHWN-2, Z, ZW-2	Types TW, UF	Types RHW, THHW, THW, THWN, XHHW, USE	Types TBS, SA, SIS, THHN, THHW, THW-2, THWN-2, RHH, RHW-2, USE-2, XHH, XHHW, XHHW-2, XHWN, XHWN-2, ZHHN	
	COPPER			ALUMINUM OR COPPER-CLAD ALUMINUM			
18*	—	—	14	—	—	—	—
16*	—	—	18	—	—	—	—
14*	15	20	25	—	—	—	—
12*	20	25	30	15	20	25	12*
10*	30	35	40	25	30	35	10*
8	40	50	55	35	40	45	8
6	55	65	75	40	50	55	6
4	70	85	95	55	65	75	4
3	85	100	115	65	75	85	3
2	95	115	130	75	90	100	2
1	110	130	145	85	100	115	1

(continued)

Table 310.16 Ampacities of Insulated Conductors with Not More Than Three Current-Carrying Conductors in Raceway, Cable, or Earth (Directly Buried) — *Continued*

Size AWG or kcmil	Temperature Rating of Conductor [See Table 310.4(A).]						Size AWG or kcmil
	60°C (140°F)	75°C (167°F)	90°C (194°F)	60°C (140°F)	75°C (167°F)	90°C (194°F)	
	Types TW, UF	Types RHW, THW, THHW, THWN, XHHW, USE, ZW	Types TBS, SA, SIS, FEP, FEPB, MI, PFA, RHH, RHW-2, THHN, THHW, THW-2, THWN-2, USE-2, XHH, XHHW, XHHW-2, XHWN, XHWN-2, Z, ZW-2	Types TW, UF	Types RHW, THW, THHW, THWN, XHHW, USE	Types TBS, SA, SIS, THHN, THHW, THW-2, THWN-2, RHH, RHW-2, USE-2, XHH, XHHW, XHHW-2, XHWN, XHWN-2, XHHN	
	COPPER			**ALUMINUM OR COPPER-CLAD ALUMINUM**			
1/0	125	150	170	100	120	135	1/0
2/0	145	175	195	115	135	150	2/0
3/0	165	200	225	130	155	175	3/0
4/0	195	230	260	150	180	205	4/0
250	215	255	290	170	205	230	250
300	240	285	320	195	230	260	300
350	260	310	350	210	250	280	350
400	280	335	380	225	270	305	400
500	320	380	430	260	310	350	500
600	350	420	475	285	340	385	600
700	385	460	520	315	375	425	700
750	400	475	535	320	385	435	750
800	410	490	555	330	395	445	800
900	435	520	585	355	425	480	900

Wiring Methods and Conductors

Size AWG or kcmil	60°C (140°F) Types TW, UF	75°C (167°F) Types RHW, THW, THHW, XHHW, USE, ZW	90°C (194°F) Types TBS, SA, SIS, FEP, FEPB, MI, PFA, RHH, RHW-2, THHN, THHW, THW-2, THWN-2, USE-2, XHH, XHHW, XHHW-2, XHWN, XHWN-2, Z, ZW-2	60°C (140°F) Types TW, UF	75°C (167°F) Types RHW, THW, THHW, THWN, XHHW, USE	90°C (194°F) Types TBS, SA, SIS, THHN, THHW, THW-2, THWN-2, RHH, RHW-2, USE-2, XHH, XHHW, XHHW-2, XHWN, XHWN-2, XHHN	Size AWG or kcmil
	COPPER			ALUMINUM OR COPPER-CLAD ALUMINUM			
1000	455	545	615	375	445	500	1000
1250	495	590	665	405	485	545	1250
1500	525	625	705	435	520	585	1500
1750	545	650	735	455	545	615	1750
2000	555	665	750	470	560	630	2000

Notes:
1. Section 310.15(B) shall be referenced for ampacity correction factors where the ambient temperature is other than 30°C (86°F).
2. Section 310.15(C)(1) shall be referenced for more than three current-carrying conductors.
3. Section 310.16 shall be referenced for conditions of use.
*Section 240.4(D) shall be referenced for conductor overcurrent protection limitations, except as modified elsewhere in the *Code*.

CHAPTER 16
BRANCH CIRCUITS

INTRODUCTION

This chapter contains requirements from Article 210 — Branch Circuits. The requirements in Article 210 have been divided into three chapters of this *Pocket Guide*. This chapter covers Part I (general provisions) and Part II (branch-circuit ratings). Some of the general requirements pertain to multiwire branch circuits, ground-fault circuit interrupters, required branch circuits, and arc-fault circuit interrupters. Part II covers the branch-circuit ratings for conductors, equipment, and outlets.

Chapter 17 contains the receptacle outlet requirements in Article 210, and Chapter 19 contains the lighting outlet provisions.

ARTICLE 210
Branch Circuits

PART I. GENERAL PROVISIONS

210.1 Scope. This article provides the general requirements for branch circuits.

210.3 Other Articles for Specific-Purpose Branch Circuits. Table 210.3 lists references for specific equipment and applications not located in Chapters 5, 6, and 7 that amend or supplement the requirements of this article.

210.4 Multiwire Branch Circuits.

(A) General. Branch circuits recognized by this article shall be permitted as multiwire circuits. A multiwire circuit shall be permitted to be considered as multiple circuits. All conductors of a multiwire branch circuit shall originate from the same panelboard or similar distribution equipment.

(B) Disconnecting Means. Each multiwire branch circuit shall be provided with a means that will simultaneously disconnect all ungrounded conductors at the point where the branch circuit originates.

Informational Note: See 240.15(B) for information on the use of single-pole circuit breakers as the disconnecting means.

(C) Line-to-Neutral Loads. Multiwire branch circuits shall supply only line-to-neutral loads.

Exception No. 1: A multiwire branch circuit that supplies only one utilization equipment.

Exception No. 2: Where all ungrounded conductors of the multiwire branch circuit are opened simultaneously by the branch-circuit overcurrent device.

(D) Grouping. The ungrounded and grounded circuit conductors of each multiwire branch circuit shall be grouped in accordance with 200.4(B).

210.5 Identification for Branch Circuits.

(A) Grounded Conductor. The grounded conductor of a branch circuit shall be identified in accordance with 200.6.

Branch Circuits 210.8

(B) Equipment Grounding Conductor. The equipment grounding conductor shall be identified in accordance with 250.119.

(C) Identification of Ungrounded Conductors. Ungrounded conductors shall be identified in accordance with 210.5(C)(1) or (2), as applicable.

(2) Branch Circuits Supplied from Direct-Current Systems. Where a branch circuit is supplied from a dc system operating at more than 60 volts, each ungrounded conductor of 4 AWG or larger shall be identified by polarity at all termination, connection, and splice points by marking tape, tagging, or other approved means; each ungrounded conductor of 6 AWG or smaller shall be identified by polarity at all termination, connection, and splice points in compliance with 210.5(C)(2)(a) and (b). The identification methods utilized for conductors originating within each branch-circuit panelboard or similar branch-circuit distribution equipment shall be documented in a manner that is readily available or shall be permanently posted at each branch-circuit panelboard or similar branch-circuit distribution equipment.

210.7 Multiple Branch Circuits. Where two or more branch circuits supply devices or equipment on the same yoke or mounting strap, a means to simultaneously disconnect the ungrounded supply conductors shall be provided at the point at which the branch circuits originate.

210.8 Ground-Fault Circuit-Interrupter Protection for Personnel. Ground-fault circuit-interrupter protection for personnel shall be provided as required in 210.8(A) through (F). The ground-fault circuit interrupter shall be installed in a readily accessible location.

For the purposes of this section, when determining the distance from receptacles the distance shall be measured as the shortest path the supply cord of an appliance connected to the receptacle would follow without piercing a floor, wall, ceiling, or fixed barrier, or the shortest path without passing through a window.

(A) Dwelling Units. All 125-volt through 250-volt receptacles installed in the locations specified in 210.8(A)(1) through (A)(11) and supplied by single-phase branch circuits rated 150 volts or less to ground shall have ground-fault circuit-interrupter protection for personnel.

(1) Bathrooms
(2) Garages and also accessory buildings that have a floor located at or below grade level not intended as habitable rooms and limited to storage areas, work areas, and areas of similar use

(3) Outdoors

Exception to (3): Receptacles that are not readily accessible and are supplied by a branch circuit dedicated to electric snow-melting, deicing, or pipeline and vessel heating equipment shall be permitted to be installed in accordance with 426.28 or 427.22, as applicable.

(4) Crawl spaces — at or below grade level
(5) Basements

Exception to (5): A receptacle supplying only a permanently installed fire alarm or burglar alarm system shall not be required to have ground-fault circuit-interrupter protection.

Receptacles installed under the exception to 210.8(A)(5) shall not be considered as meeting the requirements of 210.52(G).

(6) Kitchens — where the receptacles are installed to serve the countertop surfaces
(7) Sinks — where receptacles are installed within 1.8 m (6 ft) from the top inside edge of the bowl of the sink
(8) Boathouses
(9) Bathtubs or shower stalls — where receptacles are installed within 1.8 m (6 ft) of the outside edge of the bathtub or shower stall
(10) Laundry areas
(11) Indoor damp and wet locations

(C) Crawl Space Lighting Outlets. GFCI protection shall be provided for lighting outlets not exceeding 120 volts installed in crawl spaces.

(D) Specific Appliances. Unless GFCI protection is provided in accordance with 422.5(B)(3) through (B)(5), the outlets supplying the appliances specified in 422.5(A) shall have GFCI protection in accordance with 422.5(B)(1) or (B)(2).

Where the appliance is a vending machine as specified in 422.5(A)(5) and GFCI protection is not provided in accordance with 422.5(B)(3) or (B)(4), branch circuits supplying vending machines shall have GFCI protection in accordance with 422.5(B)(1) or (B)(2).

(E) Equipment Requiring Servicing. GFCI protection shall be provided for the receptacles required by 210.63.

210.11 Branch Circuits Required. Branch circuits for lighting and for appliances, including motor-operated appliances, shall be provided to supply the loads calculated in accordance with 220.10. In addition,

Branch Circuits 210.12

branch circuits shall be provided for specific loads not covered by 220.10 where required elsewhere in this *Code* and for dwelling unit loads as specified in 210.11(C).

210.12 Arc-Fault Circuit-Interrupter Protection. Arc-fault circuit-interrupter protection shall be provided as required in 210.12(A), (B), (C), and (D). The arc-fault circuit interrupter shall be installed in a readily accessible location.

(A) Dwelling Units. All 120-volt, single-phase, 15- and 20-ampere branch circuits supplying outlets or devices installed in dwelling unit kitchens, family rooms, dining rooms, living rooms, parlors, libraries, dens, bedrooms, sunrooms, recreation rooms, closets, hallways, laundry areas, or similar rooms or areas shall be protected by any of the means described in 210.12(A)(1) through (A)(6):

(1) A listed combination-type arc-fault circuit interrupter installed to provide protection of the entire branch circuit

(2) A listed branch/feeder-type AFCI installed at the origin of the branch-circuit in combination with a listed outlet branch-circuit-type arc-fault circuit interrupter installed at the first outlet box on the branch circuit. The first outlet box in the branch circuit shall be marked to indicate that it is the first outlet of the circuit.

(3) A listed supplemental arc protection circuit breaker installed at the origin of the branch circuit in combination with a listed outlet branch-circuit-type arc-fault circuit interrupter installed at the first outlet box on the branch circuit where all of the following conditions are met:

   a. The branch-circuit wiring shall be continuous from the branch-circuit overcurrent device to the outlet branch-circuit arc-fault circuit interrupter.
   b. The maximum length of the branch-circuit wiring from the branch-circuit overcurrent device to the first outlet shall not exceed 15.2 m (50 ft) for a 14 AWG conductor or 21.3 m (70 ft) for a 12 AWG conductor.
   c. The first outlet box in the branch circuit shall be marked to indicate that it is the first outlet of the circuit.

(4) A listed outlet branch-circuit-type arc-fault circuit interrupter installed at the first outlet on the branch circuit in combination with a listed branch-circuit overcurrent protective device where all of the following conditions are met:

a. The branch-circuit wiring shall be continuous from the branch-circuit overcurrent device to the outlet branch-circuit arc-fault circuit interrupter.
b. The maximum length of the branch-circuit wiring from the branch-circuit overcurrent device to the first outlet shall not exceed 15.2 m (50 ft) for a 14 AWG conductor or 21.3 m (70 ft) for a 12 AWG conductor.
c. The first outlet box in the branch circuit shall be marked to indicate that it is the first outlet of the circuit.
d. The combination of the branch-circuit overcurrent device and outlet branch-circuit AFCI shall be identified as meeting the requirements for a system combination-type AFCI and shall be listed as such.

(5) If metal raceway, metal wireways, metal auxiliary gutters, or Type MC, or Type AC cable meeting the applicable requirements of 250.118, with metal boxes, metal conduit bodies, and metal enclosures are installed for the portion of the branch circuit between the branch-circuit overcurrent device and the first outlet, it shall be permitted to install a listed outlet branch-circuit-type AFCI at the first outlet to provide protection for the remaining portion of the branch circuit.

(6) Where a listed metal or nonmetallic conduit or tubing or Type MC cable is encased in not less than 50 mm (2 in.) of concrete for the portion of the branch circuit between the branch-circuit overcurrent device and the first outlet, it shall be permitted to install a listed outlet branch-circuit-type AFCI at the first outlet to provide protection for the remaining portion of the branch circuit.

Exception: AFCI protection shall not be required for an individual branch circuit supplying a fire alarm system installed in accordance with 760.41(B) or 760.121(B). The branch circuit shall be installed in a metal raceway, metal auxiliary gutter, steel-armored cable, Type MC or Type AC, meeting the applicable requirements of 250.118, with metal boxes, conduit bodies, and enclosures.

Informational Note No. 1: For information on combination-type and branch/feeder-type arc-fault circuit interrupters, see UL 1699-2011, *Standard for Arc-Fault Circuit Interrupters*. For information on outlet branch-circuit-type arc-fault circuit interrupters, see UL Subject 1699A, *Outline of Investigation for Outlet Branch Circuit Arc-Fault Circuit-Interrupters*.

Branch Circuits 210.18

Informational Note No. 2: See 29.6.3(5) of *NFPA 72-2019, National Fire Alarm and Signaling Code,* for information related to secondary power-supply requirements for smoke alarms installed in dwelling units.

Informational Note No. 3: See 760.41(B) and 760.121(B) for power-supply requirements for fire alarm systems.

(D) Branch Circuit Extensions or Modifications — Dwelling Units, Dormitory Units, and Guest Rooms and Guest Suites. Where branch circuit wiring for any of the areas specified in 210.12(A), (B), or (C) is modified, replaced, or extended, the branch circuit shall be protected by one of the following:

(1) By any of the means described in 210.12(A)(1) through (A)(6)
(2) A listed outlet branch-circuit-type AFCI located at the first receptacle outlet of the existing branch circuit

Exception: AFCI protection shall not be required where the extension of the existing branch circuit conductors is not more than 1.8 m (6 ft) and does not include any additional outlets or devices, other than splicing devices. This measurement shall not include the conductors inside an enclosure, cabinet, or junction box.

210.15 Reconditioned Equipment. The following shall not be reconditioned:

(1) Equipment that provides ground-fault circuit-interrupter protection for personnel
(2) Equipment that provides arc-fault circuit-interrupter protection

PART II. BRANCH-CIRCUIT RATINGS

210.18 Rating. Branch circuits recognized by this article shall be rated in accordance with the maximum permitted ampere rating or setting of the overcurrent device. The rating for other than individual branch circuits shall be 15, 20, 30, 40, and 50 amperes. Where conductors of higher ampacity are used for any reason, the ampere rating or setting of the specified overcurrent device shall determine the circuit rating.

210.19 Conductors — Minimum Ampacity and Size.

(A) Branch Circuits Not More Than 600 Volts.

(1) General. Branch-circuit conductors shall have an ampacity not less than the larger of 210.19(A)(1)(a) or (A)(1)(b) and comply with 110.14(C) for equipment terminations.

(a) Where a branch circuit supplies continuous loads or any combination of continuous and noncontinuous loads, the minimum branch-circuit conductor size shall have an ampacity not less than the noncontinuous load plus 125 percent of the continuous load in accordance with 310.14

(b) The minimum branch-circuit conductor size shall have an ampacity not less than the maximum load to be served after the application of any adjustment or correction factors in accordance with 310.15.

(2) Branch Circuits with More than One Receptacle. Conductors of branch circuits supplying more than one receptacle for cord-and-plug-connected portable loads shall have an ampacity of not less than the rating of the branch circuit.

(3) Household Ranges and Cooking Appliances. Branch-circuit conductors supplying household ranges, wall-mounted ovens, counter-mounted cooking units, and other household cooking appliances shall have an ampacity not less than the rating of the branch circuit and not less than the maximum load to be served. For ranges of 8¾ kW or more rating, the minimum branch-circuit rating shall be 40 amperes.

Exception No. 1: Conductors tapped from a branch circuit not exceeding 50 amperes supplying electric ranges, wall-mounted electric ovens, and counter-mounted electric cooking units shall have an ampacity of not less than 20 amperes and shall be sufficient for the load to be served. These tap conductors include any conductors that are a part of the leads supplied with the appliance that are smaller than the branch-circuit conductors. The taps shall not be longer than necessary for servicing the appliance.

Exception No. 2: The neutral conductor of a 3-wire branch circuit supplying a household electric range, a wall-mounted oven, or a counter-mounted cooking unit shall be permitted to be smaller than the ungrounded conductors where the maximum demand of a range of 8¾ kW or more rating has been calculated according to Column C of Table 220.55, but such conductor shall have an ampacity of not less than 70 percent of the branch-circuit rating and shall not be smaller than 10 AWG.

Branch Circuits   210.21

(4) Other Loads. Branch-circuit conductors that supply loads other than those specified in 210.3 and other than cooking appliances as covered in 210.19(A)(3) shall have an ampacity sufficient for the loads served and shall not be smaller than 14 AWG.

210.20 Overcurrent Protection. Branch-circuit conductors and equipment shall be protected by overcurrent protective devices that have a rating or setting that complies with 210.20(A) through (D).

(A) Continuous and Noncontinuous Loads. Where a branch circuit supplies continuous loads or any combination of continuous and noncontinuous loads, the rating of the overcurrent device shall not be less than the noncontinuous load plus 125 percent of the continuous load.

(B) Conductor Protection. Conductors shall be protected in accordance with 240.4. Flexible cords and fixture wires shall be protected in accordance with 240.5.

(D) Outlet Devices. The rating or setting shall not exceed that specified in 210.21 for outlet devices.

210.21 Outlet Devices. Outlet devices shall have an ampere rating that is not less than the load to be served and shall comply with 210.21(A) and (B).

(B) Receptacles.

(1) Single Receptacle on an Individual Branch Circuit. A single receptacle installed on an individual branch circuit shall have an ampere rating not less than that of the branch circuit.

(2) Total Cord-and-Plug-Connected Load. Where connected to a branch circuit supplying two or more receptacles or outlets, a receptacle shall not supply a total cord-and-plug-connected load in excess of the maximum specified in Table 210.21(B)(2).

Table 210.21(B)(2) Maximum Cord-and-Plug-Connected Load to Receptacle

Circuit Rating (Amperes)	Receptacle Rating (Amperes)	Maximum Load (Amperes)
15 or 20	15	12
20	20	16
30	30	24

Table 210.21(B)(3) Receptacle Ratings for Various Size Circuits

Circuit Rating (Amperes)	Receptacle Rating (Amperes)
15	Not over 15
20	15 or 20
30	30
40	40 or 50
50	50

(3) Receptacle Ratings. Where connected to a branch circuit supplying two or more receptacles or outlets, receptacle ratings shall conform to the values listed in Table 210.21(B)(3), or, where rated higher than 50 amperes, the receptacle rating shall not be less than the branch-circuit rating.

(4) Range Receptacle Rating. The ampere rating of a range receptacle shall be permitted to be based on a single range demand load as specified in Table 220.55.

210.22 Permissible Loads, Individual Branch Circuits. An individual branch circuit shall be permitted to supply any load for which it is rated, but in no case shall the load exceed the branch-circuit ampere rating.

210.23 Permissible Loads, Multiple-Outlet Branch Circuits. In no case shall the load exceed the branch-circuit ampere rating. A branch circuit supplying two or more outlets or receptacles shall supply only the loads specified according to its size as specified in 210.23(A) through (D) and as summarized in 210.24 and Table 210.24.

(A) 15- and 20-Ampere Branch Circuits. A 15- or 20-ampere branch circuit shall be permitted to supply lighting units or other utilization equipment, or a combination of both, and shall comply with 210.23(A)(1) and (A)(2).

Exception: The small-appliance branch circuits, laundry branch circuits, and bathroom branch circuits required in a dwelling unit(s) by 210.11 (C)(1), (C)(2), and (C)(3) shall supply only the receptacle outlets specified in that section.

(1) Cord-and-Plug-Connected Equipment Not Fastened in Place. The rating of any one cord-and-plug-connected utilization equipment not fastened in place shall not exceed 80 percent of the branch-circuit ampere rating.

Branch Circuits 210.25

(2) Utilization Equipment Fastened in Place. The total rating of utilization equipment fastened in place, other than luminaires, shall not exceed 50 percent of the branch-circuit ampere rating where lighting units, cord-and-plug-connected utilization equipment not fastened in place, or both, are also supplied.

(B) 30-Ampere Branch Circuits. A 30-ampere branch circuit shall be permitted to supply fixed lighting units with heavy-duty lampholders in other than a dwelling unit(s) or utilization equipment in any occupancy. A rating of any one cord-and-plug-connected utilization equipment shall not exceed 80 percent of the branch-circuit ampere rating.

(C) 40- and 50-Ampere Branch Circuits. A 40- or 50-ampere branch circuit shall be permitted to supply cooking appliances that are fastened in place in any occupancy. In other than dwelling units, such circuits shall be permitted to supply fixed lighting units with heavy-duty lampholders, infrared heating units, or other utilization equipment.

(D) Branch Circuits Larger Than 50 Amperes. Branch circuits larger than 50 amperes shall supply only nonlighting outlet loads.

210.24 Branch-Circuit Requirements — Summary. The requirements for circuits that have two or more outlets or receptacles, other than the receptacle circuits of 210.11(C)(1), (C)(2), and (C)(3), are summarized in Table 210.24. This table provides only a summary of minimum requirements. See 210.19, 210.20, and 210.21 for the specific requirements applying to branch circuits.

210.25 Branch Circuits in Buildings with More Than One Occupancy.

(A) Dwelling Unit Branch Circuits. Branch circuits in each dwelling unit shall supply only loads within that dwelling unit or loads associated only with that dwelling unit.

(B) Common Area Branch Circuits. Branch circuits installed for lighting, central alarm, signal, communications, or other purposes for public or common areas of a two-family dwelling, a multifamily dwelling, or a multi-occupancy building shall not be supplied from equipment that supplies an individual dwelling unit or tenant space.

PART III. REQUIRED OUTLETS

(See Chapter 17 for required receptacles and Chapter 19 for required lighting outlets.)

Table 210.24 Summary of Branch-Circuit Requirements

Circuit Rating	15 A	20 A	30 A	40 A	50 A
Conductors (min. size):¹					
Circuit wires¹	14	12	10	8	6
Taps	14	14	14	12	12
Fixture wires and cords — see 240.5					
Overcurrent Protection	**15 A**	**20 A**	**30 A**	**40 A**	**50 A**
Outlet devices:					
Lampholders permitted	Any type	Any type	Heavy duty	Heavy duty	Heavy duty
Receptacle rating²	15 max. A	15 or 20 A	30 A	40 or 50 A	50 A
Maximum Load	**15 A**	**20 A**	**30 A**	**40 A**	**50 A**
Permissible load	See 210.23(A)	See 210.23(A)	See 210.23(B)	See 210.23(C)	See 210.23(C)

¹These gauges are for copper conductors.
²For receptacle rating of cord-connected electric-discharge luminaires, see 410.62(C).

CHAPTER 17
RECEPTACLE PROVISIONS

INTRODUCTION

This chapter contains requirements from Article 210 — Branch Circuits, and Article 406 — Receptacles, Cord Connectors, and Attachment Plugs (Caps). The requirements from Article 210 covered in this chapter include 210.50 through 210.64. These sections cover receptacle outlets for both general and specific locations. General locations include kitchens, family rooms, dining rooms, living rooms, bedrooms, and similar rooms and areas. Specific locations include bathrooms, hallways, basements and garages, laundry areas, and kitchen countertop areas.

Article 406 covers the rating, type, and installation of receptacles, cord connectors, and attachment plugs (cord caps). Some of the topics covered are receptacle rating and type, general installation requirements, receptacle mounting, receptacles in damp or wet locations, and connecting the receptacle's grounding terminal to the box. This article also contains requirements on replacing receptacles installed in existing dwelling units.

PART III. REQUIRED OUTLETS

210.50 Receptacle Outlets. Receptacle outlets shall be installed as specified in 210.52 through 210.65.

(A) Cord Pendants. A cord connector that is supplied by a permanently connected cord pendant shall be considered a receptacle outlet.

(B) Cord Connections. A receptacle outlet shall be installed wherever flexible cords with attachment plugs are used. Where flexible cords are permitted to be permanently connected, receptacles shall be permitted to be omitted for such cords.

(C) Appliance Receptacle Outlets. Appliance receptacle outlets installed in a dwelling unit for specific appliances, such as laundry equipment, shall be installed within 1.8 m (6 ft) of the intended location of the appliance.

210.52 Dwelling Unit Receptacle Outlets. This section provides requirements for 125-volt, 15- and 20-ampere receptacle outlets. The receptacles required by this section shall be in addition to any receptacle that is as follows:

(1) Part of a luminaire or appliance, or
(2) Controlled by a listed wall-mounted control device in accordance with 210.70(A)(1), Exception No. 1, or
(3) Located within cabinets or cupboards, or
(4) Located more than 1.7 m (5½ ft) above the floor

Permanently installed electric baseboard heaters equipped with factory-installed receptacle outlets or outlets provided as a separate assembly by the manufacturer shall be permitted as the required outlet or outlets for the wall space utilized by such permanently installed heaters. Such receptacle outlets shall not be connected to the heater circuits.

Informational Note: Listed baseboard heaters include instructions that may not permit their installation below receptacle outlets.

(A) General Provisions. In every kitchen, family room, dining room, living room, parlor, library, den, sunroom, bedroom, recreation room, or similar room or area of dwelling units, receptacle outlets shall be installed in accordance with the general provisions specified in 210.52 (A)(1) through (A)(4).

(1) Spacing. Receptacles shall be installed such that no point measured horizontally along the floor line of any wall space is more than 1.8 m (6 ft) from a receptacle outlet.

Receptacle Provisions 210.52

(2) Wall Space. As used in this section, a wall space shall include the following:

(1) Any space 600 mm (2 ft) or more in width (including space measured around corners) and unbroken along the floor line by doorways and similar openings, fireplaces, and fixed cabinets that do not have countertops or similar work surfaces
(2) The space occupied by fixed panels in walls, excluding sliding panels
(3) The space afforded by fixed room dividers, such as freestanding bar-type counters or railings

(3) Floor Receptacles. Receptacle outlets in or on floors shall not be counted as part of the required number of receptacle outlets unless located within 450 mm (18 in.) of the wall.

(4) Countertop and Similar Work Surface Receptacle Outlets. Receptacles installed for countertop and similar work surfaces as specified in 210.52(C) shall not be considered as the receptacle outlets required by 210.52(A).

(B) Small Appliances.

(1) Receptacle Outlets Served. In the kitchen, pantry, breakfast room, dining room, or similar area of a dwelling unit, the two or more 20-ampere small-appliance branch circuits required by 210.11(C)(1) shall serve all wall and floor receptacle outlets covered by 210.52(A), all countertop outlets covered by 210.52(C), and receptacle outlets for refrigeration equipment.

Exception No. 1: In addition to the required receptacles specified by 210.52, switched receptacles supplied from a general-purpose 15- or 20-ampere branch circuit as required in 210.70(A)(1), Exception No. 1, shall be permitted.

Exception No. 2: In addition to the required receptacles specified by 210.52, a receptacle outlet to serve a specific appliance shall be permitted to be supplied from an individual branch circuit rated 15 amperes or greater.

(2) No Other Outlets. The two or more small-appliance branch circuits specified in 210.52(B)(1) shall have no other outlets.

Exception No. 1: A receptacle installed solely for the electrical supply to and support of an electric clock in any of the rooms specified in 210.52(B)(1).

Exception No. 2: Receptacles installed to provide power for supplemental equipment and lighting on gas-fired ranges, ovens, or counter-mounted cooking units.

(3) Kitchen Receptacle Requirements. Receptacles installed in a kitchen to serve countertop surfaces shall be supplied by not fewer than two small-appliance branch circuits, either or both of which shall also be permitted to supply receptacle outlets in the same kitchen and in other rooms specified in 210.52(B)(1). Additional small-appliance branch circuits shall be permitted to supply receptacle outlets in the kitchen and other rooms specified in 210.52(B)(1). No small-appliance branch circuit shall serve more than one kitchen.

(C) Countertops and Work Surfaces. In kitchens, pantries, breakfast rooms, dining rooms, and similar areas of dwelling units, receptacle outlets for countertop and work surfaces that are 300 mm (12 in.) or wider shall be installed in accordance with 210.52(C)(1)through (C)(3) and shall not be considered as the receptacle outlets required by 210.52(A).

For the purposes of this section, where using multioutlet assemblies, each 300 mm (12 in.) of multioutlet assembly containing two or more receptacles installed in individual or continuous lengths shall be considered to be one receptacle outlet.

(1) Wall Spaces. Receptacle outlets shall be installed so that no point along the wall line is more than 600 mm (24 in.) measured horizontally from a receptacle outlet in that space.

Exception: Receptacle outlets shall not be required directly behind a range, counter-mounted cooking unit, or sink in the installation described in Figure 210.52(C)(1).

(2) Island and Peninsular Countertops and Work Surfaces. Receptacle outlets shall be installed in accordance with 210.52(C)(2)(a) and (C)(2)(b).

(a) At least one receptacle outlet shall be provided for the first 0.84 m² (9 ft²), or fraction thereof, of the countertop or work surface. A receptacle outlet shall be provided for every additional 1.7 m² (18 ft²), or fraction thereof, of the countertop or work surface.

(b) At least one receptacle outlet shall be located within 600 mm (2 ft) of the outer end of a peninsular countertop or work surface. Additional required receptacle outlets shall be permitted to be located as determined by the installer, designer, or building owner. The location of the receptacle outlets shall be in accordance with 210.52(C)(3).

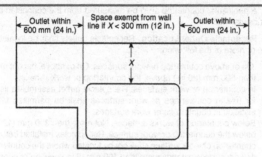

Range, counter-mounted cooking unit, or sink extending from face of counter

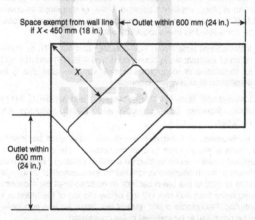

Range, counter-mounted cooking unit, or sink mounted in corner

Figure 210.52(C)(1) Determination of Area Behind a Range, Counter-Mounted Cooking Unit, or Sink.

A peninsular countertop shall be measured from the connected perpendicular wall.

(3) Receptacle Outlet Location. Receptacle outlets shall be located in one or more of the following:

(1) On or above countertop or work surfaces: On or above, but not more than 500 mm (20 in.) above, the countertop or work surface.
(2) In countertop or work surfaces: Receptacle outlet assemblies listed for use in countertops or work surfaces shall be permitted to be installed in countertops or work surfaces.
(3) Below countertop or works surfaces: Not more than 300 mm (12 in.) below the countertop or work surface. Receptacles installed below a countertop or work surface shall not be located where the countertop or work surface extends more than 150 mm (6 in.) beyond its support base.

Receptacle outlets rendered not readily accessible by appliances fastened in place, appliance garages, sinks, or rangetops as covered in 210.52(C)(1), Exception, or appliances occupying assigned spaces shall not be considered as these required outlets.

Informational Note No. 1: See 406.5(E) and 406.5(G) for installation of receptacles in countertops and 406.5(F) and 406.5(G) for installation of receptacles in work surfaces. See 380.10 for installation of multioutlet assemblies.

Informational Note No. 2: See Annex J and ANSI/ICC A117.1-2009, *Standard on Accessible and Usable Buildings and Facilities*.

(D) Bathrooms. At least one receptacle outlet shall be installed in bathrooms within 900 mm (3 ft) of the outside edge of each basin. The receptacle outlet shall be located on a wall or partition that is adjacent to the basin or basin countertop, located on the countertop, or installed on the side or face of the basin cabinet. In no case shall the receptacle be located more than 300 mm (12 in.) below the top of the basin or basin countertop. Receptacle outlet assemblies listed for use in countertops shall be permitted to be installed in the countertop.

(E) Outdoor Outlets. Outdoor receptacle outlets shall be installed in accordance with 210.52(E)(1) through (E)(3).

(1) One-Family and Two-Family Dwellings. For a one-family dwelling and each unit of a two-family dwelling that is at grade level, at least one receptacle outlet readily accessible from grade and not more than 2.0 m (6½ ft) above grade level shall be installed at the front and back of the dwelling.

(2) Multifamily Dwellings. For each dwelling unit of a multifamily dwelling where the dwelling unit is located at grade level and provided with individual exterior entrance/egress, at least one receptacle outlet readily accessible from grade and not more than 2.0 m (6½ ft) above grade level shall be installed.

(3) Balconies, Decks, and Porches. Balconies, decks, and porches that are within 102 mm (4 in.) horizontally of the dwelling unit shall have at least one receptacle outlet accessible from the balcony, deck, or porch. The receptacle outlet shall not be located more than 2.0 m (6½ ft) above the balcony, deck, or porch walking surface.

(F) Laundry Areas. In dwelling units, at least one receptacle outlet shall be installed in areas designated for the installation of laundry equipment.

Exception No. 1: A receptacle for laundry equipment shall not be required in a dwelling unit of a multifamily building where laundry facilities are provided on the premises for use by all building occupants.

Exception No. 2: A receptacle for laundry equipment shall not be required in other than one-family dwellings where laundry facilities are not to be installed or permitted.

(G) Basements, Garages, and Accessory Buildings. For one- and two- family dwellings, and multifamily dwellings, at least one receptacle outlet shall be installed in the areas specified in 210.52(G)(1) through (G)(3). These receptacles shall be in addition to receptacles required for specific equipment.

(1) Garages. In each attached garage and in each detached garage with electric power, at least one receptacle outlet shall be installed in each vehicle bay and not more than 1.7 m (5½ ft) above the floor.

Exception: Garage spaces not attached to an individual dwelling unit of a multifamily dwelling shall not require a receptacle outlet in each vehicle bay.

(2) Accessory Buildings. In each accessory building with electric power.

(3) Basements. In each separate unfinished portion of a basement.

(H) Hallways. In dwelling units, hallways of 3.0 m (10 ft) or more in length shall have at least one receptacle outlet.

As used in this subsection, the hallway length shall be considered the length along the centerline of the hallway without passing through a doorway.

(I) Foyers. Foyers that are not part of a hallway in accordance with 210.52(H) and that have an area that is greater than 5.6 m² (60 ft²) shall have a receptacle(s) located in each wall space 900 mm (3 ft) or more in width. Doorways, door-side windows that extend to the floor, and similar openings shall not be considered wall space.

210.63 Equipment Requiring Servicing. A 125-volt, single-phase, 15- or 20-ampere-rated receptacle outlet shall be installed at an accessible location within 7.5 m (25 ft) of the equipment as specified in 210.63(A) and (B).

(A) Heating, Air-Conditioning, and Refrigeration Equipment. The required receptacle outlet shall be located on the same level as the heating, air-conditioning, and refrigeration equipment. The receptacle outlet shall not be connected to the load side of the equipment's branch-circuit disconnecting means.

Exception: A receptacle outlet shall not be required at one- and two-family dwellings for the service of evaporative coolers.

(B) Other Electrical Equipment. In other than one- and two-family dwellings, a receptacle outlet shall be located as specified in 210.63(B)(1) and (B)(2).

(1) Indoor Service Equipment. The required receptacle outlet shall be located within the same room or area as the service equipment.

(2) Indoor Equipment Requiring Dedicated Equipment Spaces. Where equipment, other than service equipment, requires dedicated equipment space as specified in 110.26(E), the required receptacle outlet shall be located within the same room or area as the electrical equipment and shall not be connected to the load side of the equipment's branch-circuit disconnecting means.

ARTICLE 406
Receptacles, Cord Connectors, and Attachment Plugs (Caps)

406.2 Definitions. The definitions in this section shall apply only within this article.

Receptacle Provisions 406.3

Outlet Box Hood. A housing shield intended to fit over a faceplate for flush-mounted wiring devices, or an integral component of an outlet box or of a faceplate for flush-mounted wiring devices. The hood does not serve to complete the electrical enclosure; it reduces the risk of water coming in contact with electrical components within the hood, such as attachment plugs, current taps, surge protective devices, direct plug-in transformer units, or wiring devices.

406.3 Receptacle Rating and Type.

(A) Receptacles. Receptacles shall be listed and marked with the manufacturer's name or identification and voltage and ampere ratings. Receptacles shall not be permitted to be reconditioned.

(B) Rating. Receptacles and cord connectors shall be rated not less than 15 amperes, 125 volts, or 15 amperes, 250 volts, and shall be of a type not suitable for use as lampholders.

(C) Receptacles for Aluminum Conductors. Receptacles rated 20 amperes or less and designed for the direct connection of aluminum conductors shall be marked CO/ALR.

(D) Isolated Ground Receptacles. Receptacles incorporating an isolated equipment grounding conductor connection intended for the reduction of electromagnetic interference as permitted in 250.146(D) shall be identified by an orange triangle located on the face of the receptacle.

(1) Isolated Equipment Grounding Conductor Required. Receptacles so identified shall be used only with equipment grounding conductors that are isolated in accordance with 250.146(D).

(2) Installation in Nonmetallic Boxes. Isolated ground receptacles installed in nonmetallic boxes shall be covered with a nonmetallic faceplate.

Exception: Where an isolated ground receptacle is installed in a nonmetallic box, a metal faceplate shall be permitted if the box contains a feature or accessory that permits the connection of the faceplate to the equipment grounding conductor.

(E) Controlled Receptacle Marking. All nonlocking-type, 125-volt, 15- and 20-ampere receptacles that are controlled by an automatic control device, or that incorporate control features that remove power from the receptacle for the purpose of energy management or building automation, shall be permanently marked with the symbol shown in Figure 406.3(E) and the word "controlled."

For receptacles controlled by an automatic control device, the marking shall be located on the receptacle face and visible after installation.

In both cases where a multiple receptacle device is used, the required marking of the word "controlled" and symbol shall denote which contact device(s) are controlled.

(F) Receptacle with USB Charger. A 125-volt 15- or 20-ampere receptacle that additionally provides Class 2 power shall be listed and constructed such that the Class 2 circuitry is integral with the receptacle.

406.4 General Installation Requirements. Receptacle outlets shall be located in branch circuits in accordance with Part III of Article 210. General installation requirements shall be in accordance with 406.4(A) through (F).

(A) Grounding Type. Except as provided in 406.4(D), receptacles installed on 15- and 20-ampere branch circuits shall be of the grounding type. Grounding-type receptacles shall be installed only on circuits of the voltage class and current for which they are rated, except as provided in 210.21(B)(1) for single receptacles or Table 210.21(B)(2) and Table 210.21(B)(3) for two or more receptacles.

(B) To Be Grounded. Receptacles and cord connectors that have equipment grounding conductor contacts shall have those contacts connected to an equipment grounding conductor.

Exception No. 2: Replacement receptacles as permitted by 406.4(D).

(C) Methods of Grounding. The equipment grounding conductor contacts of receptacles and cord connectors shall be connected to the equipment grounding conductor of the circuit supplying the receptacle or cord connector.

The branch-circuit wiring method shall include or provide an equipment grounding conductor to which the equipment grounding conductor contacts of the receptacle or cord connector are connected.

406.5 Receptacle Mounting. Receptacles shall be mounted in identified boxes or assemblies. The boxes or assemblies shall be securely fastened in place unless otherwise permitted elsewhere in this *Code*. Screws used for the purpose of attaching receptacles to a box shall be of the type provided with a listed receptacle, or shall be machine screws having 32 threads per inch or part of listed assemblies or systems, in accordance with the manufacturer's instructions.

Receptacle Provisions 406.5

(A) Boxes That Are Set Back. Receptacles mounted in boxes that are set back from the finished surface as permitted in 314.20 shall be installed such that the mounting yoke or strap of the receptacle is held rigidly at the finished surface.

(B) Boxes That Are Flush. Receptacles mounted in boxes that are flush with the finished surface or project therefrom shall be installed such that the mounting yoke or strap of the receptacle is held rigidly against the box or box cover.

(C) Receptacles Mounted on Covers. Receptacles mounted to and supported by a cover shall be held rigidly against the cover by more than one screw or shall be a device assembly or box cover listed and identified for securing by a single screw.

(D) Position of Receptacle Faces. After installation, receptacle faces shall be flush with or project from faceplates of insulating material and shall project a minimum of 0.4 mm (0.015 in.) from metal faceplates.

Exception: Listed kits or assemblies encompassing receptacles and nonmetallic faceplates that cover the receptacle face, where the plate cannot be installed on any other receptacle, shall be permitted.

(E) Receptacles in Countertops. Receptacle assemblies for installation in countertop surfaces shall be listed for countertop applications. Where receptacle assemblies for countertop applications are required to provide ground-fault circuit-interrupter protection for personnel in accordance with 210.8, such assemblies shall be permitted to be listed as GFCI receptacle assemblies for countertop applications.

(F) Receptacles in Work Surfaces. Receptacle assemblies and GFCI receptacle assemblies listed for work surface or countertop applications shall be permitted to be installed in work surfaces.

(G) Receptacle Orientation.

(1) Countertop and Work Surfaces. Receptacles shall not be installed in a face-up position in or on countertop surfaces or work surfaces unless listed for countertop or work surface applications.

(2) Under Sinks. Receptacles shall not be installed in a face-up position in the area below a sink.

(H) Receptacles in Seating Areas and Other Similar Surfaces. In seating areas or similar surfaces, receptacles shall not be installed in a face-up position unless the receptacle is any of the following:

(1) Part of an assembly listed as a furniture power distribution unit
(2) Part of an assembly listed either as household furnishings or as commercial furnishings
(3) Listed either as a receptacle assembly for countertop applications or as a GFCI receptacle assembly for countertop applications
(4) Installed in a listed floor box

406.6 Receptacle Faceplates (Cover Plates). Receptacle faceplates shall be installed so as to completely cover the opening and seat against the mounting surface.

Receptacle faceplates mounted inside a box having a recess-mounted receptacle shall effectively close the opening and seat against the mounting surface.

(A) Thickness of Metal Faceplates. Metal faceplates shall be of ferrous metal not less than 0.76 mm (0.030 in.) in thickness or of nonferrous metal not less than 1.02 mm (0.040 in.) in thickness.

(B) Grounding. Metal faceplates shall be grounded.

(C) Faceplates of Insulating Material. Faceplates of insulating material shall be noncombustible and not less than 2.54 mm (0.10 in.) in thickness but shall be permitted to be less than 2.54 mm (0.10 in.) in thickness if formed or reinforced to provide adequate mechanical strength.

(D) Receptacle Faceplate (Cover Plates) with Integral Night Light and/or USB Charger. A flush device cover plate that additionally provides a night light and/or Class 2 output connector(s) shall be listed and constructed such that the night light and/or Class 2 circuitry is integral with the flush device cover plate.

406.7 Attachment Plugs, Cord Connectors, and Flanged Surface Devices. All attachment plugs, cord connectors, and flanged surface devices (inlets and outlets) shall be listed and marked with the manufacturer's name or identification and voltage and ampere ratings. Attachment plugs, cord connectors, and flanged surface devices shall not be permitted to be reconditioned.

(A) Construction of Attachment Plugs and Cord Connectors. Attachment plugs and cord connectors shall be constructed so that there are no exposed current-carrying parts except the prongs, blades, or pins. The cover for wire terminations shall be a part that is essential for the operation of an attachment plug or connector (dead-front construction).

Receptacle Provisions 406.9

(B) Connection of Attachment Plugs. Attachment plugs shall be installed so that their prongs, blades, or pins are not energized unless inserted into an energized receptacle or cord connectors. No receptacle shall be installed so as to require the insertion of an energized attachment plug as its source of supply.

406.9 Receptacles in Damp or Wet Locations.

(A) Damp Locations. A receptacle installed outdoors in a location protected from the weather or in other damp locations shall have an enclosure for the receptacle that is weatherproof when the receptacle is covered (attachment plug cap not inserted and receptacle covers closed).

An installation suitable for wet locations shall also be considered suitable for damp locations.

A receptacle shall be considered to be in a location protected from the weather where located under roofed open porches, canopies, marquees, and the like, and will not be subjected to a beating rain or water runoff. All 15- and 20-ampere, 125- and 250-volt nonlocking receptacles shall be a listed weather-resistant type.

(B) Wet Locations.

(1) Receptacles of 15 and 20 Amperes in a Wet Location. Receptacles of 15 and 20 amperes, 125 and 250 volts installed in a wet location shall have an enclosure that is weatherproof whether or not the attachment plug cap is inserted. An outlet box hood installed for this purpose shall be listed and shall be identified as "extra-duty." Other listed products, enclosures, or assemblies providing weatherproof protection that do not utilize an outlet box hood need not be marked "extra duty."

All 15- and 20-ampere, 125- and 250-volt nonlocking-type receptacles shall be listed and so identified as the weather-resistant type.

(2) Other Receptacles. All other receptacles installed in a wet location shall comply with 406.9(B)(2)(a) or (B)(2)(b).

(a) A receptacle installed in a wet location, where the product intended to be plugged into it is not attended while in use, shall have an enclosure that is weatherproof with the attachment plug cap inserted or removed.

(b) A receptacle installed in a wet location where the product intended to be plugged into it will be attended while in use (e.g., portable tools) shall have an enclosure that is weatherproof when the attachment plug is removed.

(C) Bathtub and Shower Space. Receptacles shall not be installed within a zone measured 900 mm (3 ft) horizontally and 2.5 m (8 ft) vertically from the top of the bathtub rim or shower stall threshold. The identified zone is all-encompassing and shall include the space directly over the tub or shower stall.

Exception: In bathrooms with less than the required zone the receptacle(s) shall be permitted to be installed opposite the bathtub rim or shower stall threshold on the farthest wall within the room.

(E) Flush Mounting with Faceplate. The enclosure for a receptacle installed in an outlet box flush-mounted in a finished surface shall be made weatherproof by means of a weatherproof faceplate assembly that provides a watertight connection between the plate and the finished surface.

406.11 Connecting Receptacle Grounding Terminal to Box. The connection of the receptacle grounding terminal shall comply with 250.146.

406.12 Tamper-Resistant Receptacles. All 15- and 20-ampere, 125- and 250-volt nonlocking-type receptacles in the areas specified in 406.12(1) through (8) shall be listed tamper-resistant receptacles.

(1) Dwelling units, including attached and detached garages and accessory buildings to dwelling units, and common areas of multifamily dwellings specified in 210.52 and 550.13

Exception to (1), (2), (3), (4), (5), (6), (7), and (8): Receptacles in the following locations shall not be required to be tamper resistant:

(1) Receptacles located more than 1.7 m (5½ ft) above the floor
(2) Receptacles that are part of a luminaire or appliance
(3) A single receptacle, or a duplex receptacle for two appliances, located within the dedicated space for each appliance that, in normal use, is not easily moved from one place to another and that is cord-and-plug-connected in accordance with 400.10(A)(6), (A)(7), or (A)(8)
(4) Nongrounding receptacles used for replacements as permitted in 406.4(D)(2)(a)

CHAPTER 18
SWITCHES

INTRODUCTION

This chapter contains requirements from Article 404 — Switches. Unlike the disconnect switches covered in Chapter 8, this chapter covers general-use snap switches. These switches are generally installed in device boxes or on box covers and are commonly used for local control of lighting or other outlets. General-use snap switches can also be used as disconnecting means for smaller motors, air-conditioning equipment, and appliances. The disconnecting means requirements in Articles 404, 422, 430, and 440 provide the limitations on this use of general-use snap switches. Requirements from Article 404 also cover general-use snap switches (single-pole, three-way, four-way, double-pole, etc.), dimmers, and similar control switches. The *Code* contains few rules on the actual placement of switches that control lighting load, and those rules are in Article 210 (see Chapter 19). Many modern switches that react to movement or a room being occupied require power for operation, and the requirements to have a circuit available to power these devices is covered in Article 404.

ARTICLE 404
Switches

PART I. INSTALLATION

404.2 Switch Connections.

(A) Three-Way and Four-Way Switches. Three-way and four-way switches shall be wired so that all switching is done only in the ungrounded circuit conductor. Where in metal raceways or metal-armored cables, wiring between switches and outlets shall be in accordance with 300.20(A).

Exception: Switch loops shall not require a grounded conductor.

(B) Grounded Conductors. Switches or circuit breakers shall not disconnect the grounded conductor of a circuit.

(C) Switches Controlling Lighting Loads. The grounded circuit conductor for the controlled lighting circuit shall be installed at the location where switches control lighting loads that are supplied by a grounded general-purpose branch circuit serving bathrooms, hallways, stairways, and habitable rooms or occupiable spaces as defined in the applicable building code. Where multiple switch locations control the same lighting load such that the entire floor area of the room or space is visible from the single or combined switch locations, the grounded circuit conductor shall only be required at one location. A grounded conductor shall not be required to be installed at lighting switch locations under any of the following conditions:

(1) Where conductors enter the box enclosing the switch through a raceway, provided that the raceway is large enough for all contained conductors, including a grounded conductor
(2) Where the box enclosing the switch is accessible for the installation of an additional or replacement cable without removing finish materials
(3) Where snap switches with integral enclosures comply with 300.15(E)
(4) Where lighting in the area is controlled by automatic means
(5) Where a switch controls a receptacle load

The grounded conductor shall be extended to any switch location as necessary and shall be connected to switching devices that require line-to-neutral voltage to operate the electronics of the switch in the standby mode and shall meet the requirements of 404.22.

Switches 404.9

Exception: The connection requirement shall become effective on January 1, 2020. It shall not apply to replacement or retrofit switches installed in locations prior to local adoption of 404.2(C) and where the grounded conductor cannot be extended without removing finish materials. The number of electronic control switches on a branch circuit shall not exceed five, and the number connected to any feeder on the load side of a system or main bonding jumper shall not exceed 25. For the purpose of this exception, a neutral busbar, in compliance with 200.2(B) and to which a main or system bonding jumper is connected shall not be limited as to the number of electronic lighting control switches connected.

404.4 Damp or Wet Locations.

(A) Surface-Mounted Switch or Circuit Breaker. A surface-mounted switch or circuit breaker shall be enclosed in a weatherproof enclosure or cabinet that complies with 312.2.

(B) Flush-Mounted Switch or Circuit Breaker. A flush-mounted switch or circuit breaker shall be equipped with a weatherproof cover.

(C) Switches in Tub or Shower Spaces. Switches shall not be installed within tub or shower spaces unless installed as part of a listed tub or shower assembly.

404.9 General-Use Snap Switches, Dimmers, and Control Switches.

(A) Faceplates. Faceplates provided for snap switches, dimmers, and control switches mounted in boxes and other enclosures shall be installed so as to completely cover the opening and, where the switch is flush mounted, seat against the finished surface.

(B) Grounding. Snap switches, dimmers, and control switches shall be connected to an equipment grounding conductor and shall provide a means to connect metal faceplates to the equipment grounding conductor, whether or not a metal faceplate is installed. Metal faceplates shall be bonded to the equipment grounding conductor. Snap switches, dimmers, control switches, and metal faceplates shall be connected to an equipment grounding conductor using either of the following methods:

(1) The switch is mounted with metal screws to a metal box or metal cover that is connected to an equipment grounding conductor or to a nonmetallic box with integral means for connecting to an equipment grounding conductor.
(2) An equipment grounding conductor or equipment bonding jumper is connected to an equipment grounding termination of the snap switch.

Exception No. 1 to (B): Where no means exists within the enclosure for bonding to the equipment grounding conductor, or where the wiring method does not include or provide an equipment grounding conductor, a snap switch without a connection to an equipment grounding conductor shall be permitted for replacement purposes only. A snap switch wired under the provisions of this exception and located within 2.5 m (8 ft) vertically, or 1.5 m (5 ft) horizontally, of ground or exposed grounded metal objects shall be provided with a faceplate of nonconducting noncombustible material with nonmetallic attachment screws, unless the switch mounting strap or yoke is nonmetallic or the circuit is protected by a ground-fault circuit interrupter.

Exception No. 2 to (B): Listed kits or listed assemblies shall not be required to be bonded to an equipment grounding conductor if all of the following conditions are met:

(1) The device is provided with a nonmetallic faceplate, and the device is designed such that no metallic faceplate replaces the one provided.
(2) The device does not have mounting means to accept other configurations of faceplates.
(3) The device is equipped with a nonmetallic yoke.
(4) All parts of the device that are accessible after installation of the faceplate are manufactured of nonmetallic materials.

Exception No. 3 to (B): A snap switch with integral nonmetallic enclosure complying with 300.15(E) shall be permitted without a bonding connection to an equipment grounding conductor.

404.10 Mounting of General-Use Snap Switches, Dimmers, and Control Switches.

(A) Surface Type. General-use snap switches, dimmers, and control switches used with open wiring on insulators shall be mounted on insulating material that separates the conductors at least 13 mm (½ in.) from the surface wired over.

(B) Box Mounted. Flush-type general-use snap switches, dimmers, and control switches mounted in boxes that are set back of the finished surface as permitted in 314.20 shall be installed so that the extension plaster ears are seated against the surface. Flush-type devices mounted in boxes that are flush with the finished surface or project from it shall be installed so that the mounting yoke or strap of the device is seated against the box. Screws used for the purpose of attaching a device to a

Switches

box shall be of the type provided with a listed device, or shall be machine screws having 32 threads per inch or part of listed assemblies or systems, in accordance with the manufacturer's instructions.

404.14 Rating and Use of Switches. Switches shall be listed and used within their ratings. Switches of the types covered in 404.14(A) through (E) shall be limited to the control of loads as specified accordingly. Switches used to control cord-and-plug-connected loads shall be limited as covered in 404.14(F).

(A) Alternating-Current General-Use Snap Switch. This form of switch shall only be used on ac circuits and used for controlling the following:

(1) Resistive and inductive loads not exceeding the ampere rating of the switch at the voltage applied
(2) Tungsten-filament lamp loads not exceeding the ampere rating of the switch at 120 volts
(3) Electric discharge lamp loads not exceeding the marked ampere and voltage rating of the switch
(4) Motor loads not exceeding 80 percent of the ampere rating of the switch at its rated voltage
(5) Electronic ballasts, self-ballasted lamps, compact fluorescent lamps, and LED lamp loads with their associated drivers, not exceeding 20 amperes and not exceeding the ampere rating of the switch at the voltage applied

(C) CO/ALR Snap Switches. Snap switches directly connected to aluminum conductors and rated 20 amperes or less shall be marked CO/ALR.

(E) Dimmer and Electronic Control Switches. General-use dimmer switches shall be used only to control permanently installed incandescent luminaires unless listed for the control of other loads and installed accordingly. Other electronic control switches, such as timing switches and occupancy sensors, shall be used to control permanently connected loads. They shall be marked by their manufacturer with their current and voltage ratings and used for loads that do not exceed their ampere rating at the voltage applied.

(F) Cord-and-Plug-Connected Loads. Where a snap switch or control device is used to control cord-and-plug-connected equipment on a general-purpose branch circuit, each snap switch or control device controlling receptacle outlets or cord connectors that are supplied by permanently connected cord pendants shall be rated at not less than the rating of the maximum permitted ampere rating or setting of the overcurrent device protecting the receptacles or cord connectors, as provided in 210.21(B).

> Informational Note: See 210.50(A) and 400.10(A)(1) for equivalency to a receptacle outlet of a cord connector that is supplied by a permanently connected cord pendant.

Exception: Where a snap switch or control device is used to control not more than one receptacle on a branch circuit, the switch or control device shall be permitted to be rated at not less than the rating of the receptacle.

404.22 Electronic Control Switches. Electronic control switches shall be listed. Electronic control switches shall not introduce current on the equipment grounding conductor during normal operation. The requirement to not introduce current on the equipment grounding conductor shall take effect on January 1, 2020.

Exception: Electronic control switches that introduce current on the equipment grounding conductor shall be permitted for applications covered by 404.2(C), Exception. Electronic control switches that introduce current on the equipment grounding conductor shall be listed and marked for use in replacement or retrofit applications only.

CHAPTER 19

LIGHTING REQUIREMENTS

INTRODUCTION

This chapter contains requirements from Article 210 — Branch Circuits; Article 410 — Luminaires, Lampholders, and Lamps; and Article 411 — Low-Voltage Lighting. This chapter covers the last section of Article 210, which provides requirements for lighting outlets and switches for dwelling units.

Article 410 covers luminaires, portable luminaires, lampholders, pendants, incandescent filament lamps, arc lamps, electric-discharge lamps, decorative lighting products, lighting accessories for temporary seasonal and holiday use, portable flexible lighting products, and the wiring and equipment forming part of such products and lighting installations. Contained in this *Pocket Guide* are eight of the fifteen parts that are included in Article 410: Part I General; Part II Luminaire Locations; Part III Provisions at Luminaire Outlet Boxes, Canopies, and Pans; Part IV Luminaire Supports; Part V Grounding; Part VI Wiring of Luminaires; Part X Special Provisions for Flush and Recessed Luminaires; and Part XIV Lighting Track.

Article 411 covers lighting systems operating at 30 volts or less and their associated components. While these systems have historically been used for landscape and other outdoor accent uses, they have become extremely popular for interior display and task lighting. Under counter low-voltage LED systems are used extensively in residential kitchens.

ARTICLE 210
Branch Circuits

PART III. REQUIRED OUTLETS

210.70 Lighting Outlets Required. Lighting outlets shall be installed where specified in 210.70(A), (B), and (C).

(A) Dwelling Units. In dwelling units, lighting outlets shall be installed in accordance with 210.70(A)(1), (A)(2), and (A)(3).

(1) Habitable Rooms. At least one lighting outlet controlled by a listed wall-mounted control device shall be installed in every habitable room, kitchen, and bathroom. The wall-mounted control device shall be located near an entrance to the room on a wall.

Exception No. 1: In other than kitchens and bathrooms, one or more receptacles controlled by a listed wall-mounted control device shall be permitted in lieu of lighting outlets.

Exception No. 2: Lighting outlets shall be permitted to be controlled by occupancy sensors that are (1) in addition to listed wall-mounted control devices or (2) located at a customary wall switch location and equipped with a manual override that will allow the sensor to function as a wall switch.

(2) Additional Locations. Additional lighting outlets shall be installed in accordance with the following:

(1) At least one lighting outlet controlled by a listed wall-mounted control device shall be installed in hallways, stairways, attached garages, and detached garages with electric power.
(2) For dwelling units, attached garages, and detached garages with electric power, at least one lighting outlet controlled by a listed wall-mounted control device shall be installed to provide illumination on the exterior side of outdoor entrances or exits with grade-level access. A vehicle door in a garage shall not be considered as an outdoor entrance or exit.
(3) Where one or more lighting outlet(s) are installed for interior stairways, there shall be a listed wall-mounted control device at each floor level and landing level that includes an entryway to control the lighting outlet(s) where the stairway between floor levels has six risers or more.

Exception to (A)(2)(1), (A)(2)(2), and (A)(2)(3): In hallways, in stairways, and at outdoor entrances, remote, central, or automatic control of lighting shall be permitted.

Lighting Requirements

(4) Lighting outlets controlled in accordance with 210.70(A)(2)(3) shall not be controlled by use of listed wall-mounted control devices unless they provide the full range of dimming control at each location.

(B) Guest Rooms or Guest Suites. In hotels, motels, or similar occupancies, guest rooms or guest suites shall have at least one lighting outlet controlled by a listed wall-mounted control device installed in every habitable room and bathroom.

Exception No. 1: In other than bathrooms and kitchens where provided, one or more receptacles controlled by a listed wall-mounted control device shall be permitted in lieu of lighting outlets.

Exception No. 2: Lighting outlets shall be permitted to be controlled by occupancy sensors that are (1) in addition to listed wall-mounted control devices or (2) located at a customary wall switch location and equipped with a manual override that allows the sensor to function as a wall switch.

(C) All Occupancies. For attics and underfloor spaces, utility rooms, and basements, at least one lighting outlet containing a switch or controlled by a wall switch or listed wall-mounted control device shall be installed where these spaces are used for storage or contain equipment requiring servicing. A point of control shall be at each entry that permits access to the attic and underfloor space, utility room, or basement. Where a lighting outlet is installed for equipment requiring service, the lighting outlet shall be installed at or near the equipment.

ARTICLE 410
Luminaires, Lampholders, and Lamps

PART I. GENERAL

410.1 Scope. This article covers luminaires, portable luminaires, lampholders, pendants, incandescent filament lamps, arc lamps, electric-discharge lamps, decorative lighting products, lighting accessories for temporary seasonal and holiday use, portable flexible lighting products, and the wiring and equipment forming part of such products and lighting installations.

410.2 Definition. The definition in this section shall apply only within this article.

Clothes Closet Storage Space. The volume bounded by the sides and back closet walls and planes extending from the closet floor vertically to a height of 1.8 m (6 ft) or to the highest clothes-hanging rod and parallel to the walls at a horizontal distance of 600 mm (24 in.) from the sides and back of the closet walls, respectively, and continuing vertically to the closet ceiling parallel to the walls at a horizontal distance of 300 mm (12 in.) or the width of the shelf, whichever is greater; for a closet that permits access to both sides of a hanging rod, this space includes the volume below the highest rod extending 300 mm (12 in.) on either side of the rod on a plane horizontal to the floor extending the entire length of the rod. See Figure 410.2.

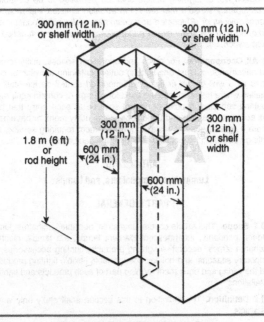

Figure 410.2 Clothes Closet Storage Space.

Lighting Requirements

410.6 Listing Required. All luminaires, lampholders, and retrofit kits shall be listed.

410.7 Reconditioned Equipment. Luminaires, lampholders, and retrofit kits shall not be permitted to be reconditioned. If a retrofit kit is installed in a luminaire in accordance with the installation instructions, the retrofitted luminaire shall not be considered reconditioned.

410.8 Inspection. Luminaires shall be installed such that the connections between the luminaire conductors and the circuit conductors can be inspected without requiring the disconnection of any part of the wiring unless the luminaires are connected by attachment plugs and receptacles.

PART II. LUMINAIRE LOCATIONS

410.10 Luminaires in Specific Locations.

(A) Wet and Damp Locations. Luminaires installed in wet or damp locations shall be installed such that water cannot enter or accumulate in wiring compartments, lampholders, or other electrical parts. All luminaires installed in wet locations shall be marked, "Suitable for Wet Locations." All luminaires installed in damp locations shall be marked "Suitable for Wet Locations" or "Suitable for Damp Locations."

(D) Bathtub and Shower Areas. A luminaire installed in a bathtub or shower area shall meet all of the following requirements:

(1) No parts of cord-connected luminaires, chain-, cable-, or cord-suspended luminaires, lighting track, pendants, or ceiling-suspended (paddle) fans shall be located within a zone measured 900 mm (3 ft) horizontally and 2.5 m (8 ft) vertically from the top of the bathtub rim or shower stall threshold. This zone is all-encompassing and includes the space directly over the tub or shower stall.

(2) Luminaires located within the actual outside dimension of the bathtub or shower to a height of 2.5 m (8 ft) vertically from the top of the bathtub rim or shower threshold shall be marked suitable for damp locations or marked suitable for wet locations. Luminaires located where subject to shower spray shall be marked suitable for wet locations.

410.11 Luminaires Near Combustible Material. Luminaires shall be constructed, installed, or equipped with shades or guards so that combustible material is not subjected to temperatures in excess of 90°C (194°F).

410.12 Luminaires over Combustible Material. Lampholders installed over highly combustible material shall be of the unswitched type. Unless an individual switch is provided for each luminaire, lampholders shall be located at least 2.5 m (8 ft) above the floor or shall be located or guarded so that the lamps cannot be readily removed or damaged.

410.16 Luminaires in Clothes Closets.

(A) Luminaire Types Permitted. Only luminaires of the following types shall be permitted in a clothes closet:

(1) Surface-mounted or recessed incandescent or LED luminaires with completely enclosed light sources
(2) Surface-mounted or recessed fluorescent luminaires
(3) Surface-mounted fluorescent or LED luminaires identified as suitable for installation within the clothes closet storage space

(B) Luminaire Types Not Permitted. Incandescent luminaires with open or partially enclosed lamps and pendant luminaires or lampholders shall not be permitted.

(C) Location. The minimum clearance between luminaires installed in clothes closets and the nearest point of a clothes closet storage space shall be as follows:

(1) 300 mm (12 in.) for surface-mounted incandescent or LED luminaires with a completely enclosed light source installed on the wall above the door or on the ceiling.
(2) 150 mm (6 in.) for surface-mounted fluorescent luminaires installed on the wall above the door or on the ceiling.
(3) 150 mm (6 in.) for recessed incandescent or LED luminaires with a completely enclosed light source installed in the wall or the ceiling.
(4) 150 mm (6 in.) for recessed fluorescent luminaires installed in the wall or the ceiling.
(5) Surface-mounted fluorescent or LED luminaires shall be permitted to be installed within the clothes closet storage space where identified for this use.

410.18 Space for Cove Lighting. Coves shall have adequate space and shall be located so that lamps and equipment can be properly installed and maintained.

Lighting Requirements

PART III. PROVISIONS AT LUMINAIRE OUTLET BOXES, CANOPIES, AND PANS

410.22 Outlet Boxes to Be Covered. In a completed installation, each outlet box shall be provided with a cover unless covered by means of a luminaire canopy, lampholder, receptacle that covers the box or is provided with a faceplate, or similar device.

410.23 Covering of Combustible Material at Outlet Boxes. Any combustible wall or ceiling finish exposed between the edge of a luminaire canopy or pan and an outlet box having a surface area of 1160 mm² (180 in.²) or more shall be covered with noncombustible material.

410.24 Connection of Electric-Discharge and LED Luminaires.

(A) Independent of the Outlet Box. Electric-discharge and LED luminaires supported independently of the outlet box shall be connected to the branch circuit through metal raceway, nonmetallic raceway, Type MC cable, Type AC cable, Type MI cable, nonmetallic sheathed cable, or by flexible cord as permitted in 410.62(B) or 410.62(C).

(B) Access to Boxes. Electric-discharge and LED luminaires surface mounted over concealed outlet, pull, or junction boxes and designed not to be supported solely by the outlet box shall be provided with suitable openings in the back of the luminaire to provide access to the wiring in the box.

PART IV. LUMINAIRE SUPPORTS

410.30 Supports.

(A) General. Luminaires and lampholders shall be securely supported. A luminaire that weighs more than 3 kg (6 lb) or exceeds 400 mm (16 in.) in any dimension shall not be supported by the screw shell of a lampholder.

(B) Metal or Nonmetallic Poles Supporting Luminaires. Metal or nonmetallic poles shall be permitted to be used to support luminaires and as a raceway to enclose supply conductors, provided the following conditions are met:

(1) A pole shall have a handhole not less than 50 mm × 100 mm (2 in. × 4 in.) with a cover suitable for use in wet locations to provide access to the supply terminations within the pole or pole base.

Exception No. 1: No handhole shall be required in a pole 2.5 m (8 ft) or less in height abovegrade where the supply wiring method continues without splice or pull point, and where the interior of the pole and any splices are accessible by removing the luminaire.

Exception No. 2: No handhole shall be required in a pole 6.0 m (20 ft) or less in height abovegrade that is provided with a hinged base.

(2) Where raceway risers or cable is not installed within the pole, a threaded fitting or nipple shall be brazed, welded, or attached to the pole opposite the handhole for the supply connection.

(3) A metal pole shall be provided with an equipment grounding terminal as follows:

   a. A pole with a handhole shall have the equipment grounding terminal accessible from the handhole.
   b. A pole with a hinged base shall have the equipment grounding terminal accessible within the base.

Exception to (3): No grounding terminal shall be required in a pole 2.5 m (8 ft) or less in height abovegrade where the supply wiring method continues without splice or pull, and where the interior of the pole and any splices are accessible by removing the luminaire.

(4) A metal pole with a hinged base shall have the hinged base and pole bonded together.
(5) Metal raceways or other equipment grounding conductors shall be bonded to the metal pole with an equipment grounding conductor recognized by 250.118 and sized in accordance with 250.122.
(6) Conductors in vertical poles used as raceway shall be supported as provided in 300.19.

410.36 Means of Support.

(A) Luminaires Supported By Outlet Boxes. Luminaires shall be permitted to be supported by outlet boxes or fittings installed as required by 314.23. The installation shall comply with the following requirements:

(1) The outlet boxes or fittings shall comply with 314.27(A)(1) and 314.27(A)(2).
(2) Luminaires shall be permitted to be supported in accordance with 314.27(E).
(3) Outlet boxes complying with 314.27(E) shall be considered lighting outlets as required by 210.70(A), (B), and (C).

(B) Suspended Ceilings. Framing members of suspended ceiling systems used to support luminaires shall be securely fastened to each other and shall be securely attached to the building structure at appropriate intervals. Luminaires shall be securely fastened to the ceiling framing member by mechanical means such as bolts, screws, or rivets. Listed clips identified for use with the type of ceiling framing member(s) and luminaire(s) shall also be permitted.

(C) Luminaire Studs. Luminaire studs that are not a part of outlet boxes, hickeys, tripods, and crowfeet shall be made of steel, malleable iron, or other material suitable for the application.

(D) Insulating Joints. Insulating joints that are not designed to be mounted with screws or bolts shall have an exterior metal casing, insulated from both screw connections.

(E) Raceway Fittings. Raceway fittings used to support a luminaire(s) shall be capable of supporting the weight of the complete fixture assembly and lamp(s).

(G) Trees. Outdoor luminaires and associated equipment shall be permitted to be supported by trees.

Informational Note No. 1: See 225.26 for restrictions for support of overhead conductors.

Informational Note No. 2: See 300.5(D) for protection of conductors.

PART V. GROUNDING

410.40 General. Luminaires and lighting equipment shall be connected to an equipment grounding conductor as required in Article 250 and Part V of this article.

410.42 Luminaire(s) with Exposed Conductive Parts. Exposed conductive parts that are accessible to unqualified persons shall be connected to an equipment grounding conductor or be separated from all live parts and other conducting surfaces by a listed system of double insulation.

Small isolated parts, such as mounting screws, clips, and decorative bands on glass spaced at least 38 mm (1½ in.) from lamp terminals, shall not require connection to an equipment grounding conductor.

Portable luminaires with a polarized attachment plug shall not require connection to an equipment grounding conductor.

410.44 Methods of Grounding. Luminaires and equipment shall be mechanically connected to an equipment grounding conductor as specified in 250.118 and sized in accordance with 250.122.

Exception No. 1: Replacement luminaires shall be permitted to connect an equipment grounding conductor in the same manner as replacement receptacles in compliance with 250.130(C). The luminaire shall then comply with 410.42.

Exception No. 2: Where no equipment grounding conductor exists at the outlet, replacement luminaires that are GFCI protected or do not have exposed conductive parts shall not be required to be connected to an equipment grounding conductor.

410.46 Equipment Grounding Conductor Attachment. Luminaires with exposed metal parts shall be provided with a means for connecting an equipment grounding conductor.

PART VI. WIRING OF LUMINAIRES

410.48 Luminaire Wiring — General. Wiring on or within luminaires shall be neatly arranged and shall not be exposed to physical damage. Excess wiring shall be avoided. Conductors shall be arranged so that they are not subjected to temperatures above those for which they are rated.

410.50 Polarization of Luminaires. Luminaires shall be wired so that the screw shells of lampholders are connected to the same luminaire or circuit conductor or terminal. The grounded conductor, where connected to a screw shell lampholder, shall be connected to the screw shell.

410.56 Protection of Conductors and Insulation.

(A) Properly Secured. Conductors shall be secured in a manner that does not tend to cut or abrade the insulation.

(B) Protection Through Metal. Conductor insulation shall be protected from abrasion where it passes through metal.

(C) Luminaire Stems. Splices and taps shall not be located within luminaire arms or stems.

(D) Splices and Taps. No unnecessary splices or taps shall be made within or on a luminaire.

> Informational Note: For approved means of making connections, see 110.14.

410.64 Luminaires as Raceways. Luminaires shall not be used as a raceway for circuit conductors unless they comply with 410.64(A), (B), or (C).

(A) Listed. Luminaires listed and marked for use as a raceway shall be permitted to be used as a raceway.

(B) Through-Wiring. Luminaires identified for through-wiring, as permitted by 410.21, shall be permitted to be used as a raceway.

(C) Luminaires Connected Together. Luminaires designed for end-to-end connection to form a continuous assembly, or luminaires connected together by recognized wiring methods, shall be permitted to contain the conductors of a 2-wire branch circuit, or one multiwire branch circuit, supplying the connected luminaires and shall not be required to be listed as a raceway. One additional 2-wire branch circuit separately supplying one or more of the connected luminaires shall also be permitted.

410.68 Feeder and Branch-Circuit Conductors and Ballasts. Feeder and branch-circuit conductors within 75 mm (3 in.) of a ballast, LED driver, power supply, or transformer shall have an insulation temperature rating not lower than 90°C (194°F), unless supplying a luminaire marked as suitable for a different insulation temperature.

410.70 Combustible Shades and Enclosures. Air space shall be provided between lamps and shades or other enclosures of combustible material.

PART X. SPECIAL PROVISIONS FOR FLUSH AND RECESSED LUMINAIRES

410.110 General. Luminaires installed in recessed cavities in walls or ceilings, including suspended ceilings, shall comply with 410.115 through 410.122.

410.115 Temperature.

(A) Combustible Material. Luminaires shall be installed so that adjacent combustible material will not be subjected to temperatures in excess of 90°C (194°F).

(B) Recessed Incandescent Luminaires. Incandescent luminaires shall have thermal protection and shall be identified as thermally protected.

Exception No. 1: Thermal protection shall not be required in a recessed luminaire identified for use and installed in poured concrete.

Exception No. 2: Thermal protection shall not be required in a recessed luminaire whose design, construction, and thermal performance characteristics are equivalent to a thermally protected luminaire and are identified as inherently protected.

410.116 Clearance and Installation.

(A) Clearance.

(1) Non-Type IC. A recessed luminaire that is not identified for contact with insulation shall have all recessed parts spaced not less than 13 mm (1/2 in.) from combustible materials. The points of support and the trim finishing off the openings in the ceiling, wall, or other finished surface shall be permitted to be in contact with combustible materials.

(2) Type IC. A recessed luminaire that is identified for contact with insulation, Type IC, shall be permitted to be in contact with combustible materials at recessed parts, points of support, and portions passing through or finishing off the opening in the building structure.

(B) Installation. Thermal insulation shall not be installed above a recessed luminaire or within 75 mm (3 in.) of the recessed luminaire's enclosure, wiring compartment, ballast, transformer, LED driver, or power supply unless the luminaire is identified as Type IC for insulation contact.

410.117 Wiring.

(A) General. Conductors that have insulation suitable for the temperature encountered shall be used.

(B) Circuit Conductors. Branch-circuit conductors that have an insulation suitable for the temperature encountered shall be permitted to terminate in the luminaire.

(C) Tap Conductors. Tap conductors of a type suitable for the temperature encountered shall be permitted to run from the luminaire terminal connection to an outlet box placed at least 300 mm (1 ft) from the luminaire. Such tap conductors shall be in suitable raceway or Type AC or MC cable of at least 450 mm (18 in.) but not more than 1.8 m (6 ft) in length.

410.118 Access to Other Boxes.
Luminaires recessed in ceilings, floors, or walls shall not be used to access outlet, pull, or junction boxes or conduit bodies, unless the box or conduit body is an integral part of the listed luminaire.

PART XIV. LIGHTING TRACK

410.151 Installation.

(A) Lighting Track. Lighting track shall be permanently installed and permanently connected to a branch circuit. Only lighting track fittings shall be installed on lighting track. Lighting track fittings shall not be equipped with general-purpose receptacles.

(B) Connected Load. The connected load on lighting track shall not exceed the rating of the track. Lighting track shall be supplied by a branch circuit having a rating not more than that of the track. The load calculation in 220.43(B) shall not be required to limit the length of track on a single branch circuit, and it shall not be required to limit the number of luminaires on a single track.

(C) Locations Not Permitted. Lighting track shall not be installed in the following locations:

(1) Where likely to be subjected to physical damage
(2) In wet or damp locations
(6) Where concealed
(7) Where extended through walls or partitions
(8) Less than 1.5 m (5 ft) above the finished floor except where protected from physical damage or track operating at less than 30 volts rms open-circuit voltage
(9) Where prohibited by 410.10(D)

(D) Support. Fittings identified for use on lighting track shall be designed specifically for the track on which they are to be installed. They shall be securely fastened to the track, shall maintain polarization and connections to the equipment grounding conductor, and shall be designed to be suspended directly from the track.

410.153 Heavy-Duty Lighting Track.
Heavy-duty lighting track is lighting track identified for use exceeding 20 amperes. Each fitting attached to a heavy-duty lighting track shall have individual overcurrent protection.

410.154 Fastening.
Lighting track shall be securely mounted so that each fastening is suitable for supporting the maximum weight of luminaires that can be installed. Unless identified for supports at greater intervals, a single section 1.2 m (4 ft) or shorter in length shall have two supports, and, where installed in a continuous row, each individual section of not more than 1.2 m (4 ft) in length shall have one additional support.

410.155 Construction Requirements.

(B) Grounding. Lighting track shall be grounded in accordance with Article 250, and the track sections shall be securely coupled to maintain continuity of the circuitry, polarization, and grounding throughout.

CHAPTER 20
APPLIANCES

INTRODUCTION

This chapter contains requirements from Article 422 — Appliances. Article 422 covers electric appliances used in any occupancy. Installation requirements cover branch-circuit ratings and overcurrent protection. Flexible-cord requirements are provided for electrically operated kitchen waste disposers, built-in dishwashers and trash compactors, and wall-mounted ovens and counter-mounted cooking units. Appliance disconnecting specifications are covered in Part III of Article 422.

269

ARTICLE 422
Appliance

PART I. GENERAL

422.5 Ground-Fault Circuit-Interrupter (GFCI) Protection for Personnel.

(A) General. Appliances identified in 422.5(A)(1) through (A)(7) rated 150 volts or less to ground and 60 amperes or less, single- or 3-phase, shall be provided with Class A GFCI protection for personnel. Multiple Class A GFCI protective devices shall be permitted but shall not be required.

(1) Automotive vacuum machines
(2) Drinking water coolers and bottle fill stations
(3) Cord-and-plug-connected high-pressure spray washing machines
(4) Tire inflation machines
(5) Vending machines
(6) Sump pumps
(7) Dishwashers

The GFCI shall be readily accessible, listed, and located in one or more of the following locations:

(1) Within the branch-circuit overcurrent device
(2) A device or outlet within the supply circuit
(3) An integral part of the attachment plug
(4) Within the supply cord not more than 300 mm (12 in.) from the attachment plug
(5) Factory installed within the appliance

422.6 Listing Required. All appliances supplied by 50 volts or higher shall be listed.

PART II. INSTALLATION

422.10 Branch Circuits. This section specifies the requirements for branch circuits capable of carrying appliance current without overheating under the conditions specified.

(A) Individual Branch Circuits. The ampacities of branch-circuit conductors shall not be less than the marked rating of the appliance or the marked rating of an appliance having combined loads.

Appliances 422.11

The ampacities of branch-circuit conductors for motor-operated appliances not having a marked rating shall be in accordance with Part II of Article 430.

The branch-circuit rating for an appliance that is a continuous load, other than a motor-operated appliance, shall not be less than 125 percent of the marked rating, or not less than 100 percent of the marked rating if the branch-circuit device and its assembly are listed for continuous loading at 100 percent of its rating.

Branch circuits and branch-circuit conductors for household ranges and cooking appliances shall be permitted to be in accordance with Table 220.55 and shall be sized in accordance with 210.19(A)(3).

(B) Branch Circuits Supplying Two or More Loads. For branch circuits supplying appliance and other loads, the rating shall be determined in accordance with 210.23.

422.11 Overcurrent Protection. Appliances shall be protected against overcurrent in accordance with 422.11(A) through (G) and 422.10.

(A) Branch-Circuit Overcurrent Protection. Branch circuits shall be protected in accordance with 240.4.

If a protective device rating is marked on an appliance, the branch-circuit overcurrent device rating shall not exceed the protective device rating marked on the appliance.

(B) Household-Type Appliances with Surface Heating Elements. Household-type appliances with surface heating elements having a maximum demand of more than 60 amperes calculated in accordance with Table 220.55 shall have their power supply subdivided into two or more circuits, each of which shall be provided with overcurrent protection rated at not over 50 amperes.

(E) Single Non–Motor-Operated Appliance. If the branch circuit supplies a single non–motor-operated appliance, the rating of overcurrent protection shall comply with the following:

(1) Not exceed that marked on the appliance.
(2) Not exceed 20 amperes if the overcurrent protection rating is not marked and the appliance is rated 13.3 amperes or less; or
(3) Not exceed 150 percent of the appliance rated current if the overcurrent protection rating is not marked and the appliance is rated over 13.3 amperes. Where 150 percent of the appliance rating does not correspond to a standard overcurrent device ampere rating, the next higher standard rating shall be permitted.

(G) Motor-Operated Appliances. Motors of motor-operated appliances shall be provided with overload protection in accordance with Part III of Article 430. Hermetic refrigerant motor-compressors in air-conditioning or refrigerating equipment shall be provided with overload protection in accordance with Part VI of Article 440. Where appliance overcurrent protective devices that are separate from the appliance are required, data for selection of these devices shall be marked on the appliance. The minimum marking shall be that specified in 430.7 and 440.4.

422.12 Central Heating Equipment. Central heating equipment other than fixed electric space-heating equipment shall be supplied by an individual branch circuit.

Exception No. 1: Auxiliary equipment, such as a pump, valve, humidifier, or electrostatic air cleaner directly associated with the heating equipment, shall be permitted to be connected to the same branch circuit.

Exception No. 2: Permanently connected air-conditioning equipment shall be permitted to be connected to the same branch circuit.

422.13 Storage-Type Water Heaters. The branch-circuit overcurrent device and conductors for fixed storage-type water heaters that have a capacity of 450 L (120 gal) or less shall be sized not smaller than 125 percent of the rating of the water heater.

422.15 Central Vacuum Outlet Assemblies.

(A) Listed central vacuum outlet assemblies shall be permitted to be connected to a branch circuit in accordance with 210.23(A).

(B) The ampacity of the connecting conductors shall not be less than the ampacity of the branch circuit conductors to which they are connected.

(C) Accessible non–current-carrying metal parts of the central vacuum outlet assembly likely to become energized shall be connected to an equipment grounding conductor in accordance with 250.110. Incidental metal parts such as screws or rivets installed into or on insulating material shall not be considered likely to become energized.

422.16 Flexible Cords.

(A) General. Flexible cord shall be permitted as follows:

(1) To connect appliances to facilitate their frequent interchange or to prevent the transmission of noise or vibration or
(2) To facilitate the removal or disconnection of appliances that are fastened in place, where the fastening means and mechanical

Appliances 422.16

connections are specifically designed to permit ready removal for maintenance or repair and the appliance is intended or identified for flexible cord connection.

(B) Specific Appliances.

(1) Electrically Operated In-Sink Waste Disposers. Electrically operated in-sink waste disposers shall be permitted to be cord-and-plug-connected with a flexible cord identified as suitable in the installation instructions of the appliance manufacturer where all of the following conditions are met:

(1) The length of the cord shall not be less than 450 mm (18 in.) and not over 900 mm (36 in.).
(2) Receptacles shall be located to protect against physical damage to the flexible cord.
(3) The receptacle shall be accessible.
(4) The flexible cord shall have an equipment grounding conductor and be terminated with a grounding-type attachment plug.

Exception: A listed appliance distinctly marked to identify it as protected by a system of double insulation shall not be required to be terminated with a grounding-type attachment plug.

(2) Built-in Dishwashers and Trash Compactors. Built-in dishwashers and trash compactors shall be permitted to be cord-and-plug-connected with a flexible cord identified as suitable for the purpose in the installation instructions of the appliance manufacturer where all of the following conditions are met:

(1) For a trash compactor, the length of the cord shall be 0.9 m to 1.2 m (3 ft to 4 ft) measured from the face of the attachment plug to the plane of the rear of the appliance.
(2) For a built-in dishwasher, the length of the cord shall be 0.9 m to 2.0 m (3 ft to 6.5 ft) measured from the face of the attachment plug to the plane of the rear of the appliance.
(3) Receptacles shall be located to protect against physical damage to the flexible cord.
(4) The receptacle for a trash compactor shall be located in the space occupied by the appliance or adjacent thereto.
(5) The receptacle for a built-in dishwasher shall be located in the space adjacent to the space occupied by the dishwasher. Where the flexible cord passes through an opening, it shall be protected against damage by a bushing, grommet, or other approved means.

(6) The receptacle shall be accessible.
(7) The flexible cord shall have an equipment grounding conductor and be terminated with a grounding-type attachment plug.

Exception: A listed appliance distinctly marked to identify it as protected by a system of double insulation shall not be required to be terminated with a grounding-type attachment plug.

(3) Wall-Mounted Ovens and Counter-Mounted Cooking Units. Wall-mounted ovens and counter-mounted cooking units complete with provisions for mounting and for making electrical connections shall be permitted to be permanently connected or cord-and-plug-connected with a flexible cord identified as suitable for the purpose in the installation instructions of the appliance manufacturer.

A separable connector or a plug and receptacle combination in the supply line to an oven or cooking unit shall be identified for the temperature of the space in which it is located.

(4) Range Hoods and Microwave Oven/Range Hood Combinations. Range hoods and over-the-range microwave ovens with integral range hoods shall be permitted to be cord-and-plug-connected with a flexible cord identified as suitable for use on range hoods in the installation instructions of the appliance manufacturer, where all of the following conditions are met:

(1) The length of the cord is not less than 450 mm (18 in.) and not over 1.2 m (4 ft).
(2) Receptacles are located to protect against physical damage to the flexible cord.
(3) The receptacle is supplied by an individual branch circuit.
(4) The receptacle shall be accessible.
(5) The flexible cord shall have an equipment grounding conductor and be terminated with a grounding-type attachment plug.

Exception: A listed appliance distinctly marked to identify it as protected by a system of double insulation shall not be required to be terminated with a grounding-type attachment plug.

422.18 Support of Ceiling-Suspended (Paddle) Fans. Ceiling-suspended (paddle) fans shall be supported independently of an outlet box or by one of the following:

(1) A listed outlet box or listed outlet box system identified for the use and installed in accordance with 314.27(C)

Appliances   422.31

(2) A listed outlet box system, a listed locking support and mounting receptacle, and a compatible factory installed attachment fitting designed for support, identified for the use and installed in accordance with 314.27(E)

422.19 Space for Conductors. The combined volume of the canopy of ceiling-suspended (paddle) fans and outlet box shall provide sufficient space so that conductors and their connecting devices are capable of being installed in accordance with 314.16.

422.20 Outlet Boxes to Be Covered. In a completed installation, each outlet box shall be provided with a cover unless covered by means of a ceiling-suspended (paddle) fan canopy.

422.21 Covering of Combustible Material at Outlet Boxes. Any combustible ceiling finish that is exposed between the edge of a ceiling-suspended (paddle) fan canopy or pan and an outlet box and that has a surface area of 1160 mm² (180 in.²) or more shall be covered with noncombustible material.

422.22 Utilizing Separable Attachment Fittings. Appliances shall be permitted to utilize listed locking support and mounting receptacles in combination with compatible attachment fittings utilized within their ratings and used in accordance with 314.27(E).

PART III. DISCONNECTING MEANS

422.30 General. A means shall be provided to simultaneously disconnect each appliance from all ungrounded conductors in accordance with the following sections of Part III. If an appliance is supplied by more than one branch circuit or feeder, these disconnecting means shall be grouped and identified as being the multiple disconnecting means for the appliance. Each disconnecting means shall simultaneously disconnect all ungrounded conductors that it controls.

422.31 Disconnection of Permanently Connected Appliances.

(A) Rated at Not over 300 Volt-Amperes or ⅛ Horsepower. For permanently connected appliances rated at not over 300 volt-amperes or ⅛ hp, the branch-circuit overcurrent device shall be permitted to serve as the disconnecting means where the switch or circuit breaker is within sight from the appliance or be capable of being locked in the open position in compliance with 110.25.

(B) Appliances Rated over 300 Volt-Amperes. For permanently connected appliances rated over 300 volt-amperes, the branch-circuit switch or circuit breaker shall be permitted to serve as the disconnecting means where the switch or circuit breaker is within sight from the appliance or be capable of being locked in the open position in compliance with 110.25.

(C) Motor-Operated Appliances Rated over ⅛ Horsepower. The disconnecting means shall comply with 430.109 and 430.110. For permanently connected motor-operated appliances with motors rated over ⅛ hp, the disconnecting means shall be within sight from the appliance or be capable of being locked in the open position in compliance with 110.25.

Exception: If an appliance of more than ⅛ hp is provided with a unit switch that complies with 422.34(A), (B), (C), or (D), the switch or circuit breaker serving as the other disconnecting means shall be permitted to be out of sight from the appliance.

422.33 Disconnection of Cord-and-Plug-Connected or Attachment Fitting–Connected Appliances.

(A) Separable Connector or an Attachment Plug (or Attachment Fitting) and Receptacle. For cord-and-plug- (or attachment fitting–) connected appliances, an accessible separable connector or an accessible plug (or attachment fitting) and receptacle combination shall be permitted to serve as the disconnecting means. The attachment fitting shall be a factory installed part of the appliance and suitable for disconnection of the appliance. Where the separable connector or plug (or attachment fitting) and receptacle combination are not accessible, cord-and-plug-connected or attachment fitting-and-plug-connected appliances shall be provided with disconnecting means in accordance with 422.31.

(B) Connection at the Rear Base of a Range. For cord-and-plug-connected household electric ranges, an attachment plug and receptacle connection at the rear base of a range, accessible from the front by removal of a drawer, shall meet the intent of 422.33(A).

(C) Rating. The rating of a receptacle or of a separable connector shall not be less than the rating of any appliance connected thereto.

422.34 Unit Switch(es) as Disconnecting Means. A unit switch(es) with a marked-off position that is a part of an appliance and disconnects all ungrounded conductors shall be permitted as the disconnecting means required by this article where other means for disconnection are provided in occupancies specified in 422.34(A) through (D).

(A) Multifamily Dwellings. In multifamily dwellings, the other disconnecting means shall be within the dwelling unit, or on the same floor as the dwelling unit in which the appliance is installed, and shall be permitted to control lamps and other appliances.

(B) Two-Family Dwellings. In two-family dwellings, the other disconnecting means shall be permitted either inside or outside of the dwelling unit in which the appliance is installed. In this case, an individual switch or circuit breaker for the dwelling unit shall be permitted and shall also be permitted to control lamps and other appliances.

(C) One-Family Dwellings. In one-family dwellings, the service disconnecting means shall be permitted to be the other disconnecting means.

422.35 Switch and Circuit Breaker to Be Indicating. Switches and circuit breakers used as disconnecting means shall be of the indicating type.

CHAPTER 21

ELECTRIC SPACE-HEATING EQUIPMENT

INTRODUCTION

This chapter contains requirements from Article 424 — Fixed Electric Space-Heating Equipment. Article 424 covers fixed electric equipment used for space heating. Heating equipment covered in this article includes heating cable, unit heaters, central systems, and other approved fixed electric space-heating equipment. Contained in this *Pocket Guide* are four of the nine parts that are included in Article 424: Part I General, Part II Installation, Part III Control and Protection of Fixed Electric Space-Heating Equipment, and Part V Electric Space-Heating Cables. Room air-conditioning equipment is covered in Article 440 (see Chapter 22).

ARTICLE 424
Fixed Electric Space-Heating Equipment

PART I. GENERAL

424.4 Branch Circuits.

(A) Branch-Circuit Requirements. An individual branch circuit shall be permitted to supply any volt-ampere or wattage rating of fixed electric space-heating equipment for which the branch circuit is rated.

Branch circuits supplying two or more outlets for fixed electric space-heating equipment shall be rated not over 30 amperes. In other than a dwelling unit, fixed infrared heating equipment shall be permitted to be supplied from branch circuits rated not over 50 amperes.

(B) Branch-Circuit Sizing. The branch-circuit conductors for fixed electric space-heating equipment and any associated motors shall be sized not smaller than 125 percent of the load.

PART II. INSTALLATION

424.9 General. Factory-installed receptacle outlets that are part of a permanently installed electric baseboard heater, or outlets provided as a separate listed assembly of an electric baseboard heater, shall be permitted in lieu of a receptacle outlet(s) that is required by 210.52. Such receptacle outlets shall not be connected to the baseboard heater circuits.

> Informational Note: Listed baseboard heaters include instructions that may not permit their installation below receptacle outlets.

424.12 Locations.

(A) Exposed to Physical Damage. Where subject to physical damage, fixed electric space-heating equipment shall be protected in an approved manner.

(B) Damp or Wet Locations. Heaters and related equipment installed in damp or wet locations shall be listed for such locations and shall be constructed and installed so that water or other liquids cannot enter or accumulate in or on wired sections, electrical components, or ductwork.

> Informational Note No. 1: See 110.11 for equipment exposed to deteriorating agents.

Electric Space-Heating Equipment 424.19

Informational Note No. 2: See 680.27(C) for pool deck areas.

424.13 Spacing from Combustible Materials. Fixed electric space-heating equipment shall be installed to provide the required spacing between the equipment and adjacent combustible material, unless it is listed to be installed in direct contact with combustible material.

PART III. CONTROL AND PROTECTION OF FIXED ELECTRIC SPACE-HEATING EQUIPMENT

424.19 Disconnecting Means. Means shall be provided to simultaneously disconnect the heater, motor controller(s), and supplementary overcurrent protective device(s) of all fixed electric space-heating equipment from all ungrounded conductors. Where heating equipment is supplied by more than one source, feeder, or branch circuit, the disconnecting means shall be grouped and identified as having multiple disconnecting means. Each disconnecting means shall simultaneously disconnect all ungrounded conductors that it controls. The disconnecting means specified in 424.19(A) and (B) shall have an ampere rating not less than 125 percent of the total load of the motors and the heaters and shall be capable of being locked in the open position in compliance with 110.25.

(A) Heating Equipment with Supplementary Overcurrent Protection. The disconnecting means for fixed electric space-heating equipment with supplementary overcurrent protection shall be within sight from the supplementary overcurrent protective device(s), on the supply side of these devices, if fuses, and, in addition, shall comply with either 424.19(A)(1) or (A)(2).

(1) Heater Containing No Motor Rated over ⅛ Horsepower. The disconnecting means provided shall be within sight from the motor controller(s) and the heater, or shall be lockable as specified in 424.19, or shall be a unit switch complying with 424.19(C).

(2) Heater Containing a Motor(s) Rated over ⅛ Horsepower. The disconnecting means required by 424.19 shall be permitted to serve as the required disconnecting means for both the motor controller(s) and heater under either of the following conditions:

(1) Where the disconnecting means is in sight from the motor controller(s) and the heater and complies with Part IX of Article 430.
(2) Where a motor(s) of more than ⅛ hp and the heater are provided with a single unit switch that complies with 422.34(A), (B), (C), or

(D), the disconnecting means shall be permitted to be out of sight from the motor controller.

(B) Heating Equipment Without Supplementary Overcurrent Protection.

(1) Without Motor or with Motor Not over ⅛ Horsepower. For fixed electric space-heating equipment without a motor rated over ⅛ hp, the branch-circuit switch or circuit breaker shall be permitted to serve as the disconnecting means where the switch or circuit breaker is within sight from the heater or is capable of being locked in the open position in compliance with 110.25.

(2) Over ⅛ Horsepower. For motor-driven electric space-heating equipment with a motor rated over ⅛ hp, a disconnecting means shall be located within sight from the motor controller or shall be permitted to comply with the requirements in 424.19(A)(2).

(C) Unit Switch(es) as Disconnecting Means. A unit switch(es) with a marked "off" position that is part of a fixed heater and disconnects all ungrounded conductors shall be permitted as the disconnecting means required by this article where other means for disconnection are provided in the types of occupancies in 424.19(C)(1) through (C)(4).

(1) Multifamily Dwellings. In multifamily dwellings, the other disconnecting means shall be within the dwelling unit, or on the same floor as the dwelling unit in which the fixed heater is installed, and shall also be permitted to control general-purpose circuits and appliance circuits.

(2) Two-Family Dwellings. In two-family dwellings, the other disconnecting means shall be permitted either inside or outside of the dwelling unit in which the fixed heater is installed. In this case, an individual switch or circuit breaker for the dwelling unit shall be permitted and shall also be permitted to control general-purpose circuits and appliance circuits.

(3) One-Family Dwellings. In one-family dwellings, the service disconnecting means shall be permitted to be the other disconnecting means.

424.20 Thermostatically Controlled Switching Devices.

(A) Serving as Both Controllers and Disconnecting Means. Thermostatically controlled switching devices and combination thermostats and manually controlled switches shall be permitted to serve as both controllers and disconnecting means, provided they meet all of the following conditions:

Electric Space-Heating Equipment   424.38

(1) Provided with a marked "off" position
(2) Directly open all ungrounded conductors when manually placed in the "off" position
(3) Designed so that the circuit cannot be energized automatically after the device has been manually placed in the "off" position
(4) Located as specified in 424.19
(5) Located in an accessible location

(B) Thermostats That Do Not Directly Interrupt All Ungrounded Conductors. Thermostats that do not directly interrupt all ungrounded conductors and thermostats that operate remote-control circuits shall not be required to meet the requirements of 424.20(A). These devices shall not be permitted as the disconnecting means.

424.21 Switch and Circuit Breaker to Be Indicating. Switches and circuit breakers used as disconnecting means shall be of the indicating type.

424.22 Overcurrent Protection.

(A) Branch-Circuit Devices. Electric space-heating equipment, other than motor-operated equipment required to have additional overcurrent protection by Parts III and IV of Article 430 or Parts III and VI of Article 440, shall be permitted to be protected against overcurrent where supplied by one of the branch circuits in Part II of Article 210.

PART V. ELECTRIC SPACE-HEATING CABLES

424.35 Marking of Heating Cables. Each unit shall be marked with the identifying name or identification symbol, catalog number, and ratings in volts and watts or in volts and amperes.

424.36 Clearances of Wiring in Ceilings. Wiring located above heated ceilings shall be spaced not less than 50 mm (2 in.) above the heated ceiling. The ampacity of conductors shall be calculated on the basis of an assumed ambient temperature of not less than 50°C (122°F), applying the correction factors in accordance with 310.15(B)(1). If this wiring is located above thermal insulation having a minimum thickness of 50 mm (2 in.), it shall be subject to the ambient correction in accordance with 310.15(B)(1).

424.38 Area Restrictions.

(A) Extending Beyond the Room or Area. Heating cables shall be permitted to extend beyond the room or area in which they originate unless prohibited by 424.38(B).

(B) Uses Not Permitted. Heating cables shall not be installed as follows:

(1) In closets, other than as noted in 424.38(C)
(2) Over the top of walls where the wall intersects the ceiling
(3) Over partitions that extend to the ceiling, unless they are isolated single runs of embedded cable
(4) Under or through walls
(5) Over cabinets whose clearance from the ceiling is less than the minimum horizontal dimension of the cabinet to the nearest cabinet edge that is open to the room or area
(6) In tub and shower walls
(7) Under cabinets or similar built-ins having no clearance to the floor

(C) In Closet Ceilings as Low-Temperature Heat Sources to Control Relative Humidity. The provisions of 424.38(B) shall not prevent the use of cable in closet ceilings as low-temperature heat sources to control relative humidity, provided they are used only in those portions of the ceiling that are unobstructed to the floor.

424.39 Clearance from Other Objects and Openings. Heating elements of cables installed in ceilings shall be separated at least 200 mm (8 in.) from the edge of outlet boxes and junction boxes that are to be used for mounting surface luminaires. A clearance of not less than 50 mm (2 in.) shall be provided from recessed luminaires and their trims, ventilating openings, and other such openings in room surfaces. No heating cable shall be covered by any ceiling surface-mounted equipment.

424.40 Splices. The length of heating cable shall only be altered using splices identified in the manufacturer's instructions.

424.41 Ceiling Installation of Heating Cables on Dry Board, in Plaster, and on Concrete.

(A) In Walls. Cables shall not be installed in walls unless it is necessary for an isolated single run of cable to be installed down a vertical surface to reach a dropped ceiling.

(B) Adjacent Runs. Adjacent runs of heating cable shall be installed in accordance with the manufacturer's instructions.

(C) Surfaces to Be Applied. Heating cables shall be applied only to gypsum board, plaster lath, or other fire-resistant material. With metal lath or other electrically conductive surfaces, a coat of plaster or other means employed in accordance with the heating cable manufacturer's

Electric Space-Heating Equipment 424.42

instructions shall be applied to completely separate the metal lath or conductive surface from the cable.

Informational Note: See also 424.41(F).

(D) Splices. All heating cables, the splice between the heating cable and nonheating leads, and 75-mm (3-in.) minimum of the nonheating lead at the splice shall be embedded in plaster or dry board in the same manner as the heating cable.

(E) Ceiling Surface. The entire ceiling surface shall have a finish of thermally noninsulating sand plaster that has a nominal thickness of 13 mm (½ in.), or other noninsulating material identified as suitable for this use and applied according to specified thickness and directions.

(F) Secured. Cables shall be secured by means of approved stapling, tape, plaster, nonmetallic spreaders, or other approved means either at intervals not exceeding 400 mm (16 in.) or at intervals not exceeding 1.8 m (6 ft) for cables identified for such use. Staples or metal fasteners that straddle the cable shall not be used with metal lath or other electrically conductive surfaces.

(G) Dry Board Installations. In dry board installations, the entire ceiling below the heating cable shall be covered with gypsum board not exceeding 13 mm (½ in.) thickness. The void between the upper layer of gypsum board, plaster lath, or other fire-resistant material and the surface layer of gypsum board shall be completely filled with thermally conductive, nonshrinking plaster or other approved material or equivalent thermal conductivity.

(H) Free from Contact with Conductive Surfaces. Cables shall be kept free from contact with metal or other electrically conductive surfaces.

(I) Joists. In dry board applications, cable shall be installed parallel to the joist, leaving a clear space centered under the joist of 65 mm (2½ in.) (width) between centers of adjacent runs of cable. A surface layer of gypsum board shall be mounted so that the nails or other fasteners do not pierce the heating cable.

(J) Crossing Joists. Cables shall cross joists only at the ends of the room unless the cable is required to cross joists elsewhere in order to satisfy the manufacturer's instructions regarding clearance from ceiling penetrations and luminaires.

424.42 Finished Ceilings. Finished ceilings shall not be covered with decorative panels or beams constructed of materials that have thermal insulating properties, such as wood, fiber, or plastic. Finished ceilings shall be permitted to be covered with paint, wallpaper, or other approved surface finishes.

424.43 Installation of Nonheating Leads of Cables.

(A) Free Nonheating Leads. Free nonheating leads of cables shall be installed in accordance with Chapter 3 wiring methods, or other listed means, from the junction box to a location within the ceiling.

(B) Leads in Junction Box. Not less than 150 mm (6 in.) of free nonheating lead shall be within the junction box. The marking of the leads shall be visible in the junction box.

(C) Excess Leads. Excess leads of heating cables shall not be cut but shall be secured to the underside of the ceiling and embedded in plaster or other approved material, leaving only a length sufficient to reach the junction box with not less than 150 mm (6 in.) of free lead within the box.

424.44 Installation of Cables in Concrete or Poured Masonry Floors.

(A) Adjacent Runs. Adjacent runs of heating cable shall be installed in accordance with the manufacturer's instructions.

(B) Secured in Place. Cables shall be secured in place by nonmetallic frames or spreaders or other approved means while the concrete or other finish is applied.

(C) Leads Protected. Leads shall be protected where they leave the floor by rigid metal conduit, intermediate metal conduit, rigid nonmetallic conduit, electrical metallic tubing, or by other approved means.

(D) Bushings or Approved Fittings. Bushings or approved fittings shall be used where the leads emerge within the floor slab.

(E) Ground-Fault Circuit-Interrupter Protection. In addition to the requirements in 210.8, ground-fault circuit-interrupter protection for personnel shall be provided for cables installed in electrically heated floors of bathrooms, kitchens, and in hydromassage bathtub locations.

424.45 Installation of Cables Under Floor Coverings.

(A) Identification. Heating cables for installation under floor covering shall be identified as suitable for installation under floor covering.

(B) Expansion Joints. Heating cables shall not be installed where they bridge expansion joints unless provided with expansion and contraction fittings applicable to the manufacture of the cable.

Electric Space-Heating Equipment 424.45

(C) Connection to Conductors. Heating cables shall be connected to branch-circuit and supply wiring by wiring methods described in the installation instructions.

(D) Anchoring. Heating cables shall be positioned and secured in place under the floor covering, in accordance with the manufacturer's instructions.

(E) Ground-Fault Circuit-Interrupter Protection. In addition to the requirements in 210.8, ground-fault circuit-interrupter protection for personnel shall be provided.

(F) Grounding Braid or Sheath. Grounding means, such as copper braid, metal sheath, or other approved means, shall be provided as part of the heated length.

CHAPTER 22

AIR-CONDITIONING AND REFRIGERATING EQUIPMENT

INTRODUCTION

This chapter contains requirements from Article 440 — Air-Conditioning and Refrigerating Equipment. Article 440 covers electric motor–driven air-conditioning and refrigerating equipment, including their branch circuits and controllers. Provided in Article 440 are special considerations necessary for circuits supplying hermetic refrigerant motor-compressors and for any air-conditioning or refrigerating equipment that is supplied from a branch circuit that supplies a hermetic refrigerant motor-compressor.

The requirements of Article 440 recognize the operational characteristics of the motors associated with air-conditioning and refrigeration equipment, much of which contains multiple motors. Motors that are sealed and immersed in the refrigerant (i.e., hermetically sealed) are the unique feature of the equipment covered in Article 440. However, the approach to overcurrent protection takes on a two-parts: one device sized to accommodate the starting current that protects against the effects of short-circuits and line-to-ground faults, and a device that is responsive to motor or equipment operating current that protects against overloads.

Installers of the equipment covered by Article 440 need to be aware of the markings on the equipment that in most cases will provide specific guidance on maximum size and type of overcurrent protective devices in the supply circuit, the minimum size for circuit conductors, and the short-circuit current rating of internal motor control and protection components. These markings are essential to providing a *Code*-compliant and operationally safe installation.

Contained in this *Pocket Guide* are six of the seven parts that are included in Article 440. The six parts are: Part I General, Part II Disconnecting Means, Part III Branch-Circuit Short-Circuit and Ground-Fault Protection, Part IV Branch-Circuit Conductors, Part VI Motor-Compressor and Branch-Circuit Overload Protection, and Part VII Provisions for Room Air Conditioners.

ARTICLE 440

Air-Conditioning and Refrigerating Equipment

PART I. GENERAL

440.2 Definitions. The definitions in this section shall apply only within this article.

Branch-Circuit Selection Current. The value in amperes to be used instead of the rated-load current in determining the ratings of motor branch-circuit conductors, disconnecting means, controllers, and branch-circuit short-circuit and ground-fault protective devices wherever the running overload protective device permits a sustained current greater than the specified percentage of the rated-load current. The value of branch-circuit selection current will always be equal to or greater than the marked rated-load current.

Leakage-Current Detector-Interrupter (LCDI). A device provided in a power supply cord or cord set that senses leakage current flowing between or from the cord conductors and interrupts the circuit at a predetermined level of leakage current.

Rated-Load Current. The current of a hermetic refrigerant motor-compressor resulting when it is operated at the rated load, rated voltage, and rated frequency of the equipment it serves.

440.3 Other Articles.

(A) Article 430. These provisions are in addition to, or amendatory of, the provisions of Article 430 and other articles in this *Code*, which apply except as modified in this article.

(B) Articles 422, 424, or 430. The rules of Articles 422, 424, or 430, as applicable, shall apply to air-conditioning and refrigerating equipment that does not incorporate a hermetic refrigerant motor-compressor. This equipment includes devices that employ refrigeration compressors driven by conventional motors, furnaces with air-conditioning evaporator coils installed, fan-coil units, remote forced air-cooled condensers, remote commercial refrigerators, and so forth.

(C) Article 422. Equipment such as room air conditioners, household refrigerators and freezers, drinking water coolers, and beverage dispensers shall be considered appliances, and Article 422 shall also apply.

Air-Conditioning and Refrigerating Equipment 440.9

440.6 Ampacity and Rating. The size of conductors for equipment covered by this article shall be selected from Table 310.16 through Table 310.19 or calculated in accordance with 310.14 as applicable. The required ampacity of conductors and rating of equipment shall be determined according to 440.6(A) and 440.6(B).

(A) Hermetic Refrigerant Motor-Compressor. For a hermetic refrigerant motor-compressor, the rated-load current marked on the nameplate of the equipment in which the motor-compressor is employed shall be used in determining the rating or ampacity of the disconnecting means, the branch-circuit conductors, the controller, the branch-circuit short-circuit and ground-fault protection, and the separate motor overload protection. Where no rated-load current is shown on the equipment nameplate, the rated-load current shown on the compressor nameplate shall be used.

Exception No. 1: Where so marked, the branch-circuit selection current shall be used instead of the rated-load current to determine the rating or ampacity of the disconnecting means, the branch-circuit conductors, the controller, and the branch-circuit short-circuit and ground-fault protection.

(B) Multimotor Equipment. For multimotor equipment employing a shaded-pole or permanent split-capacitor-type fan or blower motor, the full-load current for such motor marked on the nameplate of the equipment in which the fan or blower motor is employed shall be used instead of the horsepower rating to determine the ampacity or rating of the disconnecting means, the branch-circuit conductors, the controller, the branch-circuit short-circuit and ground-fault protection, and the separate overload protection. This marking on the equipment nameplate shall not be less than the current marked on the fan or blower motor nameplate.

440.8 Single Machine. An air-conditioning or refrigerating system shall be considered to be a single machine under the provisions of 430.87, Exception No. 1, and 430.112, Exception. The motors shall be permitted to be located remotely from each other.

440.9 Grounding and Bonding. Where equipment is installed outdoors on a roof, an equipment grounding conductor of the wire type shall be installed in outdoor portions of metallic raceway systems that use compression-type fittings.

PART II. DISCONNECTING MEANS

440.11 General. Part II is intended to require disconnecting means capable of disconnecting air-conditioning and refrigerating equipment, including motor-compressors and controllers from the circuit conductors.

440.12 Rating and Interrupting Capacity.

(A) Hermetic Refrigerant Motor-Compressor. A disconnecting means serving a hermetic refrigerant motor-compressor shall be selected on the basis of the nameplate rated-load current or branch-circuit selection current, whichever is greater, and locked-rotor current, respectively, of the motor-compressor as follows.

(1) Ampere Rating. The ampere rating shall be at least 115 percent of the nameplate rated-load current or branch-circuit selection current, whichever is greater.

Exception: A listed unfused motor circuit switch, without fuseholders, having a horsepower rating not less than the equivalent horsepower determined in accordance with 440.12(A)(2) shall be permitted to have an ampere rating less than 115 percent of the specified current.

440.13 Cord-Connected Equipment. For cord-connected equipment such as room air conditioners, household refrigerators and freezers, drinking water coolers, and beverage dispensers, a separable connector or an attachment plug and receptacle shall be permitted to serve as the disconnecting means.

Informational Note: For room air conditioners, see 440.63.

440.14 Location. Disconnecting means shall be located within sight from, and readily accessible from the air-conditioning or refrigerating equipment. The disconnecting means shall be permitted to be installed on or within the air-conditioning or refrigerating equipment.

The disconnecting means shall not be located on panels that are designed to allow access to the air-conditioning or refrigeration equipment or to obscure the equipment nameplate(s).

Exception No. 2: Where an attachment plug and receptacle serve as the disconnecting means in accordance with 440.13, their location shall be accessible but shall not be required to be readily accessible.

PART III. BRANCH-CIRCUIT SHORT-CIRCUIT AND GROUND-FAULT PROTECTION

440.21 General. Part III specifies devices intended to protect the branch-circuit conductors, control apparatus, and motors in circuits supplying hermetic refrigerant motor-compressors against overcurrent due to short circuits and ground faults. They are in addition to or amendatory of Article 240.

440.22 Application and Selection.

(A) Rating or Setting for Individual Motor-Compressor. The motor-compressor branch-circuit short-circuit and ground-fault protective device shall be capable of carrying the starting current of the motor. A protective device having a rating or setting not exceeding 175 percent of the motor-compressor rated-load current or branch-circuit selection current, whichever is greater, shall be permitted, provided that, where the protection specified is not sufficient for the starting current of the motor, the rating or setting shall be permitted to be increased but shall not exceed 225 percent of the motor rated-load current or branch-circuit selection current, whichever is greater.

(C) Protective Device Rating Not to Exceed the Manufacturer's Values. Where maximum protective device ratings shown on a manufacturer's overload relay table for use with a motor controller are less than the rating or setting selected in accordance with 440.22(A) and (B), the protective device rating shall not exceed the manufacturer's values marked on the equipment.

PART IV. BRANCH-CIRCUIT CONDUCTORS

440.31 General. Part IV and Article 310 specify ampacities of conductors required to carry the motor current without overheating under the conditions specified, except as modified in 440.6(A), Exception No. 1.

These articles shall not apply to integral conductors of motors, to motor controllers and the like, or to conductors that form an integral part of approved equipment.

440.32 Single Motor-Compressor. Branch-circuit conductors supplying a single motor-compressor shall have an ampacity not less than the greater of:

(1) 125 percent of the motor-compressor rated-load current
(2) 125 percent of the branch-circuit selection current

For a wye-start, delta-run connected motor-compressor, the selection of branch-circuit conductors between the controller and the motor-compressor shall be permitted to be based on 72 percent of either the motor-compressor rated-load current or the branch-circuit selection current, whichever is greater.

440.33 Motor-Compressor(s) With or Without Additional Motor Loads. Conductors supplying one or more motor-compressor(s) with or without an additional motor load(s) shall have an ampacity not less than the sum of each of the following:

(1) The sum of the rated-load or branch-circuit selection current, whichever is greater, of all motor-compressor(s)
(2) The sum of the full-load current rating of all other motors
(3) 25 percent of the highest motor-compressor or motor full load current in the group

Exception No. 1: Where the circuitry is interlocked so as to prevent the starting and running of a second motor-compressor or group of motor-compressors, the conductor size shall be determined from the largest motor-compressor or group of motor-compressors that is to be operated at a given time.

Exception No. 2: The branch-circuit conductors for room air conditioners shall be in accordance with Part VII of Article 440.

440.34 Combination Load. Conductors supplying a motor-compressor load in addition to other load(s) as calculated from Article 220 and other applicable articles shall have an ampacity sufficient for the other load(s) plus the required ampacity for the motor-compressor load determined in accordance with 440.33 or, for a single motor-compressor, in accordance with 440.32.

Air-Conditioning and Refrigerating Equipment 440.55

PART VI. MOTOR-COMPRESSOR AND BRANCH-CIRCUIT OVERLOAD PROTECTION

440.54 Motor-Compressors and Equipment on 15- or 20-Ampere Branch Circuits — Not Cord- and Attachment-Plug-Connected. Overload protection for motor-compressors and equipment used on 15- or 20-ampere 120-volt, or 15-ampere 208- or 240-volt single-phase branch circuits as permitted in Article 210 shall be permitted as indicated in 440.54(A) and 440.54(B).

(A) Overload Protection. The motor-compressor shall be provided with overload protection selected as specified in 440.52(A). Both the controller and motor overload protective device shall be identified for installation with the short-circuit and ground-fault protective device for the branch circuit to which the equipment is connected.

(B) Time Delay. The short-circuit and ground-fault protective device protecting the branch circuit shall have sufficient time delay to permit the motor-compressor and other motors to start and accelerate their loads.

440.55 Cord- and Attachment-Plug-Connected Motor-Compressors and Equipment on 15- or 20-Ampere Branch Circuits. Overload protection for motor-compressors and equipment that are cord- and attachment-plug-connected and used on 15- or 20-ampere 120-volt, or 15-ampere 208- or 240-volt, single-phase branch circuits as permitted in Article 210 shall be permitted as indicated in 440.55(A), (B), and (C).

(A) Overload Protection. The motor-compressor shall be provided with overload protection as specified in 440.52(A). Both the controller and the motor overload protective device shall be identified for installation with the short-circuit and ground-fault protective device for the branch circuit to which the equipment is connected.

(B) Attachment Plug and Receptacle or Cord Connector Rating. The rating of the attachment plug and receptacle or cord connector shall not exceed 20 amperes at 125 volts or 15 amperes at 250 volts.

(C) Time Delay. The short-circuit and ground-fault protective device protecting the branch circuit shall have sufficient time delay to permit the motor-compressor and other motors to start and accelerate their loads.

PART VII. PROVISIONS FOR ROOM AIR CONDITIONERS

440.60 General. Part VII shall apply to electrically energized room air conditioners that control temperature and humidity. For the purpose of Part VII, a room air conditioner (with or without provisions for heating) shall be considered as an ac appliance of the air-cooled window, console, or in-wall type that is installed in the conditioned room or that incorporates a hermetic refrigerant motor-compressor(s). Part VII covers equipment rated not over 250 volts, single phase, and the equipment shall be permitted to be cord- and attachment-plug-connected.

A room air conditioner that is rated 3-phase or rated over 250 volts shall be directly connected to a wiring method recognized in Chapter 3, and Part VII shall not apply.

440.61 Grounding. The enclosures of room air conditioners shall be connected to the equipment grounding conductor in accordance with 250.110, 250.112, and 250.114.

440.62 Branch-Circuit Requirements.

(A) Room Air Conditioner as a Single Motor Unit. A room air conditioner shall be considered as a single motor unit in determining its branch-circuit requirements where all the following conditions are met:

(1) It is cord- and attachment-plug-connected.
(2) Its rating is not more than 40 amperes and 250 volts, single phase.
(3) Total rated-load current is shown on the room air-conditioner nameplate rather than individual motor currents.
(4) The rating of the branch-circuit short-circuit and ground-fault protective device does not exceed the ampacity of the branch-circuit conductors or the rating of the receptacle, whichever is less.

(B) Where No Other Loads Are Supplied. The total marked rating of a cord- and attachment-plug-connected room air conditioner shall not exceed 80 percent of the rating of a branch circuit where no other loads are supplied.

(C) Where Lighting Units or Other Appliances Are Also Supplied. The total marked rating of a cord- and attachment-plug-connected room air conditioner shall not exceed 50 percent of the rating of a branch circuit where lighting outlets, other appliances, or general-use receptacles are also supplied. Where the circuitry is interlocked to prevent simultaneous operation of the room air conditioner and energization of other outlets on

Air-Conditioning and Refrigerating Equipment　　　　　　**440.65**

the same branch circuit, a cord- and attachment-plug-connected room air conditioner shall not exceed 80 percent of the branch-circuit rating.

440.63 Disconnecting Means. An attachment plug and receptacle or cord connector shall be permitted to serve as the disconnecting means for a single-phase room air conditioner rated 250 volts or less if (1) the manual controls on the room air conditioner are readily accessible and located within 1.8 m (6 ft) of the floor, or (2) an approved manually operable disconnecting means is installed in a readily accessible location within sight from the room air conditioner.

440.64 Supply Cords. Where a flexible cord is used to supply a room air conditioner, the length of such cord shall not exceed 3.0 m (10 ft) for a nominal, 120-volt rating or 1.8 m (6 ft) for a nominal, 208- or 240-volt rating.

440.65 Protection Devices. Single-phase cord- and plug-connected room air conditioners shall be provided with one of the following factory-installed devices:

(1) Leakage-current detector-interrupter (LCDI)
(2) Arc-fault circuit interrupter (AFCI)
(3) Heat detecting circuit interrupter (HDCI)

The protection device shall be an integral part of the attachment plug or be located in the power supply cord within 300 mm (12 in.) of the attachment plug.

CHAPTER 23

COMMUNICATIONS CIRCUITS

INTRODUCTION

Chapter 23 contains requirements from Article 800 — General Requirements for Communications Systems and Article 805 — Communications Circuits.

Article 800 covers the general requirements for communication systems. This chapter contains four of the five parts in Article 800. Part I covers general installation requirements. Grounding methods are covered in Part III, and Part IV covers installation methods within buildings. The final part, Part V, covers listing requirements for the different cable types.

Article 805 covers communications circuits and equipment, and this chapter contains four of the five parts in Article 805. Part I covers general installation requirements. Protection is covered in Part III, and Part IV covers installation methods within buildings. Part V covers listing requirements for the different cable types.

299

ARTICLE 800
General Requirements for Communications Systems

Communications Circuit. The circuit that extends service from the communications utility or service provider up to and including the customer's communications equipment.

800.21 Access to Electrical Equipment Behind Panels Designed to Allow Access. Access to electrical equipment shall not be denied by an accumulation of wires and cables that prevents removal of panels, including suspended ceiling panels.

800.24 Mechanical Execution of Work. Circuits and equipment shall be installed in a neat and workmanlike manner. Cables installed exposed on the surface of ceilings and sidewalls shall be supported by the building structure in such a manner that the cable will not be damaged by normal building use. Such cables shall be secured by hardware, including straps, staples, cable ties, hangers, or similar fittings, designed and installed so as not to damage the cable. The installation shall also conform to 300.4 and 300.11. Nonmetallic cable ties and other nonmetallic cable accessories used to secure and support cables in other spaces used for environmental air (plenums) shall be listed as having low smoke and heat release properties in accordance with 805.170(C).

800.25 Abandoned Cables. The accessible portion of abandoned cables shall be removed. Where cables are identified for future use with a tag, the tag shall be of sufficient durability to withstand the environment involved.

800.26 Spread of Fire or Products of Combustion. Installations of cables, communications raceways, cable routing assemblies in hollow spaces, vertical shafts, and ventilation or air-handling ducts shall be made so that the possible spread of fire or products of combustion will not be substantially increased. Openings around penetrations of cables, communications raceways, and cable routing assemblies through fire-resistant-rated walls, partitions, floors, or ceilings shall be firestopped using approved methods to maintain the fire resistance rating.

Communications Circuits 800.100

PART III. GROUNDING METHODS

800.100 Cable and Primary Protector Bonding and Grounding.

(A) Bonding Conductor or Grounding Electrode Conductor.

(1) Insulation. The bonding conductor or grounding electrode conductor shall be listed and shall be permitted to be insulated, covered, or bare.

(2) Material. The bonding conductor or grounding electrode conductor shall be copper or other corrosion-resistant conductive material, stranded or solid.

(3) Size. The bonding conductor or grounding electrode conductor shall not be smaller than 14 AWG. The bonding conductor or grounding electrode conductor shall have a current-carrying capacity not less than the grounded metallic cable sheath member of the communications cable, the protected conductor of the communications cable, or the outer sheath of the coaxial cable, as applicable. The bonding conductor or grounding electrode conductor shall not be required to exceed 6 AWG.

(4) Length. The bonding conductor or grounding electrode conductor shall be as short as practicable. In one- and two-family dwellings, the bonding conductor or grounding electrode conductor shall be as short as practicable, not to exceed 6.0 m (20 ft) in length.

> Informational Note: Similar bonding conductor or grounding electrode conductor length limitations applied at apartment buildings and commercial buildings help to reduce voltages that may be developed between the building's power and communications systems during lightning events.

Exception: In one- and two-family dwellings where it is not practicable to achieve an overall maximum bonding conductor or grounding electrode conductor length of 6.0 m (20 ft), a separate ground rod meeting the minimum dimensional criteria of 800.100(B)(3)(2) or (B)(3)(3) shall be driven, the bonding conductor or grounding electrode conductor shall be connected to the ground rod in accordance with 800.100(C), and the ground rod shall be connected to the power grounding electrode system in accordance with 800.100(D).

(5) Run in Straight Line. The bonding conductor or grounding electrode conductor shall be run in as straight a line as practicable.

(6) Physical Protection. Bonding conductors and grounding electrode conductors shall be protected where exposed to physical damage. Where the bonding conductor or grounding electrode conductor is installed in a metal raceway, both ends of the raceway shall be bonded to the contained conductor or to the same terminal or electrode to which the bonding conductor or grounding electrode conductor is connected.

(B) Electrode. The bonding conductor or grounding electrode conductor shall be connected in accordance with 800.100(B)(1), (B)(2), or (B)(3).

(1) In Buildings or Structures with an Intersystem Bonding Termination. If the building or structure served has an intersystem bonding termination as required by 250.94, the bonding conductor shall be connected to the intersystem bonding termination.

(2) In Buildings or Structures with Grounding Means. If an intersystem bonding termination is established, 250.94(A) shall apply. If the building or structure served has no intersystem bonding termination, the bonding conductor or grounding electrode conductor shall be connected to the nearest accessible location on one of the following:

(1) The building or structure grounding electrode system as covered in 250.50
(2) The grounded interior metal water piping system, within 1.5 m (5 ft) from its point of entrance to the building, as covered in 250.52
(3) The power service accessible means external to enclosures using the options identified in 250.94(A), Exception
(4) The nonflexible metal power service raceway
(5) The service equipment enclosure
(6) The grounding electrode conductor or the grounding electrode conductor metal enclosure of the power service
(7) The grounding electrode conductor or the grounding electrode of a building or structure disconnecting means that is connected to a grounding electrode as covered in 250.32

A bonding device intended to provide a termination point for the bonding conductor (intersystem bonding) shall not interfere with the opening of an equipment enclosure. A bonding device shall be mounted on nonremovable parts. A bonding device shall not be mounted on a door or cover even if the door or cover is nonremovable.

Communications Circuits 800.110

For purposes of this section, the mobile home service equipment or the mobile home disconnecting means located within 9.0 m (30 ft) of the exterior wall of the mobile home it serves, or at a mobile home disconnecting means connected to an electrode by a grounding electrode conductor in accordance with 250.32 and located within 9.0 m (30 ft) of the exterior wall of the mobile home it serves, shall be considered to meet the requirements of this section.

(3) In Buildings or Structures Without an Intersystem Bonding Termination or Grounding Means. If the building or structure served has no intersystem bonding termination or grounding means, as described in 800.100(B)(2), the grounding electrode conductor shall be connected to one of the following:

(1) To any one of the individual grounding electrodes described in 250.52(A)(1), (A)(2), (A)(3), or (A)(4)
(2) If the building or structure served has no intersystem bonding termination or grounding means, as described in 800.100(B)(2) or (B)(3)(1), to any one of the individual grounding electrodes described in 250.52(A)(5), (A)(7), and (A)(8)

(C) Electrode Connection. Connections to grounding electrodes shall comply with 250.70.

(D) Bonding of Electrodes. A bonding jumper not smaller than 6 AWG copper or equivalent shall be connected between the grounding electrode and power grounding electrode system at the building or structure served where separate electrodes are used.

Exception: At mobile homes as covered in 800.106.

PART IV. INSTALLATION METHODS WITHIN BUILDINGS

800.110 Raceways, Cable Routing Assemblies, and Cable Trays.

(A) Types of Raceways. Wires and cables shall be permitted to be installed in raceways that comply with either 800.110(A)(1) or (A)(2). Medium-power network-powered broadband communications cables shall not be permitted to be installed in raceways that comply with 800.110(A)(2).

(1) Raceways Recognized in Chapter 3. Wires and cables shall be permitted to be installed in any raceway included in Chapter 3. The raceways shall be installed in accordance with the requirements of Chapter 3.

(2) Communications Raceways. Wires and cables shall be permitted to be installed in plenum communications raceways, riser communications raceways, and general-purpose communications raceways selected in accordance with Table 800.154(b), listed in accordance with 800.182, and installed in accordance with 800.113 and 362.24 through 362.56, where the requirements applicable to electrical nonmetallic tubing (ENT) apply.

(3) Innerduct for Communications Wires and Cables, Coaxial Cables, or Network-Powered Broadband Communications Cables. Listed plenum communications raceways, listed riser communications raceways, and listed general-purpose communications raceways selected in accordance with Table 800.154(b) shall be permitted to be installed as innerduct in any type of listed raceway permitted in Chapter 3.

(B) Raceway Fill. The raceway fill requirements of Chapters 3 and 9 shall apply to medium-power network-powered broadband communications cables.

(C) Cable Routing Assemblies. Cables shall be permitted to be installed in plenum cable routing assemblies, riser cable routing assemblies, and general-purpose cable routing assemblies selected in accordance with Table 800.154(c), listed in accordance with 800.182, and installed in accordance with 800.110(C)(1) and (C)(2) and 800.113.

(1) Horizontal Support. Cable routing assemblies shall be supported where run horizontally at intervals not to exceed 900 mm (3 ft) and at each end or joint, unless listed for other support intervals. In no case shall the distance between supports exceed 3 m (10 ft).

(2) Vertical Support. Vertical runs of cable routing assemblies shall be supported at intervals not exceeding 1.2 m (4 ft), unless listed for other support intervals, and shall not have more than one joint between supports.

800.113 Installation of Wires, Cables, Cable Routing Assemblies, and Communications Raceways. Installation of wires, cables, cable routing assemblies, and communications raceways shall comply with 800.113(A) through (L). Installation of cable routing assemblies and communications raceways shall comply also with 800.110.

(A) Listing. Wires, cables, cable routing assemblies, and communications raceways installed in buildings shall be listed.

Communications Circuits   800.113

(G) Risers — One- and Two-Family Dwellings. The following cables, cable routing assemblies, and communications raceways shall be permitted in one- and two-family dwellings:

(1) Plenum cables, riser cables, and general-purpose cables
(2) Limited-use cables less than 6 mm (0.25 in.) in diameter
(3) Plenum, riser, and general-purpose communications raceways
(4) Plenum, riser, and general-purpose cable routing assemblies
(5) Plenum cables, riser cables, and general-purpose cables installed in:

    (a) Plenum communications raceways
    (b) Riser communications raceways
    (c) General-purpose communications raceways
    (d) Plenum cable routing assemblies
    (e) Riser cable routing assemblies
    (f) General-purpose cable routing assemblies

(K) Multifamily Dwellings. The following cables, cable routing assemblies, and communications raceways shall be permitted to be installed in multifamily dwellings in locations other than the locations covered in 800.113(B) through (G):

(1) Plenum cables, riser cables, and general-purpose cables
(2) Limited-use cables less than 6 mm (0.25 in.) in diameter in nonconcealed spaces
(3) Plenum, riser, and general-purpose communications raceways
(4) Plenum, riser, and general-purpose cable routing assemblies
(5) Communications wires, plenum cables, riser cables, and general-purpose cables installed in the following:

    (a) Plenum communications raceways
    (b) Riser communications raceways
    (c) General-purpose communications raceways

(6) Plenum cables, riser cables, and general-purpose cables installed in the following:

    (a) Plenum cable routing assemblies
    (b) Riser cable routing assemblies
    (c) General-purpose cable routing assemblies

(7) Communications wires, plenum cables, riser cables, general-purpose cables, and limited-use cables installed in raceways recognized in Chapter 3

(8) Type CMUC under-carpet communications wires and cables installed under carpet, modular flooring, and planks

(L) One- and Two-Family Dwellings. The following wires, cables, cable routing assemblies, and communications raceways shall be permitted to be installed in one- and two-family dwellings in locations other than the locations covered in 800.113(B) through (F):

(1) Plenum cables, riser cables, and general-purpose cables
(2) Limited-use cables less than 6 mm (0.25 in.) in diameter
(3) Plenum, riser, and general-purpose communications raceways
(4) Plenum, riser, and general-purpose cable routing assemblies
(5) Communications wires, plenum cables, riser cables, and general-purpose cables installed in the following:

   (a) Plenum communications raceways
   (b) Riser communications raceways
   (c) General-purpose communications raceways

(6) Plenum cables, riser cables, and general-purpose cables installed in the following:

   (a) Plenum cable routing assemblies
   (b) Riser cable routing assemblies
   (c) General-purpose cable routing assemblies

(7) Communications wires, plenum cables, riser cables, general-purpose cables, and limited-use cables installed in raceways recognized in Chapter 3
(8) Type CMUC under-carpet communications wires and cables installed under carpet, modular flooring, and planks
(9) Hybrid power and communications cable listed in accordance with 805.179(F)

800.154 Applications of Listed Communications Wires, Cables, and Raceways, and Listed Cable Routing Assemblies. Permitted and nonpermitted applications of listed communications wires, cables, coaxial cables, network-powered broadband communications system cables and raceways, and listed cable routing assemblies, shall be in accordance with one of the following:

(1) Listed communications wires and cables as indicated in Table 800.154(a)
(2) Listed communications raceways as indicated in Table 800.154(b)
(3) Listed cable routing assemblies as indicated in Table 800.154(c)

Table 800.154(a) Applications of Listed Communications Wires, Cables, and Network-Powered Broadband Communications System Cables in Buildings

Applications	Wire and Cable Type									
	Plenum	Riser	BMR	General-Purpose	BM	Limited-Use	Undercarpet	BMU, BLU	Hybrid Power and Communications Cables	Communications Wires
In ducts specifically fabricated for environmental air as described in 300.22(B)	Y	N	N	N	N	N	N	N	N	N
In metal raceway that complies with 300.22(B)	Y	Y	Y	N	Y	Y	N	N	N	Y
In other spaces used for environmental air (plenums) as described in 300.22(C)	Y	N	N	Y	N	N	N	N	N	N
In metal raceway that complies with 300.22(C)	Y	Y	N	N	Y	Y	N	N	N	Y
In plenum communications raceways	Y									N

(continued)

307

Table 800.154(a) Applications of Listed Communications Wires, Cables, and Network-Powered Broadband Communications System Cables in Buildings — *Continued*

Applications		Wire and Cable Type									
		Plenum	Riser	BMR	General-Purpose	BM	Limited-Use	Undercarpet	BMU, BLU	Hybrid Power and Communications Cables	Communications Wires
	In plenum cable routing assemblies	Y	N	N	N	N	N	N	N	N	N
	Supported by open metal cable trays	Y	N	N	N	N	N	N	N	N	N
	Supported by solid bottom metal cable trays with solid metal covers	Y	Y	Y	Y	Y	Y	N	N	N	N
In risers	In vertical runs	Y	Y	Y	N	N	N	N	N	N	N
	In metal raceways	Y	Y	Y	Y	Y	Y	N	N	N	N
	In fireproof shafts	Y	Y	Y	Y	Y	N	N	N	N	N
	In plenum communications raceways	Y	Y	N	N	N	N	N	N	N	N

	In plenum cable routing assemblies	Y	N	N	N	N	N	N	N	N
	In riser communications raceways	Y	Y	N	N	N	N	N	N	N
	In riser cable routing assemblies	Y	N	N	N	N	N	N	N	N
	In one- and two-family dwellings	Y	Y	Y	Y	Y	N	N	Y	N
Within buildings in other than air-handling spaces and risers	General	Y	Y	Y	Y	Y	N	N	N	N
	In one- and two-family dwellings	Y	Y	Y	Y	Y	N	N	Y	N
	In multifamily dwellings	Y	Y	Y	Y	Y	N	N	N	N
	In nonconcealed spaces	Y	Y	Y	Y	Y	Y	N	N	N
	Supported by cable trays	Y	N	N	N	N	N	N	N	N
	Under carpet, modular flooring, and planks	N					Y	N	N	N

(continued)

Table 800.154(a) Applications of Listed Communications Wires, Cables, and Network-Powered Broadband Communications System Cables in Buildings — *Continued*

Applications	Wire and Cable Type									
	Plenum	Riser	BMR	General-Purpose	BM	Limited-Use	Undercarpet	BMU, BLU	Hybrid Power and Communications Cables	Communications Wires
In distributing frames and cross-connect arrays	Y	Y	N	Y	N	N	N	N	N	Y
In rigid metal conduit (RMC) and intermediate metal conduit (IMC)	Y	Y	Y	Y	Y	Y	Y	Y	Y	Y
In any raceway recognized in Chapter 3	Y	Y	Y	Y	Y	Y	N	N	N	Y
In plenum communications raceways	Y	Y	N	Y	N	N	N	N	N	Y
In plenum cable routing assemblies	Y	Y	N	Y	N	N	N	N	N	Y

Wire and Cable Type

Applications	Plenum	Riser	BMR	General-Purpose	BM	Limited-Use	Undercarpet	BMU, BLU	Hybrid Power and Communications Cables	Communications Wires
In riser communications raceways	Y	Y	N	Y	N	N	N	N	N	Y
In riser cable routing assemblies	Y	Y	N	Y	N	N	N	N	N	Y
In general-purpose communications raceways	Y	Y	N	Y	N	N	N	N	N	Y
In general-purpose cable routing assemblies	Y	Y	N	Y	N	N	N	N	N	Y

Note: An "N" in the table indicates that the cable type shall not be permitted to be installed in the application. A "Y" indicates that the cable type shall be permitted to be installed in the application subject to the limitations described in 800.113. The Riser column includes all riser cables except BMR, and the General-Purpose column includes all general-purpose cables except BM.

Informational Note No. 1: Part IV of Article 800 covers installation methods within buildings. This table covers the applications of listed communications wires, cables, and raceways in buildings. See the definition of *Point of Entrance* in 800.2.

Informational Note No. 2: For information on the restrictions to the installation of communications cables in fabricated ducts, see 800.113(B).

Table 800.154(b) Applications of Listed Communications Raceways in Buildings

	Applications	Listed Communications Raceway Type		
		Plenum	Riser	General-Purpose
In ducts specifically fabricated for environmental air as described in 300.22(B)	In fabricated ducts	N	N	N
	In metal raceway that complies with 300.22(B)	N	N	N
In other spaces used for environmental air (plenums) as described in 300.22(C)	In other spaces used for environmental air	Y	N	N
	In metal raceway that complies with 300.22(C)	Y	Y	Y
	In plenum cable routing assemblies	N	N	N
	Supported by open metal cable trays	Y	N	N
	Supported by solid bottom metal cable trays with solid metal covers	Y	Y	Y
In risers	In vertical runs	Y	Y	N
	In metal raceways	Y	Y	Y
	In fireproof shafts	Y	Y	Y
	In plenum cable routing assemblies	N	N	N
	In riser cable routing assemblies	N	N	N
	In one- and two-family dwellings	Y	Y	Y

Communications Circuits 800.154

Applications		Listed Communications Raceway Type		
		Plenum	Riser	General-Purpose
Within buildings in other than air-handling spaces and risers	General	Y	Y	Y
	In one- and two-family dwellings	Y	Y	Y
	In multifamily dwellings	Y	Y	Y
	In nonconcealed spaces	Y	Y	Y
	Supported by cable trays	Y	Y	Y
	Under carpet, modular flooring, and planks	N	N	N
	In distributing frames and cross-connect arrays	Y	Y	Y
	In any raceway recognized in Chapter 3	Y	Y	Y
	In plenum cable routing assemblies	N	N	N
	In riser cable routing assemblies	N	N	N
	In general-purpose cable routing assemblies	N	N	N

Note: An "N" in the table indicates that the communications raceway type shall not be permitted to be installed in the application. A "Y" indicates that the communications raceway type shall be permitted to be installed in the application, subject to the limitations described in 800.110 and 800.113.

Table 800.154(c) Applications of Listed Cable Routing Assemblies in Buildings

	Applications	Listed Cable Routing Assembly Type		
		Plenum	Riser	General-Purpose
In ducts specifically fabricated for environmental air as described in 300.22(B)	In fabricated ducts	N	N	N
	In metal raceway that complies with 300.22(B)	N	N	N
In other spaces used for environmental air (plenums) as described in 300.22(C)	In other spaces used for environmental air	Y	N	N
	In metal raceway that complies with 300.22(C)	N	N	N
	In plenum communications raceways	N	N	N
	Supported by open metal cable trays	Y	N	N
	Supported by solid bottom metal cable trays with solid metal covers	N	N	N
In risers	In vertical runs	Y	Y	N
	In metal raceways	N	N	N
	In fireproof shafts	Y	Y	Y
	In plenum communications raceways	Y	N	N
	In riser communications raceways	N	Y	N
	In one- and two-family dwellings	Y	Y	Y

Communications Circuits — 800.154

Applications	Listed Cable Routing Assembly Type		
	Plenum	Riser	General-Purpose
General	Y	Y	Y
In one- and two-family dwellings	Y	Y	Y
In multifamily dwellings	Y	Y	Y
In nonconcealed spaces	Y	Y	Y
Supported by cable trays	Y	Y	Y
Under carpet, modular flooring, and planks	N	N	N
In distributing frames and cross-connect arrays	Y	Y	Y
In any raceway recognized in Chapter 3	N	N	N
In plenum communications raceways	N	N	N
In riser communications raceways	N	N	N
In general-purpose communications raceways	N	N	N
Within buildings in other than air-handling spaces and risers			

Note: An "N" in the table indicates that the cable routing assembly type shall not be permitted to be installed in the application. A "Y" indicates that the cable routing assembly type shall be permitted to be installed in the application subject to the limitations described in 800.113.

The permitted applications shall be subject to the installation requirements of 800.110 and 800.113.

800.179 Plenum, Riser, General-Purpose, and Limited Use Cables. Plenum, riser, general-purpose, and limited-use cables shall be listed in accordance with 800.179(A) through (D) and shall have a temperature rating of not less than 60°C (140°F). The temperature rating shall be marked on the jacket of cables that have a temperature rating exceeding 60°C (140°F). The cable voltage rating shall not be marked on the cable.

(A) Plenum Cables. Type CMP communications plenum cables, Type CATVP community antenna television plenum coaxial cables, and Type BLP network-powered broadband communication low-power plenum cables shall be listed as being suitable for use in ducts, plenums, and other spaces used for environmental air and shall also be listed as having adequate fire-resistant and low smoke-producing characteristics.

> Informational Note: One method of defining a cable that is low-smoke-producing cable and fire-resistant cable is that the cable exhibits a maximum peak optical density of 0.50 or less, an average optical density of 0.15 or less, and a maximum flame spread distance of 1.52 m (5 ft) or less when tested in accordance with NFPA 262-2019, *Standard Method of Test for Flame Travel and Smoke of Wires and Cables for Use in Air-Handling Spaces.*

(B) Riser Cables. Type CMR communications riser cables, Type CATVR community antenna television riser coaxial cables, Type BMR network-powered broadband communications medium-power riser cables, and Type BLR network-powered broadband communications low-power riser cables shall be listed as being suitable for use in a vertical run in a shaft or from floor to floor and shall also be listed as having fire-resistant characteristics capable of preventing the carrying of fire from floor to floor.

> Informational Note: One method of defining fire-resistant characteristics capable of preventing the carrying of fire from floor to floor is that the cables pass the requirements of ANSI/UL 1666-2011, *Standard Test for Flame Propagation Height of Electrical and Optical-Fiber Cable Installed Vertically in Shafts.*

(C) General-Purpose Cables. Type CM communications general-purpose cables, Type CATV community antenna television coaxial general-purpose cables, Type BM network-powered broadband communications medium-power general-purpose cables, and Type BL

Communications Circuits 800.182

network-powered broadband communications low-power general-purpose cables shall be listed as being suitable for general-purpose use, with the exception of risers and plenums, and shall also be listed as being resistant to the spread of fire.

> Informational Note: One method of defining *resistant to the spread of fire* is that the cables do not spread fire to the top of the tray in the "UL Flame Exposure, Vertical Flame Tray Test" in ANSI/UL 1685-2010, *Standard for Safety for Vertical-Tray Fire-Propagation and Smoke-Release Test for Electrical and Optical-Fiber Cables*. The smoke measurements in the test method are not applicable. Another method of defining *resistant to the spread of fire* is for the damage (char length) not to exceed 1.5 m (4 ft 11 in.) when performing the CSA "Vertical Flame Test — Cables in Cable Trays," as described in CSA C22.2 No. 0.3-09, *Test Methods for Electrical Wires and Cables*.

(D) Limited-Use Cables. Type CMX limited-use communications cables, Type CATVX limited-use community antenna television coaxial cables, and Type BLX limited-use network-powered broadband low-power cables shall be listed as being suitable for use in dwellings and for use in raceway and shall also be listed as being resistant to flame spread.

> Informational Note: One method of determining that cable is resistant to flame spread is by testing the cable to the VW-1 (vertical-wire) flame test in ANSI/UL 1581-2017, *Reference Standard for Electrical Wires, Cables and Flexible Cords*.

PART VI. LISTING REQUIREMENTS

800.182 Cable Routing Assemblies and Communications Raceways. Cable routing assemblies and communications raceways shall be listed in accordance with 800.182(A) through (C). Cable routing assemblies shall be marked in accordance with Table 800.182(a). Communications raceways shall be marked in accordance with Table 800.182(b).

> Informational Note: For information on listing requirements for both communications raceways and cable routing assemblies, see ANSI/UL 2024-5-2015, *Cable Routing Assemblies and Communications Raceways*.

(A) Plenum Cable Routing Assemblies and Plenum Communications Raceways. Plenum cable routing assemblies and plenum communications raceways shall be listed as having adequate fire-resistant and low-smoke-producing characteristics.

> Informational Note No. 1: One method of defining cable routing assemblies and communications raceways that have adequate fire-resistant and low-smoke-producing characteristics is that they exhibit a maximum flame spread index of 25 and a maximum smoke developed index of 50 when tested in accordance with ASTM E84-15a, *Standard Test Method for Surface Burning Characteristics of Building Materials*, or ANSI/UL 723-2013, *Standard Test Method for Surface Burning Characteristics of Building Materials*.

> Informational Note No. 2: Another method of defining communications raceways that have adequate fire-resistant and low-smoke-producing characteristics is that they exhibit a maximum peak optical density of 0.50 or less, an average optical density of 0.15 or less, and a maximum flame spread distance of 1.52 m (5 ft) or less when tested in accordance with NFPA 262-2015, *Standard Method of Test for Flame Travel and Smoke of Wires and Cables for Use in Air-Handling Spaces*.

> Informational Note No. 3: See 4.3.11.2.6 or 4.3.11.5.5 of NFPA 90A-2015, *Standard for the Installation of Air-Conditioning and Ventilating Systems*, for information on materials exposed to the airflow in ceiling cavity and raised floor plenums.

(B) Riser Cable Routing Assemblies and Riser Communications Raceways. Riser cable routing assemblies and riser communications raceways shall be listed as having adequate fire-resistant characteristics capable of preventing the carrying of fire from floor to floor.

> Informational Note: One method of defining fire-resistant characteristics capable of preventing the carrying of fire from floor to floor is that the cable routing assemblies and communications raceways pass the requirements of ANSI/UL 1666-2011, *Standard Test for Flame Propagation Height of Electrical and Optical-Fiber Cable Installed Vertically in Shafts*.

(C) General-Purpose Cable Routing Assemblies and General-Purpose Communications Raceways. General-purpose cable routing assemblies and general-purpose communications raceways shall be listed as being resistant to the spread of fire.

Informational Note: One method of defining resistant to the spread of fire is that the cable routing assemblies and communications raceways do not spread fire to the top of the tray in the "UL Flame Exposure, Vertical Flame Tray Test" in ANSI/UL 1685-2011, *Standard for Safety for Vertical-Tray Fire-Propagation and Smoke-Release Test for Electrical and Optical-Fiber Cables.*

ARTICLE 805
Communications Circuits

PART I. GENERAL

805.18 Installation of Equipment. Equipment electrically connected to a communications network shall be listed in accordance with 805.170.

PART III. PROTECTION

805.93 Grounding, Bonding, or Interruption of Non–Current-Carrying Metallic Sheath Members of Communications Cables. Communications cables entering the building or terminating on the outside of the building shall comply with 805.93(A) or (B).

(A) Entering Buildings. In installations where the communications cable enters a building, the metallic sheath members of the cable shall be grounded or bonded as specified in 800.100 or interrupted by an insulating joint or equivalent device. The grounding, bonding, or interruption shall be as close as practicable to the point of entrance.

(B) Terminating on the Outside of Buildings. In installations where the communications cable is terminated on the outside of the building, the metallic sheath members of the cable shall be grounded or bonded as specified in 800.100 or interrupted by an insulating joint or equivalent device. The grounding, bonding, or interruption shall be as close as practicable to the point of termination of the cable.

PART IV. INSTALLATION METHODS WITHIN BUILDINGS

805.133 Installation of Communications Wires, Cables, and Equipment. Communications wires and cables from the protector to the equipment or, where no protector is required, communications wires and cables attached to the outside or inside of the building shall comply with 805.133(A) and 805.133(B)

(A) Separation from Other Conductors.

(1) In Raceways, Cable Trays, Boxes, Cables, Enclosures, and Cable Routing Assemblies.

(a) *Other Circuits.* Communications cables shall be permitted in the same raceway, cable tray, box, enclosure, or cable routing assembly with cables of any of the following:

(1) Class 2 and Class 3 remote-control, signaling, and power-limited circuits in compliance with Article 645 or Parts I and III of Article 725
(2) Power-limited fire alarm systems in compliance with Parts I and III of Article 760
(3) Nonconductive and conductive optical fiber cables in compliance with Parts I and V of Article 770
(4) Community antenna television and radio distribution systems in compliance with Parts I and V of Article 820
(5) Low-power network-powered broadband communications circuits in compliance with Parts I and V of Article 830

(b) *Class 2 and Class 3 Circuits.* Class 1 circuits shall not be run in the same cable with communications circuits. Class 2 and Class 3 circuit conductors shall be permitted in the same listed communications cable with communications circuits.

(c) *Electric Light, Power, Class 1, Non–Power-Limited Fire Alarm, and Medium-Power Network-Powered Broadband Communications Circuits in Raceways, Compartments, and Boxes.* Communications conductors shall not be placed in any raceway, compartment, outlet box, junction box, or similar fitting with conductors of electric light, power, Class 1, non–power-limited fire alarm, or medium-power network-powered broadband communications circuits.

Exception No. 1: Section 805.133(A)(1)(c) shall not apply if all of the conductors of electric light, power, Class 1, non–power-limited fire alarm, and medium-power network-powered broadband communications circuits are separated from all of the conductors of communications circuits by a permanent barrier or listed divider.

Exception No. 2: Power conductors in outlet boxes, junction boxes, or similar fittings or compartments where such conductors are introduced solely for power supply to communications equipment. The power circuit

Communications Circuits 805.179

conductors shall be routed within the enclosure to maintain a minimum of 6 mm (¼ in.) separation from the communications circuit conductors.

(B) Support of Communications Wires and Cables. Raceways shall be used for their intended purpose. Communications wires and cables shall not be strapped, taped, or attached by any means to the exterior of any raceway as a means of support.

805.156 Dwelling Unit Communications Outlet. For new construction, a minimum of one communications outlet shall be installed within the dwelling in a readily accessible area and cabled to the service provider demarcation point.

PART V. LISTING REQUIREMENTS

805.170 Equipment. Communications equipment shall be listed as being suitable for electrical connection to a communications network.

(A) Primary Protectors. The primary protector shall consist of an arrester connected between each line conductor and ground in an appropriate mounting. Primary protector terminals shall be marked to indicate line and ground as applicable.

(B) Secondary Protectors. The secondary protector shall be listed as suitable to provide means to safely limit currents to less than the current-carrying capacity of listed indoor communications wire and cable, listed telephone set line cords, and listed communications terminal equipment having ports for external wire line communications circuits. Any overvoltage protection, arresters, or grounding connection shall be connected on the equipment terminals side of the secondary protector current-limiting means.

(C) Plenum Grade Cable Ties. Cable ties intended for use in other space used for environmental air (plenums) shall be listed as having low smoke and heat release properties.

805.179 Communications Wires and Cables. Communications wires and cables shall be listed in accordance with 805.179(A) through (F) and marked in accordance with Table 805.179 and 805.179(G). Conductors in communications cables, other than in a coaxial cable, shall be copper.

Communications wires and cables shall have a voltage rating of not less than 300 volts. The insulation for the individual conductors, other than the outer conductor of a coaxial cable, shall be rated for 300 volts

Table 805.179 Cable Markings

Cable Marking	Type
CMP	Communications plenum cable
CMR	Communications riser cable
CMG	Communications general-purpose cable
CM	Communications general-purpose cable
CMX	Communications cable, limited use
CMUC	Under-carpet communications wire and cable

Informational Note: Cable types are listed in descending order of fire performance.

minimum. The cable voltage rating shall not be marked on the cable or on the under-carpet communications wire.

Exception: Voltage markings shall be permitted where the cable has multiple listings and voltage marking is required for one or more of the listings.

(A) General Requirements. The general requirements in 805.179 shall apply.

(B) Type CMUC Undercarpet Wires and Cables. Type CMUC under-carpet communications wires and cables shall be listed as being suitable for under-carpet use and shall also be listed as being resistant to flame spread.

(C) Circuit Integrity (CI) Cable or Electrical Circuit Protective System. Cables that are used for survivability of critical circuits under fire conditions shall be listed and meet either 805.179(C)(1) or 805.179(C)(2).

(1) Circuit Integrity (CI) Cables. Circuit integrity (CI) cables specified in 805.179(A) through (D), and used for survivability of critical circuits, shall have an additional classification using the suffix "CI." In order to maintain its listed fire rating, circuit integrity (CI) cable shall only be installed in free air.

(2) Fire-Resistive Cables. Cables specified in 800.179(A) through (D) and 805.179(C)(1), that are part of an electrical circuit protective system,

Communications Circuits

shall be fire-resistive cable identified with the protective system number on the product, or on the smallest unit container in which the product is packaged, and shall be installed in accordance with the listing of the protective system.

(D) Types CMP-LP, CMR-LP, CMG-LP, and CM-LP Limited Power (LP) Cables. Types CMP-LP, CMR-LP, CMG-LP, and CM-LP communications limited power cables shall be listed as suitable for carrying power and data up to a specified current limit for each conductor without exceeding the temperature rating of the cable where the cable is installed in cable bundles in free air or installed within a raceway, cable tray, or cable routing assembly. The cables shall be marked with the suffix "-LP(XXA)," where XX designates the current limit in amperes per conductor.

(E) Communications Wires. Communications wires, such as distributing frame wire and jumper wire, shall be listed as being resistant to the spread of fire.

(F) Hybrid Power and Communications Cables. Listed hybrid power and communications cables shall be permitted where the power cable is a listed Type NM or NM-B, conforming to Part III of Article 334, and the communications cable is a listed Type CM, the jackets on the listed NM or NM-B, and listed CM cables are rated for 600 volts minimum, and the hybrid cable is listed as being resistant to the spread of fire.

(G) Optional Markings. Cables shall be permitted to be surface marked to indicate special characteristics of the cable materials.

CHAPTER 24
SEPARATE BUILDINGS OR STRUCTURES

INTRODUCTION

This chapter contains requirements from three sections of Article 225 — Outside Branch Circuits and Feeders, and from one section of Article 250 — Grounding and Bonding. This chapter covers the section pertaining to two or more buildings or structures supplied from a common service. This section's requirements cover grounding electrode(s), grounded systems, number of supplies, disconnecting means located in a separate building or structure on the same premises, and the grounding electrode conductor.

PART II. BUILDINGS OR OTHER STRUCTURES SUPPLIED BY A FEEDER(S) OR BRANCH CIRCUIT(S)

225.30 Number of Supplies. A building or other structure that is served by a branch circuit or feeder on the load side of a service disconnecting means shall be supplied by only one feeder or branch circuit unless permitted in 225.30(A) through (E). For the purpose of this section, a multiwire branch circuit shall be considered a single circuit.

Where a branch circuit or feeder originates in these additional buildings or other structures, only one feeder or branch circuit shall be permitted to supply power back to the original building or structure, unless permitted in 225.30(A) through (E).

(A) Special Conditions. Additional feeders or branch circuits shall be permitted to supply the following:

(7) Electric vehicle charging systems listed, labeled, and identified for more than a single branch circuit or feeder

225.31 Disconnecting Means. Means shall be provided for disconnecting all ungrounded conductors that supply or pass through the building or structure.

225.32 Location. The disconnecting means shall be installed either inside or outside of the building or structure served or where the conductors pass through the building or structure. The disconnecting means shall be at a readily accessible location nearest the point of entrance of the conductors. For the purposes of this section, the requirements in 230.6 shall be utilized.

Exception No. 1: For installations under single management, where documented safe switching procedures are established and maintained for disconnection, and where the installation is monitored by qualified individuals, the disconnecting means shall be permitted to be located elsewhere on the premises.

Exception No. 3: For towers or poles used as lighting standards, the disconnecting means shall be permitted to be located elsewhere on the premises.

ARTICLE 250

Grounding and Bonding

250.32 Buildings or Structures Supplied by a Feeder(s) or Branch Circuit(s).

(A) Grounding Electrode. A building(s) or structure(s) supplied by a feeder(s) or branch circuit(s) shall have a grounding electrode system and grounding electrode conductor installed in accordance with Part III of Article 250. Where there is no existing grounding electrode, the grounding electrode(s) required in 250.50 shall be installed.

Exception: A grounding electrode shall not be required where only a single branch circuit, including a multiwire branch circuit, supplies the building or structure and the branch circuit includes an equipment grounding conductor for grounding the normally non–current-carrying metal parts of equipment.

(B) Grounded Systems.

(1) Supplied by a Feeder or Branch Circuit. An equipment grounding conductor, as described in 250.118, shall be run with the supply conductors and be connected to the building or structure disconnecting means and to the grounding electrode(s). The equipment grounding conductor shall be used for grounding or bonding of equipment, structures, or frames required to be grounded or bonded. The equipment grounding conductor shall be sized in accordance with 250.122. Any installed grounded conductor shall not be connected to the equipment grounding conductor or to the grounding electrode(s).

Exception No. 1: For installations made in compliance with previous editions of this Code that permitted such connection, the grounded conductor run with the supply to the building or structure shall be permitted to serve as the ground-fault return path if all of the following requirements continue to be met:

(1) An equipment grounding conductor is not run with the supply to the building or structure.
(2) There are no continuous metallic paths bonded to the grounding system in each building or structure involved.
(3) Ground-fault protection of equipment has not been installed on the supply side of the feeder(s).

If the grounded conductor is used for grounding in accordance with the provision of this exception, the size of the grounded conductor shall not be smaller than the larger of either of the following:

(1) That required by 220.61
(2) That required by 250.122

Exception No. 2: If system bonding jumpers are installed in accordance with 250.30(A)(1), Exception No. 2, the feeder grounded circuit conductor at the building or structure served shall be connected to the equipment grounding conductors, grounding electrode conductor, and the enclosure for the first disconnecting means.

(2) Supplied by Separately Derived System.

(a) *With Overcurrent Protection.* If overcurrent protection is provided where the conductors originate, the installation shall comply with 250.32(B)(1).

(b) *Without Overcurrent Protection.* If overcurrent protection is not provided where the conductors originate, the installation shall comply with 250.30(A). If installed, the supply-side bonding jumper shall be connected to the building or structure disconnecting means and to the grounding electrode(s).

(D) Disconnecting Means Located in Separate Building or Structure on the Same Premises. Where one or more disconnecting means supply one or more additional buildings or structures under single management, and where these disconnecting means are located remote from those buildings or structures in accordance with 225.32, Exception No. 1 and No. 2, 700.12(D)(5), 701.12(D)(5), or 702.12, all of the following conditions shall be met:

(1) The connection of the grounded conductor to the grounding electrode, to normally non–current-carrying metal parts of equipment, or to the equipment grounding conductor at a separate building or structure shall not be made.
(2) An equipment grounding conductor for grounding and bonding any normally non–current-carrying metal parts of equipment, interior metal piping systems, and building or structural metal frames is run with the circuit conductors to a separate building or structure and connected to existing grounding electrode(s) required in Part III of this article, or, where there are no existing electrodes, the grounding electrode(s) required in Part III of this article shall be installed where

Separate Buildings or Structures 250.32

a separate building or structure is supplied by more than one branch circuit.

(3) The connection between the equipment grounding conductor and the grounding electrode at a separate building or structure shall be made in a junction box, panelboard, or similar enclosure located immediately inside or outside the separate building or structure.

(E) Grounding Electrode Conductor. The size of the grounding electrode conductor to the grounding electrode(s) shall not be smaller than given in 250.66, based on the largest ungrounded supply conductor. The installation shall comply with Part III of this article.

CHAPTER 25

SWIMMING POOLS, FOUNTAINS, AND SIMILAR INSTALLATIONS

INTRODUCTION

This chapter contains requirements from Article 680 — Swimming Pools, Fountains, and Similar Installations. The requirements of Article 680 cover the installation of electric wiring for (and equipment in or adjacent to) all swimming, wading, and decorative pools, fountains, hot tubs, spas, hydromassage bathtubs, whether permanently installed or storable, and electrically powered pool lifts, and also cover metallic auxiliary equipment such as pumps, filters, and so forth. Article 680 contains requirements on grounding, bonding, conductor clearances, receptacles, lighting, equipment, and ground-fault circuit-interrupter protection for personnel. Extensive use of GFCI protection and low-voltage equipment is intended to mitigate the elevated risk of electric shock in environments where users are partially or completely immersed and have body resistances lower than they would in dry locations.

More robust requirements on wiring methods and conductor sizes and material address the corrosive effects of chlorine and other treatment agents used to maintain safe water chemistry. Salt water pools have also become very popular in some areas and the elevated sodium level in the water adversely impacts many of the typical electrical materials used in locations that are noncorrosive. Article 680 requirements covering the impact of corrosive agents on electrical equipment apply only to those portions of the wiring system that are directly exposed to these caustic environments.

In this *Pocket Guide* are seven of the eight parts in Article 680: Part I General, Part II Permanently Installed Pools, Part III Storable Pools, Part IV Spas and Hot Tubs, Part V Fountains, Part VII Hydromassage Bathtubs, and Part VIII Electrically Powered Pool Lifts.

ARTICLE 680
Swimming Pools, Fountains, and Similar Installations

680.3 Approval of Equipment. All electrical equipment and products covered by this article shall be installed in compliance with this article and shall be listed.

680.4 Inspections After Installation. The authority having jurisdiction shall be permitted to require periodic inspection and testing.

680.5 Ground-Fault Circuit Interrupters. Ground-fault circuit interrupters (GFCIs) shall be self-contained units, circuit-breaker or receptacle types, or other listed types. The GFCI requirements in this article, unless otherwise noted, are in addition to the requirements in 210.8.

680.6 Bonding and Equipment Grounding. Electrical equipment shall be bonded in accordance with Part V of Article 250 and shall meet the equipment grounding requirements of Parts VI and VII of Article 250. The equipment shall be connected by the wiring methods in Chapter 3, except as modified by this article. Equipment subject to these requirements shall include the following:

(1) Through-wall lighting assemblies and underwater luminaires, other than those low-voltage lighting products listed for the application without an equipment grounding conductor
(2) All electrical equipment located within 1.5 m (5 ft) of the inside wall of the specified body of water
(3) All electrical equipment associated with the recirculating system of the specified body of water
(4) Junction boxes
(5) Transformer and power supply enclosures
(6) Ground-fault circuit interrupters
(7) Panelboards that are not part of the service equipment and that supply any electrical equipment associated with the specified body of water

680.9 Overhead Conductor Clearances. Overhead conductors shall meet the clearance requirements in this section. Where a minimum clearance from the water level is given, the measurement shall be taken from the maximum water level of the specified body of water.

(A) Power. With respect to overhead conductors and open overhead wiring, swimming pool and similar installations shall comply with the minimum clearances given in Table 680.9(A) and illustrated in Figure 680.9(A).

Table 680.9(A) Overhead Conductor Clearances

Clearance Parameters	Insulated Cables, 0–750 Volts to Ground, Supported on and Cabled Together with a Solidly Grounded Bare Messenger or Solidly Grounded Neutral Conductor		All Other Conductors Voltage to Ground			
			0 through 15 kV		Over 15 kV through 50 kV	
	m	ft	m	ft	m	ft
A. Clearance in any direction to the water level, edge of water surface, base of diving platform, or permanently anchored raft	6.9	22.5	7.5	25	8.0	27
B. Clearance in any direction to the observation stand, tower, or diving platform	4.4	14.5	5.2	17	5.5	18
C. Horizontal limit of clearance measured from inside wall of the pool	This limit shall extend to the outer edge of the structures listed in A and B of this table but not less than 3 m (10 ft).					

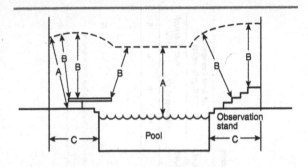

Figure 680.9(A) Clearances from Pool Structures.

Informational Note: Open overhead wiring as used in this article refers to conductor(s) not in an enclosed raceway.

(B) Communications Systems. Communications, radio, and television coaxial cables within the scope of Articles 805 through 820 shall be permitted at a height of not less than 3.0 m (10 ft) above the maximum water level of swimming and wading pools, and diving structures, observation stands, towers, or platforms.

(C) Network-Powered Broadband Communications Systems. The minimum clearances for overhead network-powered broadband communications systems conductors from pools or fountains shall comply with the provisions in Table 680.9(A) for conductors operating at 0 to 750 volts to ground.

680.11 Underground Wiring. Underground wiring shall comply with 680.11(A) through (C).

(A) Underground Wiring. Underground wiring within 1.5 m (5 ft) horizontally from the inside wall of the pool shall be permitted. The following wiring methods shall be considered suitable for the conditions in these locations:

(1) Rigid metal conduit
(2) Intermediate metal conduit

Swimming Pools, Fountains, and Similar Installations 680.21

(3) Rigid polyvinyl chloride conduit
(4) Reinforced thermosetting resin conduit
(5) Jacketed Type MC cable that is listed for burial use
(6) Liquidtight flexible nonmetallic conduit listed for direct burial use
(7) Liquidtight flexible metal conduit listed for direct burial use

(B) Wiring Under Pools. Underground wiring shall not be permitted under the pool unless this wiring is necessary to supply pool equipment permitted by this article.

(C) Minimum Cover Requirements. Minimum cover depths shall be as given in Table 300.5.

680.13 Maintenance Disconnecting Means. One or more means to simultaneously disconnect all ungrounded conductors shall be provided for all utilization equipment other than lighting. Each means shall be readily accessible and within sight from its equipment and shall be located at least 1.5 m (5 ft) horizontally from the inside walls of a pool, spa, fountain, or hot tub unless separated from the open water by a permanently installed barrier that provides a 1.5 m (5 ft) reach path or greater. This horizontal distance shall be measured from the water's edge along the shortest path required to reach the disconnect.

680.14 Wiring Methods in Corrosive Environment. Wiring methods in a corrosive environment shall be listed and identified for use in such areas. Rigid metal conduit, intermediate metal conduit, rigid polyvinyl chloride conduit, and reinforced thermosetting resin conduit shall be considered to be resistant to the corrosive environment.

PART II. PERMANENTLY INSTALLED POOLS

680.20 General. Electrical installations at permanently installed pools shall comply with the provisions of Part I and Part II of this article.

680.21 Motors.

(A) Wiring Methods. The wiring to a pool motor shall comply with 680.21(A)(1) unless modified for specific circumstances by (A)(2) or (A)(3).

(1) General. Wiring methods installed in a corrosive environment shall comply with 680.14 or shall be Type MC cable listed for that location. Wiring methods installed in these locations shall contain an insulated copper equipment grounding conductor sized in accordance with Table 250.122 but not smaller than 12 AWG.

Where installed in noncorrosive environments, the wiring methods of branch circuits shall comply with the general requirements in Chapter 3.

(2) Flexible Connections. Where necessary to employ flexible connections at or adjacent to the motor, liquidtight flexible metal or liquidtight flexible nonmetallic conduit with listed fittings shall be permitted.

(3) Cord-and-Plug Connections. Pool-associated motors shall be permitted to employ cord-and-plug connections. The flexible cord shall not exceed 900 mm (3 ft) in length. The flexible cord shall include a copper equipment grounding conductor sized in accordance with 250.122 but not smaller than 12 AWG. The cord shall terminate in a grounding-type attachment plug.

(B) Double-Insulated Pool Pumps. A listed cord-and-plug-connected pool pump incorporating an approved system of double insulation that provides a means for grounding only the internal and nonaccessible, non–current-carrying metal parts of the pump shall be connected to any wiring method recognized in Chapter 3 that is suitable for the location. Where the equipment grounding conductor of the motor circuit is connected to the equipotential bonding means in accordance with the second sentence of 680.26(B)(6)(a), the branch-circuit wiring shall comply with 680.21(A).

(C) GFCI Protection. Outlets supplying all pool motors on branch circuits rated 150 volts or less to ground and 60 amperes or less, single- or 3-phase, shall be provided with Class A ground-fault circuit-interrupter protection.

Exception: Listed low-voltage motors not requiring grounding, with ratings not exceeding the low-voltage contact limit that are supplied by listed transformers or power supplies that comply with 680.23(A)(2), shall be permitted to be installed without GFCI protection.

(D) Pool Pump Motor Replacement. Where a pool pump motor in 680.21(C) is replaced for maintenance or repair, the replacement pump motor shall be provided with ground-fault circuit-interrupter protection.

680.22 Lighting, Receptacles, and Equipment.

(A) Receptacles.

(1) Required Receptacle, Location. Where a permanently installed pool is installed, no fewer than one 125-volt, 15- or 20-ampere receptacle on a general-purpose branch circuit shall be located not less than

Swimming Pools, Fountains, and Similar Installations 680.22

1.83 m (6 ft) from, and not more than 6.0 m (20 ft) from, the inside wall of the pool. This receptacle shall be located not more than 2.0 m (6 ft 6 in.) above the floor, platform, or grade level serving the pool.

(2) Circulation and Sanitation System, Location. Receptacles that provide power for water-pump motors or for other loads directly related to the circulation and sanitation system shall be located at least 1.83 m (6 ft) from the inside walls of the pool. These receptacles shall have GFCI protection and be of the grounding type.

(3) Other Receptacles, Location. Other receptacles shall be not less than 1.83 m (6 ft) from the inside walls of a pool.

(4) GFCI Protection. All 15- and 20-ampere, single-phase, 125-volt receptacles located within 6.0 m (20 ft) of the inside walls of a pool shall be protected by a Class A ground-fault circuit interrupter. Also see 680.22(A)(5).

(5) Pool Equipment Room. At least one GFCI-protected 125-volt, 15- or 20-ampere receptacle on a general-purpose circuit shall be located within a pool equipment room, and all other receptacles supplied by branch circuits rated 150 volts or less to ground within a pool equipment room shall be GFCI protected.

(6) Measurements. In determining the dimensions in this section addressing receptacle spacings, the distance to be measured shall be the shortest path the supply cord of an appliance connected to the receptacle would follow without piercing a floor, wall, ceiling, doorway with hinged or sliding door, window opening, or other effective permanent barrier.

(B) Luminaires, Lighting Outlets, and Ceiling-Suspended (Paddle) Fans.

(1) New Outdoor Installation Clearances. In outdoor pool areas, luminaires, lighting outlets, and ceiling-suspended (paddle) fans installed above the pool or the area extending 1.5 m (5 ft) horizontally from the inside walls of the pool shall be installed at a height not less than 3.7 m (12 ft) above the maximum water level of the pool.

(2) Indoor Clearances. For installations in indoor pool areas, the clearances shall be the same as for outdoor areas unless modified as provided in this paragraph. If the branch circuit supplying the equipment is protected by a ground-fault circuit interrupter, the following equipment shall be permitted at a height not less than 2.3 m (7 ft 6 in.) above the maximum pool water level:

(1) Totally enclosed luminaires

(2) Ceiling-suspended (paddle) fans identified for use beneath ceiling structures such as provided on porches or patios

(3) Existing Installations. Existing luminaires and lighting outlets located less than 1.5 m (5 ft) measured horizontally from the inside walls of a pool shall be not less than 1.5 m (5 ft) above the surface of the maximum water level, shall be rigidly attached to the existing structure, and shall be protected by a ground-fault circuit interrupter.

(4) GFCI Protection in Adjacent Areas. Luminaires, lighting outlets, and ceiling-suspended (paddle) fans installed in the area extending between 1.5 m (5 ft) and 3.0 m (10 ft) horizontally from the inside walls of a pool shall be protected by a ground-fault circuit interrupter unless installed not less than 1.5 m (5 ft) above the maximum water level and rigidly attached to the structure adjacent to or enclosing the pool.

(5) Cord-and-Plug-Connected Luminaires. Cord-and-plug-connected luminaires shall comply with the requirements of 680.8 where installed within 4.9 m (16 ft) of any point on the water surface, measured radially.

(6) Low-Voltage Luminaires. Listed low-voltage luminaires not requiring grounding, not exceeding the low-voltage contact limit, and supplied by listed transformers or power supplies that comply with 680.23(A)(2) shall be permitted to be located less than 1.5 m (5 ft) from the inside walls of the pool.

(7) Low-Voltage Gas-Fired Luminaires, Decorative Fireplaces, Fire Pits, and Similar Equipment. Listed low-voltage gas-fired-luminaires, decorative fireplaces, fire pits, and similar equipment using low-voltage ignitors that do not require grounding, and are supplied by listed transformers or power supplies that comply with 680.23(A)(2) with outputs that do not exceed the low-voltage contact limit shall be permitted to be located less than 1.5 m (5 ft) from the inside walls of the pool. Metallic equipment shall be bonded in accordance with the requirements in 680.26(B). Transformers or power supplies supplying this type of equipment shall be installed in accordance with the requirements in 680.24. Metallic gas piping shall be bonded in accordance with the requirements in 250.104(B) and 680.26(B)(7).

(8) Measurements. In determining the dimensions in this section addressing luminaires, the distance to be measured shall be the shortest path an imaginary cord connected to the luminaire would follow without piercing a floor, wall, ceiling, doorway with hinged or sliding door, window opening, or other effective permanent barrier.

Swimming Pools, Fountains, and Similar Installations 680.23

(C) Switching Devices. Switching devices shall be located at least 1.5 m (5 ft) horizontally from the inside walls of a pool unless separated from the pool by a solid fence, wall, or other permanent barrier that provides at least a 1.5 m (5 ft) reach distance. Alternatively, a switch that is listed as being acceptable for use within 1.5 m (5 ft) shall be permitted.

(D) Other Outlets. Other outlets shall be not less than 3.0 m (10 ft) from the inside walls of the pool. Measurements shall be determined in accordance with 680.22(A)(6).

(E) Other Equipment. Other equipment with ratings exceeding the low-voltage contact limit shall be located at least 1.5 m (5 ft) horizontally from the inside walls of a pool unless separated from the pool by a solid fence, wall, or other permanent barrier.

680.23 Underwater Luminaires. This section covers all luminaires installed below the maximum water level of the pool.

(A) General.

(1) Luminaire Design, Normal Operation. The design of an underwater luminaire supplied from a branch circuit either directly or by way of a transformer or power supply meeting the requirements of this section shall be such that, where the luminaire is properly installed without a ground-fault circuit interrupter, there is no shock hazard with any likely combination of fault conditions during normal use (not relamping).

(2) Transformers and Power Supplies. Transformers and power supplies used for the supply of underwater luminaires, together with the transformer or power supply enclosure, shall be listed, labeled, and identified for swimming pool and spa use. The transformer or power supply shall incorporate either a transformer of the isolated winding type, with an ungrounded secondary that has a grounded metal barrier between the primary and secondary windings, or one that incorporates an approved system of double insulation between the primary and secondary windings.

(3) GFCI Protection, Lamping, Relamping, and Servicing. Ground-fault circuit-interrupter protection for personnel shall be installed in the branch circuit supplying luminaires operating at voltages greater than the low-voltage contact limit.

(4) Voltage Limitation. No luminaires shall be installed for operation on supply circuits over 150 volts between conductors.

(5) Location, Wall-Mounted Luminaires. Luminaires mounted in walls shall be installed with the top of the luminaire lens not less than 450 mm (18 in.) below the normal water level of the pool, unless the luminaire is listed and identified for use at lesser depths. No luminaire shall be installed less than 100 mm (4 in.) below the normal water level of the pool.

(6) Bottom-Mounted Luminaires. A luminaire facing upward shall comply with either (1) or (2):

(1) Have the lens guarded to prevent contact by any person
(2) Be listed for use without a guard

(7) Dependence on Submersion. Luminaires that depend on submersion for safe operation shall be inherently protected against the hazards of overheating when not submerged.

(8) Compliance. Compliance with these requirements shall be obtained by the use of a listed underwater luminaire and by installation of a listed ground-fault circuit interrupter in the branch circuit or a listed transformer or power supply for luminaires operating at not more than the low voltage contact limit.

(B) Wet-Niche Luminaires.

(1) Forming Shells. Forming shells shall be installed for the mounting of all wet-niche underwater luminaires and shall be equipped with provisions for conduit entries. Metal parts of the luminaire and forming shell in contact with the pool water shall be of brass or other approved corrosion-resistant metal. All forming shells used with nonmetallic conduit systems, other than those that are part of a listed low-voltage lighting system not requiring grounding, shall include provisions for terminating an 8 AWG copper conductor.

(2) Wiring Extending Directly to the Forming Shell. Conduit shall be installed from the forming shell to a junction box or other enclosure conforming to the requirements in 680.24. Conduit shall be rigid metal, intermediate metal, liquidtight flexible nonmetallic, or rigid nonmetallic.

(a) *Metal Conduit.* Metal conduit shall be listed and identified as red brass or stainless steel.

(b) *Nonmetallic Conduit.* Where a nonmetallic conduit is used, an 8 AWG insulated solid or stranded copper bonding jumper shall be installed in this conduit unless a listed low-voltage lighting system not

Swimming Pools, Fountains, and Similar Installations   680.23

requiring grounding is used. The bonding jumper shall be terminated in the forming shell, junction box or transformer enclosure, or ground-fault circuit-interrupter enclosure. The termination of the 8 AWG bonding jumper in the forming shell shall be covered with, or encapsulated in, a listed potting compound to protect the connection from the possible deteriorating effect of pool water.

(3) Equipment Grounding Provisions for Cords. Other than listed low-voltage lighting systems not requiring grounding, wet-niche luminaires that are supplied by a flexible cord or cable shall have all exposed non–current-carrying metal parts connected to an insulated copper equipment grounding conductor that is an integral part of the cord or cable. This equipment grounding conductor shall be connected to a grounding terminal in the supply junction box, transformer enclosure, or other enclosure. The equipment grounding conductor shall not be smaller than the supply conductors and not smaller than 16 AWG.

(4) Luminaire Grounding Terminations. The end of the flexible-cord jacket and the flexible-cord conductor terminations within a luminaire shall be covered with, or encapsulated in, a suitable potting compound to prevent the entry of water into the luminaire through the cord or its conductors. If present, the connection of the equipment grounding conductor within a luminaire shall be similarly treated to protect such connection from the deteriorating effect of pool water in the event of water entry into the luminaire.

(5) Luminaire Bonding. The luminaire shall be bonded to, and secured to, the forming shell by a positive locking device that ensures a low-resistance contact and requires a tool to remove the luminaire from the forming shell. Bonding shall not be required for luminaires that are listed for the application and have no non–current-carrying metal parts.

(6) Servicing. Wet-niche luminaires shall be removable from the water for inspection, relamping, or other maintenance. The forming shell location and length of cord in the forming shell shall permit personnel to place the removed luminaire on the deck or other dry location for such maintenance. The luminaire maintenance location shall be accessible without entering or going in the pool water.

In spa locations where wet-niche luminaires are installed low in the foot well of the spa, the luminaire shall only be required to reach the bench location, where the spa can be drained to make the bench location dry.

680.23

(C) Dry-Niche Luminaires.

(1) Construction. A dry-niche luminaire shall have provision for drainage of water. Other than listed low voltage luminaires not requiring grounding, a dry-niche luminaire shall have means for accommodating one equipment grounding conductor for each conduit entry.

(2) Junction Box. A junction box shall not be required but, if used, shall not be required to be elevated or located as specified in 680.24(A)(2) if the luminaire is specifically identified for the purpose.

(D) No-Niche Luminaires.
A no-niche luminaire shall meet the construction requirements of 680.23(B)(3) and be installed in accordance with the requirements of 680.23(B). Where connection to a forming shell is specified, the connection shall be to the mounting bracket.

(E) Through-Wall Lighting Assembly.
A through-wall lighting assembly shall be equipped with a threaded entry or hub, or a nonmetallic hub, for the purpose of accommodating the termination of the supply conduit. A through-wall lighting assembly shall meet the construction requirements of 680.23(B)(3) and be installed in accordance with the requirements of 680.23. Where connection to a forming shell is specified, the connection shall be to the conduit termination point.

(F) Branch-Circuit Wiring.

(1) Wiring Methods. Where branch-circuit wiring on the supply side of enclosures and junction boxes connected to conduits run to underwater luminaires are installed in corrosive environments as described in 680.2, the wiring method of that portion of the branch circuit shall be in accordance with 680.14 or shall be liquidtight flexible nonmetallic conduit. Wiring methods installed in corrosive environments as described in 680.14 shall contain an insulated copper equipment grounding conductor sized in accordance with 250.122, but not smaller than 12 AWG.

Where installed in noncorrosive environments, branch circuits shall comply with the general requirements in Chapter 3.

Exception: Where connecting to transformers or power supplies for pool lights, liquidtight flexible metal conduit shall be permitted. The length shall not exceed 1.8 m (6 ft) for any one length or exceed 3.0 m (10 ft) in total length used.

(2) Equipment Grounding. Other than listed low-voltage luminaires not requiring grounding, all through-wall lighting assemblies, wet-niche,

Swimming Pools, Fountains, and Similar Installations 680.23

dry-niche, or no-niche luminaires shall be connected to an insulated copper equipment grounding conductor installed with the circuit conductors. The equipment grounding conductor shall be installed without joint or splice except as permitted in 680.23(F)(2)(a) and (F)(2)(b). The equipment grounding conductor shall be sized in accordance with 250.122 but shall not be smaller than 12 AWG.

Exception: An equipment grounding conductor between the wiring chamber of the secondary winding of a transformer and a junction box shall be sized in accordance with the transformer secondary overcurrent protection provided.

(a) If more than one underwater luminaire is supplied by the same branch circuit, the equipment grounding conductor, installed between the junction boxes, transformer enclosures, or other enclosures in the supply circuit to wet-niche luminaires, or between the field-wiring compartments of dry-niche luminaires, shall be permitted to be terminated on grounding terminals.

(b) If the underwater luminaire is supplied from a transformer, ground-fault circuit interrupter, clock-operated switch, or a manual snap switch that is located between the panelboard and a junction box connected to the conduit that extends directly to the underwater luminaire, the equipment grounding conductor shall be permitted to terminate on grounding terminals on the transformer, ground-fault circuit interrupter, clock-operated switch enclosure, or an outlet box used to enclose a snap switch.

(3) Conductors. Conductors on the load side of a ground-fault circuit interrupter or of a transformer, used to comply with the provisions of 680.23(A)(8), shall not occupy raceways, boxes, or enclosures containing other conductors unless one of the following conditions applies:

(1) The other conductors are protected by ground-fault circuit interrupters.
(2) The other conductors are equipment grounding conductors and bonding jumpers as required per 680.23(B)(2)(b).
(3) The other conductors are supply conductors to a feed-through-type ground-fault circuit interrupter.
(4) Ground-fault circuit interrupters shall be permitted in a panelboard that contains circuits protected by other than ground-fault circuit interrupters.

680.24 Junction Boxes and Electrical Enclosures for Transformers or Ground-Fault Circuit Interrupters.

(A) Junction Boxes. A junction box connected to a conduit that extends directly to a forming shell or mounting bracket of a no-niche luminaire shall meet the requirements of this section.

(1) Construction. The junction box shall be listed, labeled, and identified as a swimming pool junction box and shall comply with the following conditions:

(1) Be equipped with threaded entries or hubs or a nonmetallic hub
(2) Be comprised of copper, brass, suitable plastic, or other approved corrosion-resistant material
(3) Be provided with electrical continuity between every connected metal conduit and the grounding terminals by means of copper, brass, or other approved corrosion-resistant metal that is integral with the box

(2) Installation. Where the luminaire operates over the low voltage contact limit, the junction box location shall comply with 680.24(A)(2)(a) and (A)(2)(b). Where the luminaire operates at the low voltage contact limit or less, the junction box location shall be permitted to comply with 680.24(A)(2)(c).

(a) *Vertical Spacing.* The junction box shall be located not less than 100 mm (4 in.), measured from the inside of the bottom of the box, above the ground level, or pool deck, or not less than 200 mm (8 in.) above the maximum pool water level, whichever provides the greater elevation.

(b) *Horizontal Spacing.* The junction box shall be located not less than 1.2 m (4 ft) from the inside wall of the pool, unless separated from the pool by a solid fence, wall, or other permanent barrier.

(c) *Flush Deck Box.* If used on a lighting system operating at the low voltage contact limit or less, a flush deck box shall be permitted if both of the following conditions are met:

(1) An approved potting compound is used to fill the box to prevent the entrance of moisture.
(2) The flush deck box is located not less than 1.2 m (4 ft) from the inside wall of the pool.

(B) Other Enclosures. An enclosure for a transformer, ground-fault circuit interrupter, or a similar device connected to a conduit that extends directly to a forming shell or mounting bracket of a no-niche luminaire shall meet the requirements of this section.

Swimming Pools, Fountains, and Similar Installations   680.24

(1) Construction. The enclosure shall be listed and labeled for the purpose and meet the following requirements:

(1) Equipped with threaded entries or hubs or a nonmetallic hub
(2) Comprised of copper, brass, suitable plastic, or other approved corrosion-resistant material
(3) Provided with an approved seal, such as duct seal at the conduit connection, that prevents circulation of air between the conduit and the enclosures
(4) Provided with electrical continuity between every connected metal conduit and the grounding terminals by means of copper, brass, or other approved corrosion-resistant metal that is integral with the box

(2) Installation.

(a) *Vertical Spacing.* The enclosure shall be located not less than 100 mm (4 in.), measured from the inside of the bottom of the box, above the ground level, or pool deck, or not less than 200 mm (8 in.) above the maximum pool water level, whichever provides the greater elevation.

(b) *Horizontal Spacing.* The enclosure shall be located not less than 1.2 m (4 ft) from the inside wall of the pool, unless separated from the pool by a solid fence, wall, or other permanent barrier.

(C) Protection. Junction boxes and enclosures mounted above the grade of the finished walkway around the pool shall not be located in the walkway unless afforded additional protection, such as by location under diving boards, adjacent to fixed structures, and the like.

(D) Grounding Terminals. Junction boxes, transformer and power-supply enclosures, and ground-fault circuit-interrupter enclosures connected to a conduit that extends directly to a forming shell or mounting bracket of a no-niche luminaire shall be provided with a number of grounding terminals that shall be no fewer than one more than the number of conduit entries.

(E) Strain Relief. The termination of a flexible cord of an underwater luminaire within a junction box, transformer or power-supply enclosure, ground-fault circuit interrupter, or other enclosure shall be provided with a strain relief.

(F) Grounding. The grounding terminals of a junction box, transformer enclosure, or other enclosure in the supply circuit to a wet-niche or no-niche luminaire and the field-wiring chamber of a dry-niche luminaire shall be connected to the equipment grounding terminal of the panelboard. This terminal shall be directly connected to the panelboard enclosure.

680.25 Feeders. These provisions shall apply to any feeder on the supply side of panelboards supplying branch circuits for pool equipment covered in Part II of this article and on the load side of the service equipment or the source of a separately derived system.

(A) Feeders. Where feeders are installed in corrosive environments as described in 680.2, Corrosive Environment, the wiring method of that portion of the feeder shall be in accordance with 680.14 or shall be liquidtight flexible nonmetallic conduit. Wiring methods installed in corrosive environments shall contain an insulated copper equipment grounding conductor sized in accordance with Table 250.122, but not smaller than 12 AWG.

Where installed in noncorrosive environments, feeders shall comply with the general requirements in Chapter 3.

(B) Aluminum Conduit. Aluminum conduit shall not be permitted in the pool area where subject to corrosion.

680.26 Equipotential Bonding.

(A) Performance. The equipotential bonding required by this section shall be installed to reduce voltage gradients in the pool area.

(B) Bonded Parts. The parts specified in 680.26(B)(1) through (B)(7) shall be bonded together using solid copper conductors, insulated covered, or bare, not smaller than 8 AWG or with rigid metal conduit of brass or other identified corrosion-resistant metal. Connections to bonded parts shall be made in accordance with 250.8. An 8 AWG or larger solid copper bonding conductor provided to reduce voltage gradients in the pool area shall not be required to be extended or attached to remote panelboards, service equipment, or electrodes.

(1) Conductive Pool Shells. Bonding to conductive pool shells shall be provided as specified in 680.26(B)(1)(a) or (B)(1)(b). Cast-in-place concrete, pneumatically applied or sprayed concrete, and concrete block with painted or plastered coatings shall all be considered conductive materials due to water permeability and porosity. Vinyl liners and fiberglass composite shells shall be considered to be nonconductive materials. Reconstructed pool shells shall also meet the requirements of this section.

(a) *Structural Reinforcing Steel.* Unencapsulated structural reinforcing steel shall be bonded together by steel tie wires or the equivalent. Where structural reinforcing steel is encapsulated in a nonconductive

Swimming Pools, Fountains, and Similar Installations 680.26

compound, a copper conductor grid shall be installed in accordance with 680.26(B)(1)(b)

(b) *Copper Conductor Grid.* A copper conductor grid shall be provided and shall comply with the following:

(1) Be constructed of minimum 8 AWG bare solid copper conductors bonded to each other at all points of crossing in accordance with 250.8 or other approved means
(2) Conform to the contour of the pool
(3) Be arranged in a 300 mm (12 in.) by 300 mm (12 in.) network of conductors in a uniformly spaced perpendicular grid pattern with a tolerance of 100 mm (4 in.)
(4) Be secured within or under the pool no more than 150 mm (6 in.) from the outer contour of the pool shell

(2) Perimeter Surfaces. The perimeter surface to be bonded shall be considered to extend for 1 m (3 ft) horizontally beyond the inside walls of the pool and shall include unpaved surfaces and other types of paving. Perimeter surfaces separated from the pool by a permanent wall or building 1.5 m (5 ft) in height or more shall require equipotential bonding only on the pool side of the permanent wall or building. Bonding to perimeter surfaces shall be provided as specified in 680.26(B)(2)(a), (B)(2)(b), or (B)(2)(c) and shall be attached to the pool reinforcing steel or copper conductor grid at a minimum of four points uniformly spaced around the perimeter of the pool. For nonconductive pool shells, bonding at four points shall not be required.

(a) *Structural Reinforcing Steel.* Structural reinforcing steel shall be bonded in accordance with 680.26(B)(1)(a).

(b) *Copper Ring.* Where structural reinforcing steel is not available or is encapsulated in a nonconductive compound, a copper conductor(s) shall be utilized where the following requirements are met:

(1) At least one minimum 8 AWG bare solid copper conductor shall be provided.
(2) The conductors shall follow the contour of the perimeter surface.
(3) Only listed splicing devices or exothermic welding shall be permitted.
(4) The required conductor shall be 450 mm to 600 mm (18 in. to 24 in.) from the inside walls of the pool.
(5) The required conductor shall be secured within or under the perimeter surface 100 mm to 150 mm (4 in. to 6 in.) below the subgrade.

(c) *Copper Grid.* Where structural reinforcing steel is not available or is encapsulated in a nonconductive compound, copper grid shall be utilized where the following requirements are met:

(1) The copper grid shall be constructed of 8 AWG solid bare copper and be arranged in accordance with 680.26(B)(1)(b)(3).
(2) The copper grid shall follow the contour of the perimeter surface extending 1 m (3 ft) horizontally beyond the inside walls of the pool.
(3) Only listed splicing devices or exothermic welding shall be permitted.
(4) The copper grid shall be secured within or under the deck or unpaved surfaces between 100 mm to 150 mm (4 in. to 6 in.) below the subgrade.

(3) Metallic Components. All metallic parts of the pool structure, including reinforcing metal not addressed in 680.26(B)(1)(a), shall be bonded. Where reinforcing steel is encapsulated with a nonconductive compound, the reinforcing steel shall not be required to be bonded.

(4) Underwater Lighting. All metal forming shells and mounting brackets of no-niche luminaires shall be bonded.

Exception: Listed low-voltage lighting systems with nonmetallic forming shells shall not require bonding.

(5) Metal Fittings. All metal fittings within or attached to the pool structure shall be bonded. Isolated parts that are not over 100 mm (4 in.) in any dimension and do not penetrate into the pool structure more than 25 mm (1 in.) shall not require bonding. Metallic pool cover anchors intended for insertion in a concrete or masonry deck surface, 25 mm (1 in.) or less in any dimension and 51 mm (2 in.) or less in length, and metallic pool cover anchors intended for insertion in a wood or composite deck surface, 51 mm (2 in.) or less in any flange dimension and 51 mm (2 in.) or less in length, shall not require bonding.

(6) Electrical Equipment. Metal parts of electrical equipment associated with the pool water circulating system, including pump motors and metal parts of equipment associated with pool covers, including electric motors, shall be bonded.

Exception: Metal parts of listed equipment incorporating an approved system of double insulation shall not be bonded.

(a) *Double-Insulated Water Pump Motors.* Where a double-insulated water pump motor is installed under the provisions of this rule, a solid 8 AWG copper conductor of sufficient length to make a bonding

Swimming Pools, Fountains, and Similar Installations 680.27

connection to a replacement motor shall be extended from the swimming pool equipotential bonding means to an accessible point in the vicinity of the pool pump motor. Where there is no connection between the swimming pool equipotential bonding means and the equipment grounding system for the premises, this bonding conductor shall be connected to the equipment grounding conductor of the motor circuit.

(b) *Pool Water Heaters.* For pool water heaters rated at more than 50 amperes and having specific instructions regarding bonding and grounding, only those parts designated to be bonded shall be bonded and only those parts designated to be grounded shall be grounded.

(7) Fixed Metal Parts. All fixed metal parts shall be bonded including, but not limited to, metal-sheathed cables and raceways, metal piping, metal awnings, metal fences, and metal door and window frames.

Exception No. 1: Those separated from the pool by a permanent barrier that prevents contact by a person shall not be required to be bonded.

Exception No. 2: Those greater than 1.5 m (5 ft) horizontally from the inside walls of the pool shall not be required to be bonded.

Exception No. 3: Those greater than 3.7 m (12 ft) measured vertically above the maximum water level of the pool, or as measured vertically above any observation stands, towers, or platforms, or any diving structures, shall not be required to be bonded.

(C) Pool Water. Where none of the bonded parts are in direct connection with the pool water, the pool water shall be in direct contact with an approved corrosion-resistant conductive surface that exposes not less than 5800 mm² (9 in.²) of surface area to the pool water at all times. The conductive surface shall be located where it is not exposed to physical damage or dislodgement during usual pool activities, and it shall be bonded in accordance with 680.26(B).

680.27 Specialized Pool Equipment.

(A) Underwater Audio Equipment. All underwater audio equipment shall be identified as underwater audio equipment.

(1) Speakers. Each speaker shall be mounted in an approved metal forming shell, the front of which is enclosed by a captive metal screen, or equivalent, that is bonded to, and secured to, the forming shell by a positive locking device that ensures a low-resistance contact and requires a tool to open for installation or servicing of the speaker. The forming shell shall be installed in a recess in the wall or floor of the pool.

(2) Wiring Methods. Rigid metal conduit of brass or other identified corrosion-resistant metal, liquidtight flexible nonmetallic conduit (LFNC), rigid polyvinyl chloride conduit, or reinforced thermosetting resin conduit shall extend from the forming shell to a listed junction box or other enclosure as provided in 680.24. Where rigid polyvinyl chloride conduit, reinforced thermosetting resin conduit, or liquidtight flexible nonmetallic conduit is used, an 8 AWG insulated solid or stranded copper bonding jumper shall be installed in this conduit. The bonding jumper shall be terminated in the forming shell and the junction box. The termination of the 8 AWG bonding jumper in the forming shell shall be covered with, or encapsulated in, a listed potting compound to protect such connection from the possible deteriorating effect of pool water.

(3) Forming Shell and Metal Screen. The forming shell and metal screen shall be of brass or other approved corrosion-resistant metal. All forming shells shall include provisions for terminating an 8 AWG copper conductor.

(B) Electrically Operated Pool Covers.

(1) Motors and Controllers. The electric motors, controllers, and wiring shall be located not less than 1.5 m (5 ft) from the inside wall of the pool unless separated from the pool by a wall, cover, or other permanent barrier. Electric motors installed below grade level shall be of the totally enclosed type. The device that controls the operation of the motor for an electrically operated pool cover shall be located such that the device operator has full view of the pool.

Exception: Motors that are part of listed systems with ratings not exceeding the low-voltage contact limit that are supplied by listed transformers or power supplies that comply with 680.23(A)(2) shall be permitted to be located less than 1.5 m (5 ft) from the inside walls of the pool.

(2) Protection. The electric motor and controller shall be connected to a branch circuit protected with ground-fault circuit-interrupter protection.

Exception: Motors that are part of listed systems with ratings not exceeding the low-voltage contact limit that are supplied by listed transformers or power supplies that comply with 680.23(A)(2).

(C) Deck Area Heating. The provisions of this section shall apply to all pool deck areas, including a covered pool, where electrically operated comfort heating units are installed within 6.0 m (20 ft) of the inside wall of the pool.

Swimming Pools, Fountains, and Similar Installations 680.32

(1) Unit Heaters. Unit heaters shall be rigidly mounted to the structure and shall be of the totally enclosed or guarded type. Unit heaters shall not be mounted over the pool or within the area extending 1.5 m (5 ft) horizontally from the inside walls of a pool.

(2) Permanently Wired Radiant Heaters. Radiant electric heaters shall be suitably guarded and securely fastened to their mounting device(s). Heaters shall not be installed over a pool or within the area extending 1.5 m (5 ft) horizontally from the inside walls of the pool and shall be mounted at least 3.7 m (12 ft) vertically above the pool deck unless otherwise approved.

(3) Radiant Heating Cables Not Permitted. Radiant heating cables embedded in or below the deck shall not be permitted.

680.28 Gas-Fired Water Heater. Circuits serving gas-fired swimming pool and spa water heaters operating at voltages above the low-voltage contact limit shall be provided with ground-fault circuit-interrupter protection for personnel.

PART III. STORABLE POOLS, STORABLE SPAS, STORABLE HOT TUBS, AND STORABLE IMMERSION POOLS

680.30 General. Electrical installations at storable pools, storable spas, storable hot tubs, or storable immersion pools shall comply with the provisions of Part I and Part III of this article.

680.31 Pumps. A cord-connected pool filter pump shall incorporate an approved system of double insulation or its equivalent and shall be provided with means for the termination of an equipment grounding conductor for the connection to the internal and nonaccessible non–current-carrying metal parts of the pump.

An equipment grounding conductor run with the power-supply conductors in the flexible cord shall terminate in a grounding-type attachment plug having a fixed grounding contact member.

Cord-connected pool filter pumps shall be provided with a ground-fault circuit interrupter that is an integral part of the attachment plug or located in the power-supply cord within 300 mm (12 in.) of the attachment plug.

680.32 Ground-Fault Circuit Interrupters Required. All electrical equipment, including power-supply cords, used with storable pools shall be protected by ground-fault circuit interrupters.

All 125-volt, 15- and 20-ampere receptacles located within 6.0 m (20 ft) of the inside walls of a storable pool, storable spa, or storable hot tub shall be protected by a ground-fault circuit interrupter. In determining these dimensions, the distance to be measured shall be the shortest path the supply cord of an appliance connected to the receptacle would follow without piercing a floor, wall, ceiling, doorway with hinged or sliding door, window opening, or other effective permanent barrier.

680.33 Luminaires. An underwater luminaire, if installed, shall be installed in or on the wall of the storable pool, storable spa, or storable hot tub. It shall comply with either 680.33(A) or (B).

(A) Within the Low Voltage Contact Limit. A luminaire shall be part of a cord-and plug connected lighting assembly. This assembly shall be listed as an assembly for the purpose and have the following construction features:

(1) No exposed metal parts
(2) A luminaire lamp that is suitable for use at the supplied voltage
(3) An impact-resistant polymeric lens, luminaire body, and transformer enclosure
(4) A transformer or power supply meeting the requirements of 680.23(A)(2) with a primary rating not over 150 V

(B) Over the Low Voltage Contact Limit But Not over 150 Volts. A lighting assembly without a transformer or power supply and with the luminaire lamp(s) operating at not over 150 volts shall be permitted to be cord-and-plug-connected where the assembly is listed as an assembly for the purpose. The installation shall comply with 680.23(A)(5), and the assembly shall have the following construction features:

(1) No exposed metal parts
(2) An impact-resistant polymeric lens and luminaire body
(3) A ground-fault circuit interrupter with open neutral conductor protection as an integral part of the assembly
(4) The luminaire lamp permanently connected to the ground-fault circuit interrupter with open-neutral protection
(5) Compliance with the requirements of 680.23(A)

680.34 Receptacle Locations. Receptacles shall not be located less than 1.83 m (6 ft) from the inside walls of a storable pool, storable spa, or storable hot tub. In determining these dimensions, the distance to be measured shall be the shortest path the supply cord of an appliance connected to the receptacle would follow without piercing a floor, wall,

Swimming Pools, Fountains, and Similar Installations 680.42

ceiling, doorway with hinged or sliding door, window opening, or other effective permanent barrier.

PART IV. SPAS, HOT TUBS, AND PERMANENTLY INSTALLED IMMERSION POOLS

680.40 General. Electrical installations at spas and hot tubs shall comply with the provisions of Part I and Part IV of this article.

680.41 Emergency Switch for Spas and Hot Tubs. A clearly labeled emergency shutoff or control switch for the purpose of stopping the motor(s) that provides power to the recirculation system and jet system shall be installed at a point readily accessible to the users and not less than 1.5 m (5 ft) away, adjacent to, and within sight of the spa or hot tub. This requirement shall not apply to one-family dwellings.

680.42 Outdoor Installations. A spa or hot tub installed outdoors shall comply with the provisions of Parts I and II of this article, except as permitted in 680.42(A) and (B), that would otherwise apply to pools installed outdoors.

(A) Flexible Connections. Listed packaged spa or hot tub equipment assemblies or self-contained spas or hot tubs utilizing a factory-installed or assembled control panel or panelboard shall be permitted to use flexible connections as covered in 680.42(A)(1) and (A)(2).

(1) Flexible Conduit. Liquidtight flexible metal conduit or liquidtight flexible nonmetallic conduit shall be permitted.

(2) Cord-and-Plug Connections. Cord-and-plug connections with a cord not longer than 4.6 m (15 ft) shall be permitted where protected by a ground-fault circuit interrupter.

(B) Bonding. Bonding by metal-to-metal mounting on a common frame or base shall be permitted. The metal bands or hoops used to secure wooden staves shall not be required to be bonded as required in 680.26.

Equipotential bonding of perimeter surfaces in accordance with 680.26(B)(2) shall not be required to be provided for spas and hot tubs where all of the following conditions apply:

(1) The spa or hot tub shall be listed, labeled, and identified as a self-contained spa for aboveground use.
(2) The spa or hot tub shall not be identified as suitable only for indoor use.

(3) The installation shall be in accordance with the manufacturer's instructions and shall be located on or above grade.

(4) The top rim of the spa or hot tub shall be at least 710 mm (28 in.) above all perimeter surfaces that are within 760 mm (30 in.), measured horizontally from the spa or hot tub. The height of nonconductive external steps for entry to or exit from the self-contained spa shall not be used to reduce or increase this rim height measurement.

Informational Note: For information regarding listing requirements for self-contained spas and hot tubs, see ANSI/UL 1563–2010, *Standard for Electric Spas, Equipment Assemblies, and Associated Equipment*.

(C) Interior Wiring to Outdoor Installations. In the interior of a dwelling unit or in the interior of another building or structure associated with a dwelling unit, any of the wiring methods recognized or permitted in Chapter 3 of this *Code* shall be permitted to be used for the connection to motor disconnecting means and the motor, heating, and control loads that are part of a self-contained spa or hot tub or a packaged spa or hot tub equipment assembly. Wiring to an underwater luminaire shall comply with 680.23 or 680.33.

680.43 Indoor Installations. A spa or hot tub installed indoors shall comply with the provisions of Parts I and II of this article except as modified by this section and shall be connected by the wiring methods of Chapter 3.

Exception No. 1: Listed spa and hot tub packaged units rated 20 amperes or less shall be permitted to be cord-and-plug-connected to facilitate the removal or disconnection of the unit for maintenance and repair.

Exception No. 2: The equipotential bonding requirements for perimeter surfaces in 680.26(B)(2) shall not apply to a listed self-contained spa or hot tub installed above a finished floor.

Exception No. 3: For a dwelling unit(s) only, where a listed spa or hot tub is installed indoors, the wiring method requirements of 680.42(C) shall also apply.

(A) Receptacles. At least one 125-volt, 15- or 20-ampere receptacle on a general-purpose branch circuit shall be located not less than 1.83 m (6 ft) from, and not exceeding 3.0 m (10 ft) from, the inside wall of the spa or hot tub.

Swimming Pools, Fountains, and Similar Installations 680.43

(1) Location. Receptacles shall be located at least 1.83 m (6 ft) measured horizontally from the inside walls of the spa or hot tub.

(2) Protection, General. Receptacles rated 125 volts and 30 amperes or less and located within 3.0 m (10 ft) of the inside walls of a spa or hot tub shall be protected by a ground-fault circuit interrupter.

(3) Protection, Spa or Hot Tub Supply Receptacle. Receptacles that provide power for a spa or hot tub shall be ground-fault circuit-interrupter protected.

(4) Measurements. In determining the dimensions in this section addressing receptacle spacings, the distance to be measured shall be the shortest path the supply cord of an appliance connected to the receptacle would follow without piercing a floor, wall, ceiling, doorway with hinged or sliding door, window opening, or other effective permanent barrier.

(B) Installation of Luminaires, Lighting Outlets, and Ceiling-Suspended (Paddle) Fans.

(1) Elevation. Luminaires, except as covered in 680.43(B)(2), lighting outlets, and ceiling-suspended (paddle) fans located over the spa or hot tub or within 1.5 m (5 ft) from the inside walls of the spa or hot tub shall comply with the clearances specified in 680.43(B)(1)(a), (B)(1)(b), and (B)(1)(c) above the maximum water level.

(a) *Without GFCI.* Where no GFCI protection is provided, the mounting height shall be not less than 3.7 m (12 ft).

(b) *With GFCI.* Where GFCI protection is provided, the mounting height shall be permitted to be not less than 2.3 m (7 ft 6 in.).

(c) *Below 2.3 m (7 ft 6 in.).* Luminaires meeting the requirements of item (1) or (2) and protected by a ground-fault circuit interrupter shall be permitted to be installed less than 2.3 m (7 ft 6 in.) over a spa or hot tub:

(1) Recessed luminaires with a glass or plastic lens, nonmetallic or electrically isolated metal trim, and suitable for use in damp locations
(2) Surface-mounted luminaires with a glass or plastic globe, a nonmetallic body, or a metallic body isolated from contact, and suitable for use in damp locations

(2) Underwater Applications. Underwater luminaires shall comply with the provisions of 680.23 or 680.33.

(C) Switches. Switches shall be located at least 1.5 m (5 ft), measured horizontally, from the inside walls of the spa or hot tub.

(D) Bonding. The following parts shall be bonded together:

(1) All metal fittings within or attached to the spa or hot tub structure
(2) Metal parts of electrical equipment associated with the spa or hot tub water circulating system, including pump motors, unless part of a listed, labeled, and identified self-contained spa or hot tub
(3) Metal raceway and metal piping that are within 1.5 m (5 ft) of the inside walls of the spa or hot tub and that are not separated from the spa or hot tub by a permanent barrier
(4) All metal surfaces that are within 1.5 m (5 ft) of the inside walls of the spa or hot tub and that are not separated from the spa or hot tub area by a permanent barrier

Exception: Small conductive surfaces not likely to become energized, such as air and water jets and drain fittings, where not connected to metallic piping, towel bars, mirror frames, and similar nonelectrical equipment, shall not be required to be bonded.

(5) Non-current-carrying metal parts of electrical devices and controls that are not associated with the spas or hot tubs and that are located within 1.5 m (5 ft) of such units

(E) Methods of Bonding. All metal parts associated with the spa or hot tub shall be bonded by any of the following methods:

(1) The interconnection of threaded metal piping and fittings
(2) Metal-to-metal mounting on a common frame or base
(3) The provisions of a solid copper bonding jumper, insulated, covered, or bare, not smaller than 8 AWG

(F) Grounding. The following equipment shall be connected to the equipment grounding conductor:

(1) All electrical equipment located within 1.5 m (5 ft) of the inside wall of the spa or hot tub
(2) All electrical equipment associated with the circulating system of the spa or hot tub

(G) Underwater Audio Equipment. Underwater audio equipment shall comply with the provisions of Part II of this article.

680.44 Protection. Except as otherwise provided in this section, the outlet(s) that supplies a self-contained spa or hot tub, a packaged spa or hot tub equipment assembly, or a field-assembled spa or hot tub shall be protected by a ground-fault circuit interrupter.

Swimming Pools, Fountains, and Similar Installations 680.45

(A) Listed Units. If so marked, a listed, labeled, and identified self-contained unit or a listed, labeled, and identified packaged equipment assembly that includes integral ground-fault circuit-interrupter protection for all electrical parts within the unit or assembly (pumps, air blowers, heaters, lights, controls, sanitizer generators, wiring, and so forth) shall be permitted without additional GFCI protection.

(B) Other Units. A field-assembled spa or hot tub rated 3 phase or rated over 250 volts or with a heater load of more than 50 amperes shall not require the supply to be protected by a ground-fault circuit interrupter.

680.45 Permanently Installed Immersion Pools. Electrical installations at permanently installed immersion pools, whether installed indoors or outdoors, shall comply with the provisions of Part I, Part II, and Part IV of this article except as modified by this section and shall be connected by the wiring methods of Chapter 3. With regard to provisions in Part IV of this article, an immersion pool shall be considered to be a spa or hot tub.

Exception No. 1: The equipotential bonding requirements in 680.26(B) shall not apply to immersion pools that incorporate no permanently installed or permanently connected electrical equipment, and that are installed with all portions located on or above a finished floor.

Exception No. 2: The equipotential bonding requirements for perimeter surfaces in 680.26(B)(2) shall not apply to nonconductive perimeter surfaces, such as steps, treads, and walking surfaces made of fiberglass composite.

(A) Cord-and-Plug Connections. To facilitate the removal or disconnection of the unit(s) for maintenance, storage, and repair, self-contained portable packaged immersion pools with integral pumps and/or heaters, including circulation heaters, rated 120 volts and 20 amperes or less shall be permitted to be cord-and-plug-connected with a cord not shorter than 1.83 m (6 ft) and not longer than 4.6 m (15 ft) and shall be protected by a ground-fault circuit interrupter. The cord shall ground all non-current-carrying metal parts of the electrical equipment. The means for grounding shall be an equipment grounding conductor run with the power-supply conductors in the flexible cord that is properly terminated in a grounding-type attachment plug having a fixed grounding contact member. If the ground-fault circuit interrupter is provided as an integral part of the cord assembly, it shall be located at the attachment plug or in the power-supply cord within 300 mm (12 in.) of the attachment plug.

(B) Storable and Portable Pumps. A cord-connected storable or portable pump utilized with, but not built-in or permanently attached as an integral part of, a permanently installed immersion pool, shall be identified for swimming pool and spa use. It shall incorporate an approved system of double insulation or its equivalent and shall be provided with means for grounding only the internal and nonaccessible non-current-carrying metal parts of the appliance. The means for grounding shall be an equipment grounding conductor run with the power-supply conductors in the flexible cord that is properly terminated in a grounding-type attachment plug having a fixed grounding contact member. Cord-connected pool filter pumps shall be provided with a ground-fault circuit interrupter that is an integral part of the attachment plug or located in the power-supply cord within 300 mm (12 in.) of the attachment plug.

(C) Heaters. Heaters used with permanently installed immersion pools shall comply with either 680.45(C)(1) or (C)(2).

(1) Permanently Installed Heaters. A permanently installed heater, including immersion heaters, circulation heaters, and combination pump-heater units, built-in or permanently attached as an integral part of a permanently installed immersion pool, rated 120 volts or 250 volts, shall be identified for swimming pool and spa use; shall be grounded and bonded; and heaters supplied by branch circuits rated 150 volts or less to ground shall be provided with Class A ground-fault circuit-interrupter protection. Permanently installed immersion heaters, rated 120 volts and 20 amperes or less or 250 volts and 30 amperes or less, single phase, are permitted to be cord-and-plug-connected with a cord not shorter than 1.83 m (6 ft) and not longer than 4.6 m (15 ft), shall be protected by a ground-fault circuit interrupter, and shall be provided with means for grounding all non-current-carrying metal parts of the appliance. The means for grounding shall be an equipment grounding conductor run with the power-supply conductors in the flexible cord that is properly terminated in a grounding-type attachment plug having a fixed grounding contact member. If the ground-fault circuit interrupter is provided as an integral part of the cord assembly, it shall be located at the attachment plug or in the power-supply cord within 300 mm (12 in.) of the attachment plug.

(2) Storable and Portable Heaters. A cord-connected storable or portable heater, including immersion heaters, circulation heaters, and combination pump-heater units, utilized with, but not permanently installed or attached as an integral part of a permanently installed immersion pool, rated 120 volts and 20 amperes or less or 250 volts and 30 amperes or

Swimming Pools, Fountains, and Similar Installations 680.51

less, single phase, shall be identified for swimming pool and spa use; shall be cord-and-plug-connected with a cord not shorter than 1.83 m (6 ft) and not longer than 4.6 m (15 ft), heaters supplied by branch circuits rated 150 volts or less to ground shall be provided with Class A ground-fault circuit-interrupter protection, and shall be provided with means for grounding all non-current-carrying metal parts of the appliance. The means for grounding shall be an equipment grounding conductor run with the power-supply conductors in the flexible cord that is properly terminated in a grounding-type attachment plug having a fixed grounding contact member. If the ground-fault circuit interrupter is provided as an integral part of the cord assembly, it shall be located at the attachment plug or in the power-supply cord within 300 mm (12 in.) of the attachment plug.

(D) Audio Equipment. Audio equipment shall not be installed in or on a permanently installed immersion pool. All audio equipment operating at greater than the low-voltage contact limit and located within 1.83 m (6 ft) from the inside walls of a permanently installed immersion pool shall be connected to the equipment grounding conductor and protected by a ground-fault circuit interrupter.

PART V. FOUNTAINS

680.50 General. Part I and Part V of this article shall apply to all permanently installed fountains as defined in 680.2. Fountains that have water common to a pool and fountains intended for recreational use by pedestrians, including splash pads, shall additionally comply with the requirements in Part II of this article. Part V does not cover self-contained, portable fountains. Portable fountains shall comply with Parts II and III of Article 422.

680.51 Luminaires, Submersible Pumps, and Other Submersible Equipment.

(A) Ground-Fault Circuit Interrupter. Luminaires, submersible pumps, and other submersible equipment, unless listed for operation at low voltage contact limit or less and supplied by a transformer or power supply that complies with 680.23(A)(2), shall be protected by a ground-fault circuit interrupter.

(B) Operating Voltage. No luminaires shall be installed for operation on supply circuits over 150 volts between conductors. Submersible pumps and other submersible equipment shall operate at 300 volts or less between conductors.

(C) Luminaire Lenses. Luminaires shall be installed with the top of the luminaire lens below the normal water level of the fountain unless listed for above-water locations. A luminaire facing upward shall comply with either (1) or (2):

(1) Have the lens guarded to prevent contact by any person
(2) Be listed for use without a guard

(D) Overheating Protection. Electrical equipment that depends on submersion for safe operation shall be protected against overheating by a low-water cutoff or other approved means when not submerged.

(E) Wiring. Equipment shall be equipped with provisions for threaded conduit entries or be provided with a suitable flexible cord. The maximum length of each exposed cord in the fountain shall be limited to 3.0 m (10 ft). Cords extending beyond the fountain perimeter shall be enclosed in approved wiring enclosures. Metal parts of equipment in contact with water shall be of brass or other approved corrosion-resistant metal.

(F) Servicing. All equipment shall be removable from the water for relamping or normal maintenance. Luminaires shall not be permanently embedded into the fountain structure such that the water level must be reduced or the fountain drained for relamping, maintenance, or inspection.

(G) Stability. Equipment shall be inherently stable or be securely fastened in place.

680.52 Junction Boxes and Other Enclosures.

(A) General. Junction boxes and other enclosures used for other than underwater installation shall comply with 680.24.

(B) Underwater Junction Boxes and Other Underwater Enclosures. Junction boxes and other underwater enclosures shall meet the requirements of 680.52(B)(1) and (B)(2).

(1) Construction.

(a) Underwater enclosures shall be equipped with provisions for threaded conduit entries or compression glands or seals for cord entry.

(b) Underwater enclosures shall be listed and rated for prolonged submersion and made of copper, brass, or other corrosion-resistant material.

Swimming Pools, Fountains, and Similar Installations 680.54

(2) Installation. Underwater enclosure installations shall comply with 680.52(B)(2)(a) and (B)(2)(b).

(a) Underwater enclosures shall be filled with a listed potting compound to prevent the entry of moisture.

(b) Underwater enclosures shall be firmly attached to the supports or directly to the fountain surface and bonded as required. Where the junction box is supported only by conduits in accordance with 314.23(E) and (F), the conduits shall be of copper, brass, stainless steel, or other corrosion-resistant metal. Where the box is fed by nonmetallic conduit, it shall have additional supports and fasteners of copper, brass, or other corrosion-resistant material.

680.54 Grounding and Bonding.

(A) Grounding. The following equipment shall be connected to the equipment grounding conductor:

(1) Other than listed low-voltage luminaires not requiring grounding, all electrical equipment located within the fountain or within 1.5 m (5 ft) of the inside wall of the fountain
(2) All electrical equipment associated with the recirculating system of the fountain
(3) Panelboards that are not part of the service equipment and that supply any electrical equipment associated with the fountain

Informational Note: See 250.122 for sizing of these conductors.

(B) Bonding. The following parts shall be bonded together and connected to an equipment grounding conductor on a branch circuit supplying the fountain:

(1) All metal piping systems associated with the fountain
(2) All metal fittings within or attached to the fountain
(3) Metal parts of electrical equipment associated with the fountain water-circulating system, including pump motors
(4) Metal raceways that are within 1.5 m (5 ft) of the inside wall or perimeter of the fountain and that are not separated from the fountain by a permanent barrier
(5) All metal surfaces that are within 1.5 m (5 ft) of the inside wall or perimeter of the fountain and that are not separated from the fountain by a permanent barrier
(6) Electrical devices and controls that are not associated with the fountain and are located less than 1.5 m (5 ft) of the inside wall or perimeter of the fountain

680.55 Methods of Grounding.

(A) Applied Provisions. The provisions of 680.21(A), 680.23(B)(3), 680.23(F)(1) and (F)(2), 680.24(F), and 680.25 shall apply.

(B) Supplied by a Flexible Cord. Electrical equipment that is supplied by a flexible cord shall have all exposed non–current-carrying metal parts grounded by an insulated copper equipment grounding conductor that is an integral part of this cord. The equipment grounding conductor shall be connected to an equipment grounding terminal in the supply junction box, transformer enclosure, power supply enclosure, or other enclosure.

680.56 Cord-and-Plug-Connected Equipment.

(A) Ground-Fault Circuit Interrupter. All electrical equipment, including power-supply cords, shall be protected by ground-fault circuit interrupters.

(B) Cord Type. Flexible cord immersed in or exposed to water shall be of a type for extra-hard usage, as designated in Table 400.4, and shall be a listed type with a "W" suffix.

(C) Sealing. The end of the flexible cord jacket and the flexible cord conductor termination within equipment shall be covered with, or encapsulated in, a suitable potting compound to prevent the entry of water into the equipment through the cord or its conductors. In addition, the ground connection within equipment shall be similarly treated to protect such connections from the deteriorating effect of water that may enter into the equipment.

(D) Terminations. Connections with flexible cord shall be permanent, except that grounding-type attachment plugs and receptacles shall be permitted to facilitate removal or disconnection for maintenance, repair, or storage of fixed or stationary equipment not located in any water-containing part of a fountain.

680.58 GFCI Protection for Adjacent Receptacle Outlets.
All 15- or 20-ampere, single-phase 125-volt through 250-volt receptacles located within 6.0 m (20 ft) of a fountain edge shall be provided with GFCI protection.

680.59 GFCI Protection for Permanently Installed Nonsubmersible Pumps.
Outlets supplying all permanently installed nonsubmersible pump motors rated 250 volts or less and 60 amperes or less, single- or 3-phase, shall be provided with ground-fault circuit-interrupter protection.

Swimming Pools, Fountains, and Similar Installations 680.74

PART VII. HYDROMASSAGE BATHTUBS

680.70 General. Hydromassage bathtubs as defined in 680.2 shall comply with Part VII of this article. They shall not be required to comply with other parts of this article.

680.71 Protection. Hydromassage bathtubs and their associated electrical components shall be on an individual branch circuit(s) and protected by a readily accessible ground-fault circuit interrupter. All 125-volt, single-phase receptacles not exceeding 30 amperes and located within 1.83 m (6 ft) measured horizontally of the inside walls of a hydromassage tub shall be protected by a ground-fault circuit interrupter.

680.72 Other Electrical Equipment. Luminaires, switches, receptacles, and other electrical equipment located in the same room, and not directly associated with a hydromassage bathtub, shall be installed in accordance with the requirements of Chapters 1 through 4 in this *Code* covering the installation of that equipment in bathrooms.

680.73 Accessibility. Hydromassage bathtub electrical equipment shall be accessible without damaging the building structure or building finish. Where the hydromassage bathtub is cord- and plug-connected with the supply receptacle accessible only through a service access opening, the receptacle shall be installed so that its face is within direct view and not more than 300 mm (1 ft) of the opening.

680.74 Bonding.

(A) General. The following parts shall be bonded together:

(1) All metal fittings within or attached to the tub structure that are in contact with the circulating water
(2) Metal parts of electrical equipment associated with the tub water circulating system, including pump and blower motors
(3) Metal-sheathed cables, metal raceways, and metal piping within 1.5 m (5 ft) of the inside walls of the tub and not separated from the tub by a permanent barrier
(4) All exposed metal surfaces that are within 1.5 m (5 ft) of the inside walls of the tub and not separated from the tub area by a permanent barrier
(5) Non–current-carrying metal parts of electrical devices and controls that are not associated with the hydromassage tubs within 1.5 m (5 ft) from such units

Exception No. 1: Small conductive surfaces not likely to become energized, such as air and water jets, supply valve assemblies, and drain fittings not connected to metallic piping, and towel bars, mirror frames, and similar nonelectrical equipment not connected to metal framing shall not be required to be bonded.

Exception No. 2: Double-insulated motors and blowers shall not be bonded.

(B) Bonding Conductor. All metal parts required to be bonded by this section shall be bonded together using a solid copper bonding jumper, insulated, covered, or bare, not smaller than 8 AWG. The bonding jumper(s) shall be required for equipotential bonding in the area of the hydromassage bathtub and shall not be required to be extended or attached to any remote panelboard, service equipment, or any electrode. In all installations a bonding jumper long enough to terminate on a replacement non-double-insulated pump or blower motor shall be provided and shall be terminated to the equipment grounding conductor of the branch circuit of the motor when a double-insulated circulating pump or blower motor is used.

PART VIII. ELECTRICALLY POWERED POOL LIFTS

680.80 General. Electrically powered pool lifts as defined in 680.2 shall comply with Part VIII of this article. Part VIII shall not be subject to the requirements of other parts of this article except where the requirements are specifically referenced.

680.81 Equipment Approval. Lifts shall be listed, labeled, and identified for swimming pool and spa use.

Exception No. 1: Lifts where the battery is removed for charging at another location and the battery is rated less than or equal to the low-voltage contact limit shall not be required to be listed or labeled.

Exception No. 2: Solar-operated or solar-recharged lifts where the solar panel is attached to the lift and the battery is rated less than or equal to 24 volts shall not be required to be listed or labeled.

Exception No. 3: Lifts that are supplied from a source not exceeding the low-voltage contact limit and supplied by listed transformers or power supplies that comply with 680.23(A)(2) shall not be required to be listed or labeled.

680.82 Protection. Pool lifts connected to premises wiring and operated above the low-voltage contact limit shall be provided with GFCI protection and comply with 680.5.

680.83 Bonding. Lifts shall be bonded in accordance with 680.26(B)(5) and (B)(7).

680.84 Switching Devices and Receptacles. Switches and switching devices that are operated above the low-voltage contact limit shall comply with 680.22(C). Receptacles for electrically powered pool lifts that are operated above the low-voltage contact limit shall comply with 680.22(A)(3) and (A)(4).

680.85 Nameplate Marking. Electrically powered pool lifts shall be provided with a nameplate giving the identifying name and model and rating in volts and amperes, or in volts and watts. If the lift is to be used on a specific frequency or frequencies, it shall be so marked. Battery-powered pool lifts shall indicate the type reference of the battery or battery pack to be used. Batteries and battery packs shall be provided with a battery type reference and voltage rating.

Exception: Nameplate ratings for battery-powered pool lifts shall only need to provide a rating in volts in addition to the identifying name and model.

CHAPTER 26

INTERACTIVE AND STAND-ALONE PHOTOVOLTAIC (PV) SYSTEMS

INTRODUCTION

This chapter contains requirements from Article 690 — Solar Photovoltaic (PV) Systems, Article 705 — Interconnected Electric Power Production Sources, and Chapter 11 of NFPA 1, *Fire Code*. Article 690 of the *NEC* provides the electrical installation requirements for photovoltaic (PV) electric power production systems that supply electric energy to a premises wiring system. Where the PV system is interconnected with another source of electric power (as is the case with many residential, commercial, institutional, and industrial installations), the requirements of Article 705 also have to be used. Article 691 — Large-Scale Photovoltaic (PV) Electric Power covers solar-fueled electric power production facilities, known as solar farms, that are constructed for providing renewable resource-fueled electric power into the transmission and distribution grid of a utility or other large scale user. This article is not covered in this *Pocket Guide*.

As the use of PV systems has expanded rapidly across all segments of building construction, so has the need to address the safety concerns of those who respond to a building with a PV system in the event of a fire or other emergency. Access to roofs is an essential part of firefighting strategy, and Chapter 11 of NFPA 1 provides setback and roof area coverage requirements intended to facilitate first responder access. Requirements for quickly reducing power or completely de-energizing conductors and equipment, referred to in Article 690 as rapid shutdown, are focused exclusively on making rooftops electrically safe for emergency personnel. Recognition that interactive inverter circuit conductors not only export power from the PV system, but can also import fault currents fed by the utility source, is the basis for rules in Article 705 covering the interconnection of sources and limiting the length of "unprotected" conductors on or in buildings and structures.

The requirements in Articles 690 and 705 of the *NEC* have seen significant expansion and modification to address the constantly changing PV technology and the lessons learned from the exponential growth of these systems over the past two decades. Better methods to control

Chapter 26

and protect PV systems are continuously evolving to keep pace with the demand for this increasingly viable and mainstream power production source.

To provide the user of this *Pocket Guide* with the necessary rules for installing stand-alone and interactive PV systems, installation requirements are provided from the eight parts of Article 690: Part I General, Part II Circuit Requirements, Part III Disconnecting Means, Part IV Wiring Methods, Part V Grounding and Bonding, Part VI Marking, Part VII Connection to Other Sources, and Part VIII Energy Storage Systems. Requirements are also provided from two parts of Article 705 (Part I General, and Part II Interactive Inverters) and from Chapter 11 (Building Services) of NFPA 1. Due to the specialized nature of the terminology used in Articles 690 and 705, several definitions that are essential to proper application of the requirements in these two articles are included. However, users are urged to review all of the definitions contained in these two articles to be completely conversant with the terms used in each. Unless specifically modified or indicated otherwise in Articles 690 and 705, the definitions in Article 100 also apply.

ARTICLE 690
Solar Photovoltaic (PV) Systems
PART I. GENERAL

[Article 100. Photovoltaic (PV) System. The total components and subsystem that, in combination, convert solar energy into electric energy for connection to a utilization load. (CMP-4)]

690.2 Definitions. The definitions in this section shall apply only within this article.

DC-to-DC Converter Output Circuit. The dc circuit conductors connected to the output of a dc combiner for dc-to-dc converter source circuits.

DC-to-DC Converter Source Circuit. Circuits between dc-to-dc converters and from dc-to-dc converters to the common connection point(s) of the dc system.

PV Output Circuit. The dc circuit conductors from two or more connected PV source circuits to their point of termination.

PV Source Circuit. The dc circuit conductors between modules and from modules to dc combiners, electronic power converters, or a dc PV system disconnecting means.

PV System DC Circuit. Any dc conductor in PV source circuits, PV output circuits, dc-to-dc converter source circuits, and dc-to-dc converter output circuits.

690.4 General Requirements.

(B) Equipment. Inverters, motor generators, PV modules, ac modules and ac module systems, dc combiners, dc-to-dc converters, rapid shutdown equipment, dc circuit controllers, and charge controllers intended for use in PV systems shall be listed or be evaluated for the application and have a field label applied.

(D) Multiple PV Systems. Multiple PV systems shall be permitted to be installed in or on a single building or structure. Where the PV systems are remotely located from each other, a directory in accordance with 705.10 shall be provided at each PV system disconnecting means.

(E) Locations Not Permitted. PV system equipment and disconnecting means shall not be installed in bathrooms.

690.6 Alternating-Current (ac) Modules and Systems.

(A) Photovoltaic Source Circuits. The requirements of Article 690 pertaining to PV source circuits shall not apply to ac modules or ac module systems. The PV source circuit, conductors, and inverters shall be considered as internal components of an ac module or ac module system.

(B) Output Circuit. The output of an ac module or ac module system shall be considered an inverter output circuit.

PART II. CIRCUIT REQUIREMENTS

690.7 Maximum Voltage. The maximum voltage of PV system dc circuits shall be the highest voltage between any two conductors of a circuit or any conductor and ground. The maximum voltage shall be used to determine the voltage and voltage to ground of circuits in the application of this *Code*. Maximum voltage shall be used for conductors, cables, equipment, working space, and other applications where voltage limits and ratings are used.

PV system dc circuits on or in buildings shall be permitted to have a maximum voltage no greater than 1000 volts. PV system dc circuits on or in one- and two-family dwellings shall be permitted to have a maximum voltage no greater than 600 volts. Where not located on or in buildings, listed dc PV equipment, rated at a maximum voltage no greater than 1500 volts, shall not be required to comply with Parts II and III of Article 490.

(A) Photovoltaic Source and Output Circuits. In a dc PV source circuit or output circuit, the maximum PV system voltage for that circuit shall be calculated in accordance with one of the following methods:

(1) The sum of the PV module–rated open-circuit voltage of the series-connected modules corrected for the lowest expected ambient temperature using the open-circuit voltage temperature coefficients in accordance with the instructions included in the listing or labeling of the module
(2) For crystalline and multicrystalline silicon modules, the sum of the PV module–rated open-circuit voltage of the series-connected modules corrected for the lowest expected ambient temperature using the correction factors provided in Table 690.7(A)
(3) For PV systems with an inverter generating capacity of 100 kW or greater, a documented and stamped PV system design, using an

Interactive and Stand-Alone Photovoltaic (PV) Systems 690.7

Table 690.7(A) Voltage Correction Factors for Crystalline and Multicrystalline Silicon Modules

Correction Factors for Ambient Temperatures Below 25°C (77°F). (Multiply the rated open-circuit voltage by the appropriate correction factor shown below.)

Ambient Temperature (°C)	Factor	Ambient Temperature (°F)
24 to 20	1.02	76 to 68
19 to 15	1.04	67 to 59
14 to 10	1.06	58 to 50
9 to 5	1.08	49 to 41
4 to 0	1.10	40 to 32
−1 to −5	1.12	31 to 23
−6 to −10	1.14	22 to 14
−11 to −15	1.16	13 to 5
−16 to −20	1.18	4 to −4
−21 to −25	1.20	−5 to −13
−26 to −30	1.21	−14 to −22
−31 to −35	1.23	−23 to −31
−36 to −40	1.25	−32 to −40

industry standard method maximum voltage calculation provided by a licensed professional electrical engineer

(B) DC-to-DC Converter Source and Output Circuits. In a dc-to-dc converter source and output circuit, the maximum voltage shall be calculated in accordance with 690.7(B)(1) or (B)(2).

(1) Single DC-to-DC Converter. For circuits connected to the output of a single dc-to-dc converter, the maximum voltage shall be determined in accordance with the instructions included in the listing or labeling of the dc-to-dc converter. If the instructions do not provide a method to determine the maximum voltage, the maximum voltage shall be the maximum rated voltage output of the dc-to-dc converter.

(2) Two or More Series-Connected DC-to-DC Converters. For circuits connected to the output of two or more series-connected dc-to-dc converters, the maximum voltage shall be determined in accordance with the instructions included in the listing or labeling of the dc-to-dc converter. If the instructions do not provide a method to determine the maximum voltage, the maximum voltage shall be the sum of the maximum rated voltage output of the dc-to-dc converters in series.

(C) Bipolar Source and Output Circuits. For monopole subarrays in bipolar systems, the maximum voltage shall be the highest voltage between the monopole subarray circuit conductors where one conductor of the monopole subarray circuit is connected to the functionally grounded reference. To prevent overvoltage in the event of a ground fault or arc fault, the monopole subarray circuits shall be isolated from ground.

690.8 Circuit Sizing and Current.

(A) Calculation of Maximum Circuit Current. The maximum current for the specific circuit shall be calculated in accordance with one of the methods in 690.8(A)(1)or(A)(2).

(1) PV System Circuits. The maximum current shall be calculated in accordance with 690.8(A)(1)(a) through (A)(1)(e).

(a) *Photovoltaic Source Circuit Currents.* The maximum current shall be as calculated in either of the following:

(1) The maximum current shall be the sum of the short-circuit current ratings of the PV modules connected in parallel multiplied by 125 percent.
(2) For PV systems with an inverter generating capacity of 100 kW or greater, a documented and stamped PV system design, using an industry standard method maximum current calculation provided by a licensed professional electrical engineer, shall be permitted. The calculated maximum current value shall be based on the highest 3-hour current average resulting from the simulated local irradiance on the PV array accounting for elevation and orientation. The current value used by this method shall not be less than 70 percent of the value calculated using 690.8(A)(1)(a)(1).

Informational Note: One industry standard method for calculating maximum current of a PV system is available from Sandia National Laboratories, reference SAND 2004-3535, *Photovoltaic Array Performance Model*. This model is used by the System Advisor Model simulation program provided by the National Renewable Energy Laboratory.

(b) *Photovoltaic Output Circuit Currents.* The maximum current shall be the sum of parallel source circuit maximum currents as calculated in 690.8(A)(1)(a).

(c) *DC-to-DC Converter Source Circuit Current.* The maximum current shall be the dc-to-dc converter continuous output current rating.

Interactive and Stand-Alone Photovoltaic (PV) Systems 690.9

(d) *DC-to-DC Converter Output Circuit Current.* The maximum current shall be the sum of parallel connected dc-to-dc converter source circuit currents as calculated in 690.8(A)(1)(c).

(e) *Inverter Output Circuit Current.* The maximum current shall be the inverter continuous output current rating.

(2) Circuits Connected to the Input of Electronic Power Converters. Where a circuit is protected with an overcurrent device not exceeding the conductor ampacity, the maximum current shall be permitted to be the rated input current of the electronic power converter input to which it is connected.

(B) Conductor Ampacity. Circuit conductors shall be sized to carry not less than the larger ampacity calculated in accordance with 690.8(B)(1) or (B)(2).

(1) Without Adjustment and Correction Factors. The maximum currents calculated in 690.8(A) multiplied by 125 percent without adjustment or correction factors.

Exception: Circuits containing an assembly, together with its overcurrent device(s), that is listed for continuous operation at 100 percent of its rating shall be permitted to be used at 100 percent of its rating.

(2) With Adjustment and Correction Factors. The maximum currents calculated in 690.8(A) with adjustment and correction factors.

(C) Systems with Multiple Direct-Current Voltages. For a PV power source that has multiple output circuit voltages and employs a common-return conductor, the ampacity of the common-return conductor shall not be less than the sum of the ampere ratings of the overcurrent devices of the individual output circuits.

(D) Sizing of Module Interconnection Conductors. Where a single overcurrent device is used to protect a set of two or more parallel-connected module circuits, the ampacity of each of the module interconnection conductors shall not be less than the sum of the rating of the single overcurrent device plus 125 percent of the short-circuit current from the other parallel-connected modules.

690.9 Overcurrent Protection.

(A) Circuits and Equipment. PV system dc circuit and inverter output conductors and equipment shall be protected against overcurrent. Circuits sized in accordance with 690.8(A)(2) are required to be protected against overcurrent with overcurrent protective devices. Each circuit shall be protected from overcurrent in accordance with 690.9(A)(1), (A)(2), or (A)(3).

(1) Circuits Where Overcurrent Protection Not Required. Overcurrent protective devices shall not be required where both of the following conditions are met:

(1) The conductors have sufficient ampacity for the maximum circuit current.
(2) The currents from all sources do not exceed the maximum overcurrent protective device rating specified for the PV module or electronic power converter.

(2) Circuits Where Overcurrent Protection Is Required on One End. A circuit conductor connected at one end to a current-limited supply, where the conductor is rated for the maximum circuit current from that supply, and also connected to sources having an available maximum circuit current greater than the ampacity of the conductor, shall be protected from overcurrent at the point of connection to the higher current source.

Informational Note: Photovoltaic system dc circuits and electronic power converter outputs powered by these circuits are current-limited and in some cases do not need overcurrent protection. Where these circuits are connected to higher current sources, such as parallel-connected PV system dc circuits, energy storage systems, or a utility service, the overcurrent device is often installed at the higher current source end of the circuit conductor.

(3) Other Circuits. Circuits that do not comply with 690.9(A)(1) or (A)(2) shall be protected with one of the following methods:

(1) Conductors not greater than 3 m (10 ft) in length and not in buildings, protected from overcurrent on one end
(2) Conductors not greater than 3 m (10 ft) in length and in buildings, protected from overcurrent on one end and in a raceway or metal clad cable
(3) Conductors protected from overcurrent on both ends
(4) Conductors not installed on or in buildings are permitted to be protected from overcurrent on one end of the circuit where the circuit complies with all of the following conditions:

   a. The conductors are installed in metal raceways or metal-clad cables, or installed in enclosed metal cable trays, or underground, or where directly entering pad-mounted enclosures.

Interactive and Stand-Alone Photovoltaic (PV) Systems 690.9

b. The conductors for each circuit terminate on one end at a single circuit breaker or a single set of fuses that limit the current to the ampacity of the conductors.
c. The overcurrent device for the conductors is an integral part of a disconnecting means or shall be located within 3 m (10 ft) of conductor length of the disconnecting means.
d. The disconnecting means for the conductors is installed outside of a building, or at a readily accessible location nearest the point of entrance of the conductors inside of a building, including installations complying with 230.6.

(B) Device Ratings. Overcurrent devices used in PV system dc circuits shall be listed for use in PV systems. Electronic devices that are listed to prevent backfeed current in PV system dc circuits shall be permitted to prevent overcurrent of conductors on the PV array side of the device. Overcurrent devices, where required, shall be rated in accordance with one of the following and permitted to be rounded up to the next higher standard size in accordance with 240.4(B):

(1) Not less than 125 percent of the maximum currents calculated in 690.8(A).
(2) An assembly, together with its overcurrent device(s), that is listed for continuous operation at 100 percent of its rating shall be permitted to be used at 100 percent of its rating.

Informational Note: Some electronic devices prevent backfeed current, which in some cases is the only source of overcurrent in PV system dc circuits.

(C) Source and Output Circuits. A single overcurrent protective device, where required, shall be permitted to protect the PV modules, dc-to-dc converters, and conductors of each source circuit or the conductors of each output circuit. Where single overcurrent protection devices are used to protect source or output circuits, all overcurrent devices shall be placed in the same polarity for all circuits within a PV system. The overcurrent devices shall be accessible but shall not be required to be readily accessible.

Informational Note: Due to improved ground-fault protection required in PV systems by 690.41(B), a single overcurrent protective device in either the positive or negative conductors of a PV system in combination with this ground-fault protection provides adequate overcurrent protection.

(D) Power Transformers. Overcurrent protection for a transformer with a source(s) on each side shall be provided in accordance with 450.3 by considering first one side of the transformer, then the other side of the transformer, as the primary.

Exception: A power transformer with a current rating on the side connected toward the interactive inverter output, not less than the rated continuous output current of the inverter, shall be permitted without overcurrent protection from the inverter.

690.10 Stand-Alone Systems. The wiring system connected to a stand-alone system shall be installed in accordance with 710.15.

690.11 Arc-Fault Circuit Protection (Direct Current). Photovoltaic systems with PV system dc circuits operating at 80 volts dc or greater between any two conductors shall be protected by a listed PV arc-fault circuit interrupter or other system components listed to provide equivalent protection. The system shall detect and interrupt arcing faults resulting from a failure in the intended continuity of a conductor, connection, module, or other system component in the PV system dc circuits.

Exception: For PV systems not installed on or in buildings, PV output circuits and dc-to-dc converter output circuits that are installed in metallic raceways or metal-clad cables, or installed in enclosed metallic cable trays, or are underground shall be permitted without arc-fault circuit protection. Detached structures whose sole purpose is to house PV system equipment shall not be considered buildings according to this exception.

690.12 Rapid Shutdown of PV Systems on Buildings. PV system circuits installed on or in buildings shall include a rapid shutdown function to reduce shock hazard for firefighters in accordance with 690.12(A) through (D).

Exception: Ground-mounted PV system circuits that enter buildings, of which the sole purpose is to house PV system equipment, shall not be required to comply with 690.12.

(A) Controlled Conductors. Requirements for controlled conductors shall apply to the following:

(1) PV system dc circuits
(2) Inverter output circuits originating from inverters located within the array boundary

Interactive and Stand-Alone Photovoltaic (PV) Systems 690.12

Informational Note: The rapid shutdown function reduces the risk of electrical shock that dc circuits in a PV system could pose for firefighters. The ac output conductors from PV systems that include inverters will either be de-energized after shutdown initiation or will remain energized by other sources such as a utility service. To prevent PV arrays with attached inverters from having energized ac conductors within the PV array(s), those circuits are also specifically controlled after shutdown initiation.

(B) Controlled Limits. The use of the term *array boundary* in this section is defined as 305 mm (1 ft) from the array in all directions. Controlled conductors outside the array boundary shall comply with 690.12(B)(1) and inside the array boundary shall comply with 690.12(B)(2).

(1) Outside the Array Boundary. Controlled conductors located outside the boundary or more than 1 m (3 ft) from the point of entry inside a building shall be limited to not more than 30 volts within 30 seconds of rapid shutdown initiation. Voltage shall be measured between any two conductors and between any conductor and ground.

(2) Inside the Array Boundary. The PV system shall comply with one of the following:

(1) A PV hazard control system listed for the purpose shall be installed in accordance with the instructions included with the listing or field labeling. Where a hazard control system requires initiation to transition to a controlled state, the rapid shutdown initiation device required in 690.12(C) shall perform this initiation.
(2) Controlled conductors located inside the boundary shall be limited to not more than 80 volts within 30 seconds of rapid shutdown initiation. Voltage shall be measured between any two conductors and between any conductor and ground.
(3) PV arrays shall have no exposed wiring methods or conductive parts and be installed more than 2.5 m (8 ft) from exposed grounded conductive parts or ground.

(C) Initiation Device. The initiation device(s) shall initiate the rapid shutdown function of the PV system. The device's "off" position shall indicate that the rapid shutdown function has been initiated for all PV systems connected to that device. For one-family and two-family dwellings an initiation device(s) shall be located at a readily accessible location outside the building.

For a single PV system, the rapid shutdown initiation shall occur by the operation of any single initiation device. Devices shall consist of at least one or more of the following:

(1) Service disconnecting means
(2) PV system disconnecting means
(3) Readily accessible switch that plainly indicates whether it is in the "off" or "on" position

Where multiple PV systems are installed with rapid shutdown functions on a single service, the initiation device(s) shall consist of not more than six switches or six sets of circuit breakers, or a combination of not more than six switches and sets of circuit breakers, mounted in a single enclosure, or in a group of separate enclosures. These initiation device(s) shall initiate the rapid shutdown of all PV systems with rapid shutdown functions on that service.

(D) Equipment. Equipment that performs the rapid shutdown functions, other than initiation devices such as listed disconnect switches, circuit breakers, or control switches, shall be listed for providing rapid shutdown protection.

PART III. DISCONNECTING MEANS

690.13 Photovoltaic System Disconnecting Means. Means shall be provided to disconnect the PV system from all wiring systems including power systems, energy storage systems, and utilization equipment and its associated premises wiring.

(A) Location. The PV system disconnecting means shall be installed at a readily accessible location. Where disconnecting means of systems above 30 V are readily accessible to unqualified persons, any enclosure door or hinged cover that exposes live parts when open shall be locked or require a tool to open.

(B) Marking. Each PV system disconnecting means shall plainly indicate whether in the open (off) or closed (on) position and be permanently marked "PV SYSTEM DISCONNECT" or equivalent. Additional markings shall be permitted based upon the specific system configuration. For PV system disconnecting means where the line and load terminals may be energized in the open position, the device shall be marked with the following words or equivalent:

Interactive and Stand-Alone Photovoltaic (PV) Systems 690.13

> **WARNING**
> ELECTRIC SHOCK HAZARD TERMINALS ON THE LINE AND LOAD SIDES MAY BE ENERGIZED IN THE OPEN POSITION

The warning sign(s) or label(s) shall comply with 110.21(B).

(C) Maximum Number of Disconnects. Each PV system disconnecting means shall consist of not more than six switches or six sets of circuit breakers, or a combination of not more than six switches and sets of circuit breakers, mounted in a single enclosure, or in a group of separate enclosures. A single PV system disconnecting means shall be permitted for the combined ac output of one or more inverters or ac modules in an interactive system.

> Informational Note: This requirement does not limit the number of PV systems connected to a service as permitted in 690.4(D). This requirement allows up to six disconnecting means to disconnect a single PV system. For PV systems where all power is converted through interactive inverters, a dedicated circuit breaker, in 705.12(B)(1), is an example of a single PV system disconnecting means.

(D) Ratings. The PV system disconnecting means shall have ratings sufficient for the maximum circuit current, available fault current, and voltage that is available at the terminals of the PV system disconnect.

(E) Type of Disconnect. The PV system disconnecting means shall simultaneously disconnect the PV system conductors that are not solidly grounded from all conductors of other wiring systems. The PV system disconnecting means or its remote operating device or the enclosure providing access to the disconnecting means shall be capable of being locked in accordance with 110.25. The PV system disconnecting means shall be one of the following:

(1) A manually operable switch or circuit breaker
(2) A connector meeting the requirements of 690.33(D)(1) or (D)(3)
(3) A pull-out switch with the required interrupting rating
(4) A remote-controlled switch or circuit breaker that is operable locally and opens automatically when control power is interrupted
(5) A device listed or approved for the intended application

> Informational Note: Circuit breakers marked "line" and "load" may not be suitable for backfeed or reverse current.

690.15 Disconnecting Means for Isolating Photovoltaic Equipment.

Disconnecting means of the type required in 690.15(D) shall be provided to disconnect ac PV modules, fuses, dc-to-dc converters, inverters, and charge controllers from all conductors that are not solidly grounded.

(A) Location. Isolating devices or equipment disconnecting means shall be installed in circuits connected to equipment at a location within the equipment, or within sight and within 3 m (10 ft) of the equipment. An equipment disconnecting means shall be permitted to be remote from the equipment where the equipment disconnecting means can be remotely operated from within 3 m (10 ft) of the equipment. Where disconnecting means of equipment operating above 30 volts are readily accessible to unqualified persons, any enclosure door or hinged cover that exposes live parts when open shall be locked or require a tool to open.

(B) Isolating Device. An isolating device shall not be required to have an interrupting rating. Where an isolating device is not rated for interrupting the circuit current, it shall be marked "Do Not Disconnect Under Load" or "Not for Current Interrupting." An isolating device shall not be required to simultaneously disconnect all current-carrying conductors of a circuit. The isolating device shall be one of the following:

(1) A mating connector meeting the requirements of 690.33 and listed and identified for use with specific equipment
(2) A finger-safe fuse holder
(3) An isolating device that requires a tool to place the device in the open (off) position
(4) An isolating device listed for the intended application

(C) Equipment Disconnecting Means. Equipment disconnecting means shall have ratings sufficient for the maximum circuit current, available fault current, and voltage that is available at the terminals. Equipment disconnecting means shall simultaneously disconnect all current-carrying conductors that are not solidly grounded to the circuit to which it is connected. Equipment disconnecting means shall be externally operable without exposing the operator to contact with energized parts and shall indicate whether in the open (off) or closed (on) position. Where not within sight or not within 3 m (10 ft) of the equipment, the disconnecting means or its remote operating device or the enclosure providing access to the disconnecting means shall be capable of being locked in accordance with 110.25. Equipment disconnecting means, where used, shall be one of the types in 690.13(E)(1) through (E)(5).

Interactive and Stand-Alone Photovoltaic (PV) Systems 690.31

Equipment disconnecting means, other than those complying with 690.33, shall be marked in accordance with the warning in 690.13(B) if the line and load terminals can be energized in the open position.

Informational Note: A common installation practice is to terminate PV source-side dc conductors in the same manner that utility source-side ac conductors are generally connected on the line side of a disconnecting means. This practice is more likely to de-energize load-side terminals, blades, and fuses when the disconnect is in the open position and no energized sources are connected to the load side of the disconnect.

(D) Type of Disconnecting Means. Where disconnects are required to isolate equipment, the disconnecting means shall be one of the following applicable types:

(1) An equipment disconnecting means in accordance with 690.15(C) shall be required to isolate dc circuits with a maximum circuit current over 30 amperes.
(2) An isolating device in accordance with 690.15(B) shall be permitted for circuits other than those covered by 690.15(D)(1).

PART IV. WIRING METHODS AND MATERIALS

690.31 Wiring Methods.

(A) Wiring Systems. All raceway and cable wiring methods included in this *Code*, other wiring systems and fittings specifically listed for use in PV arrays, and wiring as part of a listed system shall be permitted. Where wiring devices with integral enclosures are used, sufficient length of cable shall be provided to facilitate replacement.

Where PV source and output circuits operating at voltages greater than 30 volts are installed in readily accessible locations, circuit conductors shall be guarded or installed in Type MC cable or in raceway. The ampacity of 105°C (221°F) and 125°C (257°F) conductors shall be permitted to be determined by Table 690.31(A)(b). For ambient temperatures greater than 30°C (86°F), the ampacities of these conductors shall be corrected in accordance with Table 690.31(A)(a).

(B) Identification and Grouping. PV system dc circuits and Class 1 remote control, signaling, and power-limited circuits of a PV system shall be permitted to occupy the same equipment wiring enclosure, cable, or raceway. PV system dc circuits shall not occupy the same equipment

Table 690.31(A)(a) Correction Factors

Ambient Temperature (°C)	Temperature Rating of Conductor		Ambient Temperature (°F)
	105°C (221°F)	125°C (257°F)	
30	1	1	86
31–35	0.97	0.97	87–95
36–40	0.93	0.95	96–104
41–45	0.89	0.92	105–113
46–50	0.86	0.89	114–122
51–55	0.82	0.86	123–131
56–60	0.77	0.83	132–140
61–65	0.73	0.79	141–149
66–70	0.68	0.76	150–158
71–75	0.63	0.73	159–167
76–80	0.58	0.69	168–176
81–85	0.52	0.65	177–185
86–90	0.45	0.61	186–194
91–95	0.37	0.56	195–203
96–100	0.26	0.51	204–212
101–105	—	0.46	213–221
106–110	—	0.4	222–230
111–115	—	0.32	231–239
116–120	—	0.23	240–248

wiring enclosure, cable, or raceway, as other non-PV systems, or inverter output circuits, unless the PV system dc circuits are separated from other circuits by a barrier or partition. PV system circuit conductors shall be identified and grouped as required by 690.31(B)(1) and (B)(2).

Exception: PV system dc circuits utilizing multiconductor jacketed cable or metal-clad cable assemblies or listed wiring harnesses identified for the application shall be permitted to occupy the same wiring method as inverter output circuits and other non-PV systems. All conductors, harnesses, or assemblies shall have an insulation rating equal to at least the maximum circuit voltage applied to any conductor within the enclosure, cable, or raceway.

(1) Identification. PV system dc circuit conductors shall be identified at all termination, connection, and splice points by color coding, marking

Interactive and Stand-Alone Photovoltaic (PV) Systems 690.31

Table 690.31(A)(b) Ampacities of Insulated Conductors Rated Up To and Including 2000 Volts, 105°C Through 125°C (221°F Through 257°F), Not More Than Three Current-Carrying Conductors in Raceway, Cable, or Earth (Directly Buried), Based on Ambient Temperature of 30°C (86°F)

	Types	
AWG	PVC, CPE, XLPE 105°C	XLPE, EPDM 125°C
18	15	16
16	19	20
14	29	31
12	36	39
10	46	50
8	64	69
6	81	87
4	109	118
3	129	139
2	143	154
1	168	181
1/0	193	208
2/0	229	247
3/0	263	284
4/0	301	325

Informational Note: See 110.14(C) for conductor temperature limitations due to termination provisions.

tape, tagging, or other approved means. Conductors relying on other than color coding for polarity identification shall be identified by an approved permanent marking means such as labeling, sleeving, or shrink-tubing that is suitable for the conductor size. The permanent marking means for nonsolidly grounded positive conductors shall include imprinted plus signs (+) or the word POSITIVE or POS durably marked on insulation of a color other than green, white, or gray. The permanent marking means for nonsolidly grounded negative conductors shall include imprinted negative signs (−) or the word NEGATIVE or NEG durably marked on insulation of a color other than green, white, gray, or red. Only solidly grounded PV system dc circuit conductors shall be marked in accordance with 200.6.

Exception: Where the identification of the conductors is evident by spacing or arrangement, further identification shall not be required.

(2) Grouping. Where the conductors of more than one PV system occupy the same junction box or raceway with a removable cover(s), the PV system conductors of each system shall be grouped separately by cable ties or similar means at least once and shall then be grouped at intervals not to exceed 1.8 m (6 ft).

Exception: The requirement for grouping shall not apply if the circuit enters from a cable or raceway unique to the circuit that makes the grouping obvious.

(C) Cables. Type PV wire or cable and Type distributed generation (DG) cable shall be listed.

Informational Note: See UL 4703, *Standard for Photovoltaic Wire*, for PV wire and UL 3003, *Distributed Generation Cables*, for DG cable.

(1) Single-Conductor Cable. Single-conductor cable in exposed outdoor locations in PV system dc circuits within the PV array shall be permitted to be one of the following:

(1) PV wire or cable
(2) Single-conductor cable marked sunlight resistant and Type USE-2 and Type RHW-2

Exposed cables shall be supported and secured at intervals not to exceed 600 mm (24 in.) by cable ties, straps, hangers, or similar fittings listed and identified for securement and support in outdoor locations. PV wire or cable shall be permitted in all locations where RHW-2 is permitted.

Exception: PV systems meeting the requirements of 691.4 shall be permitted to have support and securement intervals as defined in the engineered design.

(2) Cable Tray. Single-conductor PV wire or cable of all sizes or distributed generation (DG) cable of all sizes, with or without a cable tray rating, shall be permitted in cable trays installed in outdoor locations, provided that the cables are supported at intervals not to exceed 300 mm (12 in.) and secured at intervals not to exceed 1.4 m (4½ ft).

Informational Note: PV wire and cable and DG cable have a nonstandard outer diameter. Table 1 of Chapter 9 contains the allowable percent of cross section of conduit and tubing for conductors and cables.

Interactive and Stand-Alone Photovoltaic (PV) Systems 690.31

(3) Multiconductor Jacketed Cables. Where part of a listed PV assembly, multiconductor jacketed cables shall be installed in accordance with the included instructions. Where not part of a listed assembly, or where not otherwise covered in this *Code*, multiconductor jacketed cables, including DG cable, shall be installed in accordance with the product listing and shall be permitted in PV systems. These cables shall be installed in accordance with the following:

(1) In raceways, where on or in buildings other than rooftops
(2) Where not in raceways, in accordance with the following:

   a. Marked sunlight resistant in exposed outdoor locations
   b. Protected or guarded, where subject to physical damage
   c. Closely follow the surface of support structures
   d. Secured at intervals not exceeding 1.8 m (6 ft)
   e. Secured within 600 mm (24 in.) of mating connectors or entering enclosures
   f. Marked direct burial, where buried in the earth

(4) Flexible Cords and Cables Connected to Tracking PV Arrays. Flexible cords and flexible cables, where connected to moving parts of tracking PV arrays, shall comply with Article 400 and shall be of a type identified as a hard service cord or portable power cable; they shall be suitable for extra-hard usage, listed for outdoor use, water resistant, and sunlight resistant. Allowable ampacities shall be in accordance with 400.5. Stranded copper PV wire shall be permitted to be connected to moving parts of tracking PV arrays in accordance with the minimum number of strands specified in Table 690.31(C)(4).

(5) Flexible, Fine-Stranded Cables. Flexible, fine-stranded cables shall be terminated only with terminals, lugs, devices, or connectors in accordance with 110.14.

Table 690.31(C)(4) Minimum PV Wire Strands

PV Wire AWG	Minimum Strands
18	17
16–10	19
8–4	49
2	130
1 AWG–1000 MCM	259

(6) Small-Conductor Cables. Single-conductor cables listed for outdoor use that are sunlight resistant and moisture resistant in sizes 16 AWG and 18 AWG shall be permitted for module interconnections where such cables meet the ampacity requirements of 400.5. Section 310.14 shall be used to determine the cable ampacity adjustment and correction factors.

(D) Direct-Current Circuits on or in Buildings. Where inside buildings, PV system dc circuits that exceed 30 volts or 8 amperes shall be contained in metal raceways, in Type MC metal-clad cable that complies with 250.118(10), or in metal enclosures.

Exception: PV hazard control systems installed in accordance with 690.12(B)(2)(1) shall be permitted to be provided with or listed for use with nonmetallic enclosure(s), nonmetallic raceway(s), and cables other than Type MC metal-clad cable(s), at the point of penetration of the surface of the building to the PV hazard control actuator.

Wiring methods on or in buildings shall comply with the additional installation requirements in 690.31(D)(1) and (D)(2).

(1) Flexible Wiring Methods. Where flexible metal conduit (FMC) smaller than metric designator 21 (trade size ¾) or Type MC cable smaller than 25 mm (1 in.) in diameter containing PV power circuit conductors is installed across ceilings or floor joists, the raceway or cable shall be protected by substantial guard strips that are at least as high as the raceway or cable. Where run exposed, other than within 1.8 m (6 ft) of their connection to equipment, these wiring methods shall closely follow the building surface or be protected from physical damage by an approved means.

(2) Marking and Labeling Required. Unless located and arranged so the purpose is evident, the following wiring methods and enclosures that contain PV system dc circuit conductors shall be marked with the wording PHOTOVOLTAIC POWER SOURCE or SOLAR PV DC CIRCUIT by means of permanently affixed labels or other approved permanent marking:

(1) Exposed raceways, cable trays, and other wiring methods
(2) Covers or enclosures of pull boxes and junction boxes
(3) Conduit bodies in which any of the available conduit openings are unused

Interactive and Stand-Alone Photovoltaic (PV) Systems 690.33

The labels or markings shall be visible after installation. All letters shall be capitalized and shall be a minimum height of 9.5 mm (⅜ in.) in white on a red background. Labels shall appear on every section of the wiring system that is separated by enclosures, walls, partitions, ceilings, or floors. Spacing between labels or markings, or between a label and a marking, shall not be more than 3 m (10 ft). Labels required by this section shall be suitable for the environment where they are installed.

(E) Bipolar Photovoltaic Systems. Where the sum, without consideration of polarity, of the voltages of the two monopole circuits exceeds the rating of the conductors and connected equipment, monopole circuits in a bipolar PV system shall be physically separated, and the electrical output circuits from each monopole circuit shall be installed in separate raceways until connected to the inverter. The disconnecting means and overcurrent protective devices for each monopole circuit output shall be in separate enclosures. All conductors from each separate monopole circuit shall be routed in the same raceway. Solidly grounded bipolar PV systems shall be clearly marked with a permanent, legible warning notice indicating that the disconnection of the grounded conductor(s) may result in overvoltage on the equipment.

Exception: Listed switchgear rated for the maximum voltage between circuits and containing a physical barrier separating the disconnecting means for each monopole circuit shall be permitted to be used instead of disconnecting means in separate enclosures.

(F) Wiring Methods and Mounting Systems. Roof-mounted PV array mounting systems shall be permitted to be held in place with an approved means other than those required by 110.13 and shall utilize wiring methods that allow any expected movement of the array.

Informational Note: Expected movement of unattached PV arrays is often included in structural calculations.

690.32 Component Interconnections. Fittings and connectors that are intended to be concealed at the time of on-site assembly, where listed for such use, shall be permitted for on-site interconnection of modules or other array components. Such fittings and connectors shall be equal to the wiring method employed in insulation, temperature rise, and short-circuit current rating, and shall be capable of resisting the effects of the environment in which they are used.

690.33 Mating Connectors. Mating connectors, other than connectors covered by 690.32, shall comply with 690.33(A) through (D).

(A) Configuration. The mating connectors shall be polarized and shall have a configuration that is noninterchangeable with receptacles in other electrical systems on the premises.

(B) Guarding. The mating connectors shall be constructed and installed so as to guard against inadvertent contact with live parts by persons.

(C) Type. The mating connectors shall be of the latching or locking type. Mating connectors that are readily accessible and that are used in circuits operating at over 30 volts dc or 15 volts ac shall require a tool for opening. Where mating connectors are not of the identical type and brand, they shall be listed and identified for intermatability, as described in the manufacturer's instructions.

(D) Interruption of Circuit. Mating connectors shall be one of the following:

(1) Rated for interrupting current without hazard to the operator
(2) A type that requires the use of a tool to open and marked "Do Not Disconnect Under Load" or "Not for Current Interrupting"
(3) Supplied as part of listed equipment and used in accordance with instructions provided with the listed connected equipment

Informational Note: Some listed equipment, such as microinverters, are evaluated to make use of mating connectors as disconnect devices even though the mating connectors are marked as "Do Not Disconnect Under Load" or "Not for Current Interrupting."

690.34 Access to Boxes. Junction, pull, and outlet boxes located behind modules or panels shall be so installed that the wiring contained in them can be rendered accessible directly or by displacement of a module(s) or panel(s) secured by removable fasteners and connected by a flexible wiring system.

PART V. GROUNDING AND BONDING

690.41 System Grounding.

(A) PV System Grounding Configurations. One or more of the following system configurations shall be employed:

(1) 2-wire PV arrays with one functionally grounded conductor
(2) Bipolar PV arrays according to 690.7(C) with a functional ground reference (center tap)
(3) PV arrays not isolated from the grounded inverter output circuit

Interactive and Stand-Alone Photovoltaic (PV) Systems 690.41

(4) Ungrounded PV arrays
(5) Solidly grounded PV arrays as permitted in 690.41(B)
(6) PV systems that use other methods that accomplish equivalent system protection in accordance with 250.4(A) with equipment listed and identified for the use

(B) Ground-Fault Protection. PV system dc circuits that exceed 30 volts or 8 amperes shall be provided with dc ground-fault protection meeting the requirements of 690.41(B)(1) and (B)(2) to reduce fire hazards.

Solidly grounded PV source circuits with not more than two modules in parallel and not on or in buildings shall be permitted without ground-fault protection.

Informational Note: Not all inverters, charge controllers, or dc-to-dc converters include ground-fault protection. Equipment that does not have ground-fault protection often includes the following statement in the manual: "Warning: This unit is not provided with a GFDI device."

(1) Ground-Fault Detection. The ground-fault protection device or system shall detect ground fault(s) in the PV system dc circuit conductors, including any functional grounded conductors, and be listed for providing PV ground-fault protection. For dc-to-dc converters not listed as providing ground-fault protection, where required, listed ground fault protection equipment identified for the combination of the dc-to-dc converter and ground-fault protection device shall be installed to protect the circuit.

Informational Note: Some dc-to-dc converters without integral ground-fault protection on their input (source) side can prevent other ground-fault protection equipment from properly functioning on portions of PV system dc circuits.

(2) Faulted Circuits. The faulted circuits shall be controlled by one of the following methods:

(1) The current-carrying conductors of the faulted circuit shall be automatically disconnected.
(2) The device providing ground-fault protection fed by the faulted circuit shall automatically cease to supply power to output circuits and interrupt the faulted PV system dc circuits from the ground reference in a functionally grounded system.

(3) Indication of Faults. Ground-fault protection equipment shall provide indication of ground faults at a readily accessible location.

Informational Note: Examples of indication include, but are not limited to, the following: remote indicator light, display, monitor, signal to a monitored alarm system, or receipt of notification by web-based services.

690.42 Point of System Grounding Connection. Systems with a ground-fault protective device in accordance with 690.41(B) shall have any current-carrying conductor-to-ground connection made by the ground-fault protective device. For solidly grounded PV systems, the dc circuit grounding connection shall be made at any single point on the PV output circuit.

690.43 Equipment Grounding and Bonding. Exposed non-current-carrying metal parts of PV module frames, electrical equipment, and conductor enclosures of PV systems shall be connected to an equipment grounding conductor in accordance with 250.134 or 250.136, regardless of voltage. Equipment grounding conductors and devices shall comply with 690.43(A) through (D).

(A) Photovoltaic Module Mounting Systems and Devices. Devices and systems used for mounting PV modules that are also used for bonding module frames shall be listed, labeled, and identified for bonding PV modules. Devices that mount adjacent PV modules shall be permitted to bond adjacent PV modules.

(B) Equipment Secured to Grounded Metal Supports. Devices listed, labeled, and identified for bonding and grounding the metal parts of PV systems shall be permitted to bond the equipment to grounded metal supports. Metallic support structures shall have identified bonding jumpers connected between separate metallic sections or shall be identified for equipment bonding and shall be connected to the equipment grounding conductor.

(C) With Circuit Conductors. Equipment grounding conductors for the PV array and support structure where installed shall be contained within the same raceway or cable or otherwise run with the PV system conductors where those circuit conductors leave the vicinity of the PV array.

(D) Bonding for Over 250 Volts. The bonding requirements contained in 250.97 shall apply only to solidly grounded PV system circuits operating over 250 volts to ground.

690.45 Size of Equipment Grounding Conductors. Equipment grounding conductors for PV system circuits shall be sized in accordance with 250.122. Where no overcurrent protective device is used in the

Interactive and Stand-Alone Photovoltaic (PV) Systems 690.47

circuit, an assumed overcurrent device rated in accordance with 690.9(B) shall be used when applying Table 250.122.

Increases in equipment grounding conductor size to address voltage drop considerations shall not be required.

690.47 Grounding Electrode System.

(A) Buildings or Structures Supporting a PV System. A building or structure(s) supporting a PV system shall utilize a grounding electrode system installed in accordance with Part III of Article 250.

PV array equipment grounding conductors shall be connected to a grounding electrode system in accordance with Part VII of Article 250. This connection shall be in addition to any other equipment grounding conductor requirements in 690.43(C). The PV array equipment grounding conductors shall be sized in accordance with 690.45. For specific PV system grounding configurations permitted in 690.41(A), one of the following conditions shall apply:

(1) For PV systems that are not solidly grounded, the equipment grounding conductor for the output of the PV system, where connected to associated distribution equipment connected to a grounding electrode system, shall be permitted to be the only connection to ground for the system.
(2) For solidly grounded PV systems, as permitted in 690.41(A)(5), the grounded conductor shall be connected to a grounding electrode system by means of a grounding electrode conductor sized in accordance with 250.166.

Informational Note: Most PV systems are functionally grounded systems rather than solidly grounded systems as defined in this *Code*. For functionally grounded PV systems with an interactive inverter output, the ac equipment grounding conductor is connected to associated grounded ac distribution equipment. This connection is most often the connection to ground for ground-fault protection and equipment grounding of the PV array.

(B) Grounding Electrodes and Grounding Electrode Conductors. Additional grounding electrodes shall be permitted to be installed in accordance with 250.52 and 250.54. Grounding electrodes shall be permitted to be connected directly to the PV module frame(s) or support structure. A grounding electrode conductor shall be sized according to 250.66. A support structure for a ground-mounted PV array shall be permitted to be considered a grounding electrode if it meets the requirements of 250.52.

PV arrays mounted to buildings shall be permitted to use the metal structural frame of the building if the requirements of 250.68(C)(2) are met.

PART VI. MARKING

690.51 Modules and AC Modules. Modules and ac modules shall be marked in accordance with their listing.

690.53 DC PV Circuits. A permanent readily visible label indicating the highest maximum dc voltage in a PV system, calculated in accordance with 690.7, shall be provided by the installer at one of the following locations:

(1) DC PV system disconnecting means
(2) PV system electronic power conversion equipment
(3) Distribution equipment associated with the PV system

690.54 Interactive System Point of Interconnection. All interactive system(s) points of interconnection with other sources shall be marked at an accessible location at the disconnecting means as a power source and with the rated ac output current and the nominal operating ac voltage.

690.55 Photovoltaic Systems Connected to Energy Storage Systems. The PV system output circuit conductors shall be marked to indicate the polarity where connected to energy storage systems.

690.56 Identification of Power Sources.

(A) Facilities with Stand-Alone Systems. Plaques or directories shall be installed in accordance with 710.10.

(B) Facilities with Utility Services and Photovoltaic Systems. Plaques or directories shall be installed in accordance with 705.10 and 712.10, as required.

(C) Buildings with Rapid Shutdown. Buildings with PV systems shall have a permanent label located at each service equipment location to which the PV systems are connected or at an approved readily visible location and shall indicate the location of rapid shutdown initiation devices. The label shall include a simple diagram of a building with a roof and shall include the following words:

SOLAR PV SYSTEM IS EQUIPPED WITH RAPID SHUTDOWN.
TURN RAPID SHUTDOWN SWITCH TO THE "OFF" POSITION
TO SHUT DOWN PV SYSTEM AND REDUCE SHOCK HAZARD
IN ARRAY.

Interactive and Stand-Alone Photovoltaic (PV) Systems   **690.56**

SOLAR PV SYSTEM EQUIPPED WITH RAPID SHUTDOWN

TURN RAPID SHUTDOWN
SWITCH TO THE
"OFF" POSITION TO
SHUT DOWN PV SYSTEM
AND REDUCE
SHOCK HAZARD
IN THE ARRAY.

SOLAR ELECTRIC
PV PANELS

Figure Informational Note Figure 690.56(C) Label for Roof-Mounted PV Systems with Rapid Shutdown.

The title "SOLAR PV SYSTEM IS EQUIPPED WITH RAPID SHUTDOWN" shall utilize capitalized characters with a minimum height of 9.5 mm (⅜ in.) in black on yellow background, and the remaining characters shall be capitalized with a minimum height of 4.8 mm (³⁄₁₆ in.) in black on white background.

Informational Note: See Informational Note Figure 690.56(C).

(1) Buildings with More Than One Rapid Shutdown Type. For buildings that have PV systems with more than one rapid shutdown type or PV systems with no rapid shutdown, a detailed plan view diagram of the roof shall be provided showing each different PV system with a dotted line around areas that remain energized after rapid shutdown is initiated.

(2) Rapid Shutdown Switch. A rapid shutdown switch shall have a label that includes the following wording located on or no more than 1 m (3 ft) from the switch:

RAPID SHUTDOWN SWITCH FOR SOLAR PV SYSTEM

The label shall be reflective, with all letters capitalized and having a minimum height of 9.5 mm (⅜ in.) in white on red background.

PART VII. CONNECTION TO OTHER SOURCES

690.59 Connection to Other Sources. PV systems connected to other sources shall be installed in accordance with Parts I and II of Article 705 and Article 712.

PART VIII. ENERGY STORAGE SYSTEMS

690.71 General. An energy storage system connected to a PV system shall be installed in accordance with Article 706.

690.72 Self-Regulated PV Charge Control. The PV source circuit shall be considered to comply with the requirements of 706.33 if:

(1) The PV source circuit is matched to the voltage rating and charge current requirements of the interconnected battery cells and,
(2) The maximum charging current multiplied by 1 hour is less than 3 percent of the rated battery capacity expressed in ampere-hours or as recommended by the battery manufacturer

ARTICLE 705
Interconnected Electric Power Production Sources

PART I. GENERAL

705.2 Definitions. The definitions in this section shall apply within this article and throughout the *Code*.

Microgrid Interconnect Device (MID). A device that enables a microgrid system to separate from and reconnect to operate in parallel with a primary power source.

> Informational Note: Microgrid controllers typically are used to measure and evaluate electrical parameters and provide the logic for the signal to initiate and complete transition processes. IEEE Std 2030.7-2017, *IEEE Standard for the Specification of Microgrid Controllers*, and IEEE Std 2030.8-2018, *IEEE Standard for the Testing of Microgrid Controllers*, provide information on microgrid controllers. IEEE Std 1547-2018, *IEEE Standard*

Interactive and Stand-Alone Photovoltaic (PV) Systems 705.12

for Interconnection and Interoperability of Distributed Energy Resources with Associated Electric Power Systems Interfaces, provides information on interconnection requirements.

Microgrid System. A premises wiring system that has generation, energy storage, and load(s), or any combination thereof, that includes the ability to disconnect from and parallel with the primary source.

Informational Note: The application of Article 705 to microgrid systems is limited by the exclusions in 90.2(B)(5) related to electric utilities. Additional information may be found in IEEE 1547, IEEE 2030.7, and IEEE 2030.8.

Power Source Output Circuit. The conductors between power production equipment and the service or distribution equipment.

705.6 Equipment Approval. All equipment shall be approved for the intended use. Interactive equipment intended to operate in parallel with electric power production sources including, but not limited to, interactive inverters, engine generators, energy storage equipment, and wind turbines shall be listed for interactive function or be evaluated for interactive function and have a field label applied, or both.

705.10 Identification of Power Sources. A permanent plaque or directory shall be installed at each service equipment location, or at an approved readily visible location. The plaque or directory shall denote the location of each power source disconnecting means for the building or structure and be grouped with other plaques or directories for other on-site sources. The plaque or directory shall be marked with the wording "CAUTION: MULTIPLE SOURCES OF POWER." Any posted diagrams shall be correctly oriented with respect to the diagram's location. The marking shall comply with 110.21(B).

Exception: Installations with multiple co-located power production sources shall be permitted to be identified as a group(s). The plaque or directory shall not be required to identify each power source individually.

705.12 Load-Side Source Connections. The output of an interconnected electric power source shall be permitted to be connected to the load side of the service disconnecting means of the other source(s) at any distribution equipment on the premises. Where distribution equipment or feeders are fed simultaneously by a primary source of electricity and one or more other power source and are capable of supplying multiple branch circuits or feeders, or both, the interconnecting equipment shall comply with 705.12(A) through (E). Where a power control system (PCS) is installed in accordance with 705.13, the setting of the PCS controller shall be considered the power-source output circuit current in 705.12(A) through (E).

(A) Dedicated Overcurrent and Disconnect. Each source interconnection of one or more power sources installed in one system shall be made at a dedicated circuit breaker or fusible disconnecting means.

(B) Bus or Conductor Ampere Rating. The power source output circuit current multiplied by 125 percent shall be used in ampacity calculations for 705.12(B)(1) through (B)(3).

(1) Feeders. Where the power source output connection is made to a feeder, the feeder shall have an ampacity greater than or equal to 125 percent of the power-source output circuit current. Where the power-source output connection is made to a feeder at a location other than the opposite end of the feeder from the primary source overcurrent device, that portion of the feeder on the load side of the power source output connection shall be protected by one of the following:

(a) The feeder ampacity shall be not less than the sum of the primary source overcurrent device and 125 percent of the power-source output circuit current.

(b) An overcurrent device at the load side of the power source connection point shall be rated not greater than the ampacity of the feeder.

(2) Taps. Where power source output connections are made at feeders, all taps shall be sized based on the sum of 125 percent of all power source(s) output circuit current(s) and the rating of the overcurrent device protecting the feeder conductors for sizing tap conductors using the calculations in 240.21(B).

(3) Busbars. One of the following methods shall be used to determine the ratings of busbars:

(1) The sum of 125 percent of the power source(s) output circuit current and the rating of the overcurrent device protecting the busbar shall not exceed the ampacity of the busbar.

   Informational Note: This general rule assumes no limitation in the number of the loads or sources applied to busbars or their locations.

(2) Where two sources, one a primary power source and the other another power source, are located at opposite ends of a busbar that contains loads, the sum of 125 percent of the power-source(s) output

circuit current and the rating of the overcurrent device protecting the busbar shall not exceed 120 percent of the ampacity of the busbar. The busbar shall be sized for the loads connected in accordance with Article 220. A permanent warning label shall be applied to the distribution equipment adjacent to the back-fed breaker from the power source that displays the following or equivalent wording:

> WARNING: POWER SOURCE OUTPUT CONNECTION—
> DO NOT RELOCATE THIS OVERCURRENT DEVICE.

The warning sign(s) or label(s) shall comply with 110.21(B).

(3) The sum of the ampere ratings of all overcurrent devices on panelboards, both load and supply devices, excluding the rating of the overcurrent device protecting the busbar, shall not exceed the ampacity of the busbar. The rating of the overcurrent device protecting the busbar shall not exceed the rating of the busbar. Permanent warning labels shall be applied to distribution equipment displaying the following or equivalent wording:

> WARNING: THIS EQUIPMENT FED BY MULTIPLE SOURCES.
> TOTAL RATING OF ALL OVERCURRENT DEVICES
> EXCLUDING MAIN SUPPLY OVERCURRENT DEVICE
> SHALL NOT EXCEED AMPACITY OF BUSBAR.

The warning sign(s) or label(s) shall comply with 110.21(B).

(4) A connection at either end of a center-fed panelboard in dwellings shall be permitted where the sum of 125 percent of the power-source(s) output circuit current and the rating of the overcurrent device protecting the busbar does not exceed 120 percent of the current rating of the busbar.

(5) Connections shall be permitted on switchgear, switchboards, and panelboards in configurations other than those permitted in 705.12(B)(3)(1) through (B)(3)(4) where designed under engineering supervision that includes available fault-current and busbar load calculations.

(6) Connections shall be permitted on busbars of panelboards that supply lugs connected to feed-through conductors. The feed-through conductors shall be sized in accordance with 705.12(B)(1). Where an overcurrent device is installed at the supply end of the feed-through conductors, the busbar in the supplying panelboard shall be permitted to be sized in accordance with 705.12(B)(3)(1) through 705.12(B)(3)(3).

(C) Marking. Equipment containing overcurrent devices in circuits supplying power to a busbar or conductor supplied from multiple sources shall be marked to indicate the presence of all sources.

(D) Suitable for Backfeed. Fused disconnects, unless otherwise marked, shall be considered suitable for backfeed. Circuit breakers not marked "line" and "load" shall be considered suitable for backfeed. Circuit breakers marked "line" and "load" shall be considered suitable for backfeed or reverse current if specifically rated.

(E) Fastening. Listed plug-in-type circuit breakers back-fed from electric power sources that are listed and identified as interactive shall be permitted to omit the additional fastener normally required by 408.36(D) for such applications.

705.13 Power Control Systems. A power control system (PCS) shall be listed and evaluated to control the output of one or more power production sources, energy storage systems (ESS), and other equipment. The PCS shall limit current and loading on the busbars and conductors supplied by the PCS.

For the circuits connected to a PCS, the PCS shall limit the current to the ampacity of the conductors or the ratings of the busbars to which it is connected in accordance with 705.13(A) through (E).

(A) Monitoring. The PCS controller shall monitor all currents within the PCS. Any busbar or conductor on the load side of the service disconnecting means that is not monitored by the PCS shall comply with 705.12. Where the PCS is connected in accordance with 705.11, the PCS shall monitor the service conductors and prevent overload of these conductors.

(B) Settings. The sum of all PCS-controlled currents plus all monitored currents from other sources of supply shall not exceed the ampacity of any busbar or conductor supplied by the power production sources. Where the PCS is connected to an overcurrent device protecting any busbar or conductor not monitored by the PCS, the setting of the PCS controller shall be set within the ratings of that overcurrent device.

(C) Overcurrent Protection. The PCS shall provide overcurrent protection either by overcurrent devices or by the PCS including the functionality as an overcurrent device in the product listing.

> Informational Note: Some PCS are listed to provide overcurrent protection.

Interactive and Stand-Alone Photovoltaic (PV) Systems 705.20

(D) Single Power Source Rating. The rating of the overcurrent device for any single power source controlled by the PCS shall not exceed the rating of the busbar or the ampacity of the conductors to which it is connected.

(E) Access to Settings. The access to settings of the PCS shall be restricted to qualified personnel in accordance with the requirements of 240.6(C).

705.16 Interrupting and Short-Circuit Current Rating. Consideration shall be given to the contribution of fault currents from all interconnected power sources for the interrupting and short-circuit current ratings of equipment on interactive systems.

705.20 Disconnecting Means, Source. Means shall be provided to disconnect power source output circuit conductors of electric power production equipment from conductors of other systems.

The disconnecting means shall comply with the following:

(1) Be one of the following types:

   (a) A manually operable switch or circuit breaker
   (b) A load-break-rated pull-out switch
   (c) A power-operated or remote-controlled switch or circuit breaker that is manually operable locally and opens automatically when control power is interrupted
   (d) A device listed or approved for the intended application

(2) Simultaneously disconnect all ungrounded conductors of the circuit
(3) Located where readily accessible
(4) Externally operable without exposed live parts
(5) Enclosures with doors or hinged covers with exposed live parts when open that require a tool to open or are lockable where readily accessible to unqualified persons
(6) Plainly indicate whether in the open (off) or closed (on) position
(7) Have ratings sufficient for the maximum circuit current, available fault current, and voltage that is available at the terminals
(8) Be marked in accordance with the warning in 690.13(B), where the line and load terminals are capable of being energized in the open position

   Informational Note: With interconnected power sources, some equipment, including switches and fuses, is likely to be energized from both directions. See 240.40.

705.25 Wiring Methods.

(A) General. All raceway and cable wiring methods included in Chapter 3 of this *Code* and other wiring systems and fittings specifically listed, intended, and identified for use with power production systems and equipment shall be permitted. Where wiring devices with integral enclosures are used, sufficient length of cable shall be provided to facilitate replacement.

(B) Flexible Cords and Cables. Flexible cords and cables, where used to connect the moving parts of a power production system or where used for ready removal for maintenance and repair, shall comply with Article 400 and shall be listed and identified as DG Cable, Distributed Generation Cable, hard service cord, or portable power cable, shall be suitable for extra-hard usage, shall be listed for outdoor use, and shall be water resistant. Cables exposed to sunlight shall be sunlight resistant. Flexible, fine-stranded cables shall be terminated only with terminals, lugs, devices, or connectors in accordance with 110.14(A).

(C) Multiconductor Cable Assemblies. Multiconductor cable assemblies used in accordance with their listings shall be permitted.

Informational Note: See UL 3003, *Distributed Generation Cables*, for additional information on DG Cable, Distributed Generation Cable. An ac module harness is one example of a multiconductor cable assembly.

705.28 Circuit Sizing and Current.

(A) Calculation of Maximum Circuit Current. Where not elsewhere required or permitted in this *Code*, the maximum current for the circuit shall be the continuous output current rating of the power production equipment.

(B) Conductor Ampacity. Where not elsewhere required or permitted in this *Code*, the circuit conductors shall be sized to carry not less than the largest of the following:

(1) The maximum currents in 705.28(A) multiplied by 125 percent without adjustment or correction factors
(2) The maximum currents in 705.28(A) with adjustment and correction factors
(3) Where connected to feeders, if smaller than the feeder conductors, the ampacity as calculated in 240.21(B) based on the over-current device protecting the feeder

Interactive and Stand-Alone Photovoltaic (PV) Systems 705.32

(C) Neutral Conductors. Neutral conductors shall be permitted to be sized in accordance with either 705.28(C)(1) or (C)(2).

(1) Single-Phase Line-to-Neutral Power Sources. Where not elsewhere required or permitted in this *Code*, the ampacity of a neutral conductor to which a single-phase line-to-neutral power source is connected shall not be smaller than the ampacity in 705.28(B).

(2) Neutral Conductor Used Solely for Instrumentation, Voltage, Detection, or Phase Detection. A power production equipment neutral conductor used solely for instrumentation, voltage detection, or phase detection shall be permitted to be sized in accordance with 250.102.

705.30 Overcurrent Protection.

(A) Circuit and Equipment. Power source output circuit conductors and equipment shall be provided with overcurrent protection. Circuits connected to more than one electrical source shall have overcurrent devices located to provide overcurrent protection from all sources.

(B) Overcurrent Device Ratings. The overcurrent devices in other than generator systems shall be sized to carry not less than 125 percent of the maximum currents as calculated in 705.28(A). The rating or setting of overcurrent devices shall be permitted in accordance with 240.4(B) and (C).

Exception: Circuits containing an assembly together with its overcurrent device(s) that is listed for continuous operation at 100 percent of its rating shall be permitted to be utilized at 100 percent of its rating.

(C) Power Transformers. Transformers with sources on each side shall be provided with overcurrent protection in accordance with 450.3. The primary shall be the side connected to the largest source of available fault current. Secondary protection shall not be required for a transformer secondary that has a current rating not less than the sum of the rated continuous output currents of the power sources connected to that secondary.

(D) Generators. Generators shall be provided with overcurrent protection in accordance with 445.12.

705.32 Ground-Fault Protection.
Where protection is installed in accordance with 230.95, the output of an interactive system shall be connected to the supply side of the ground-fault protection.

Exception: Connection shall be permitted to be made to the load side of ground-fault protection, if there is ground-fault protection for equipment from all ground-fault current sources.

705.40 Loss of Primary Source. The output of electric power production equipment shall be automatically disconnected from all ungrounded conductors of the interconnected systems when one or more of the phases to which it is connected opens. The electric power production equipment shall not be reconnected until all the phases of the interconnected system to which it is connected are restored. This requirement shall not be applicable to electric power production equipment providing power to an emergency or legally required standby system.

Exception: A listed interactive inverter shall trip or shall be permitted to automatically cease exporting power when one or more of the phases of the interconnected system opens and shall not be required to automatically disconnect all ungrounded conductors from the primary source. A listed interactive inverter shall be permitted to automatically or manually resume exporting power to the interconnected system once all phases of the source to which it is connected are restored.

Informational Note No. 1: Risks to personnel and equipment associated with the primary source could occur if an interactive electric power production source can operate as an intentional island. Special detection methods are required to determine that a primary source supply system outage has occurred and whether there should be automatic disconnection. When the primary source supply system is restored, special detection methods are typically required to limit exposure of power production sources to out-of-phase reconnection.

Informational Note No. 2: Induction-generating equipment connected on systems with significant capacitance can become self-excited upon loss of the primary source and experience severe overvoltage as a result.

Interactive power production equipment shall be permitted to operate in island mode to supply loads that have been disconnected from the electric power production and distribution network.

705.45 Unbalanced Interconnections.

(A) Single Phase. Single-phase power sources in interactive systems shall be connected to 3-phase power systems in order to limit unbalanced voltages at the point of interconnection to not more than 3 percent.

Informational Note: For interactive power sources, unbalanced voltages can be minimized by the same methods that are used for single-phase loads on a 3-phase power system. See ANSI/C84.1-2016, *Electric Power Systems and Equipment — Voltage Ratings (60 Hertz)*.

(B) Three Phase. Three-phase power sources in interactive systems shall have all phases automatically de-energized upon loss of, or unbalanced, voltage in one or more phases unless the interconnected system is designed so that significant unbalanced voltages will not result.

PART II. MICROGRID SYSTEMS

705.50 System Operation. Microgrid systems shall be permitted to disconnect from the primary source of power or other interconnected electric power production sources and operate as an isolated microgrid system operating in island mode.

705.60 Primary Power Source Connection. Connections to primary power sources that are external to the microgrid system shall comply with the requirements of 705.11, 705.12, or 705.13. Power source conductors connecting to a microgrid system, including conductors supplying distribution equipment, shall be considered as power source output conductors.

705.65 Reconnection to Primary Power Source. Microgrid systems that reconnect to primary power sources shall be provided with the necessary equipment to establish a synchronous transition.

705.70 Microgrid Interconnect Devices (MID). Microgrid interconnect devices shall comply with the following:

(1) Be required for any connection between a microgrid system and a primary power source
(2) Be evaluated for the application and have a field label applied or be listed for the application
(3) Have sufficient number of overcurrent devices located to provide overcurrent protection from all sources

Informational Note: MID functionality is often incorporated in an interactive or multimode inverter, energy storage system, or similar device identified for interactive operation.

CHAPTER 27
OPTIONAL STANDBY SYSTEMS

INTRODUCTION

This chapter contains requirements from Article 702 — Optional Standby Systems. The requirements of Article 702 cover the installation and operation of optional standby systems that are used to provide power for occupant comfort, business continuity, or both and are installed at the owner's discretion rather than by a codified or AHJ mandate to provide for occupant or emergency responder safety, which is the function of the systems covered in Articles 700 — Emergency Systems, and 701 — Legally Required Standby Systems. The requirements of Article 702 cover the capacity and rating of the system and alternate source, transfer equipment, conductor installation and routing, alternate power source connections to the premises wiring system, and disconnecting means for standby generators located outdoors. Generators that supply optional standby loads can also be used to supply emergency loads if the generator either has capacity to fully supply all loads connected to it, or equipment is installed to shed loads so that the emergency loads are given priority operational status. Additionally, a generator that is used to supply emergency loads must be able to provide the necessary level and quality of power to the emergency loads within 10 seconds of loss of normal (typically utility) power. This requirement does not apply to generators that are used to supply only optional standby loads.

Article 702 requirements are organized into two parts: Part I — General, and Part II — Wiring. In contrast to the requirements in Article 700, it is not necessary to provide operational time, system segregation, and wiring resiliency requirements for optional standby systems. This is because the impact of an interruption of its operation may be inconvenient, but an interruption does not place the occupant into the same level of peril and potential for mass panic that, for instance, could occur at a large assembly occupancy or high-rise building.

ARTICLE 702

Optional Standby Systems

PART I. GENERAL

702.1 Scope. This article applies to the installation and operation of optional standby systems.

The systems covered by this article consist of those that are permanently installed in their entirety, including prime movers, and those that are arranged for a connection to a premises wiring system from a portable alternate power supply.

702.2 Definition. The definition in this section shall apply within this article and throughout the *Code*.

Optional Standby Systems. Those systems intended to supply power to public or private facilities or property where life safety does not depend on the performance of the system. These systems are intended to supply on-site generated or stored power to selected loads either automatically or manually.

> Informational Note: Optional standby systems are typically installed to provide an alternate source of electric power for such facilities as industrial and commercial buildings, farms, and residences and to serve loads such as heating and refrigeration systems, data processing and communications systems, and industrial processes that, when stopped during any power outage, could cause discomfort, serious interruption of the process, damage to the product or process, or the like.

702.4 Capacity and Rating.

(A) Available Fault Current. Optional standby system equipment shall be suitable for the available fault current at its terminals.

(B) System Capacity.

(1) Manual Transfer Equipment. Where manual transfer equipment is used, an optional standby system shall have adequate capacity and rating for the supply of all equipment intended to be operated at one time. The user of the optional standby system shall be permitted to select the load connected to the system.

(2) Automatic Transfer Equipment. Where automatic transfer equipment is used, an optional standby system shall comply with 702.4(B)(2)(a)

Optional Standby Systems 702.5

or (B)(2)(b) in accordance with Article 220 or by another approved method.

(a) *Full Load.* The standby source shall be capable of supplying the full load that is transferred by the automatic transfer equipment.

(b) *Load Management.* Where a system is employed that will automatically manage the connected load, the standby source shall have a capacity sufficient to supply the maximum load that will be connected by the load management system.

702.5 Transfer Equipment.

(A) General. Transfer equipment shall be required for all standby systems subject to the requirements of this article and for which an electric utility supply is either the normal or standby source. Transfer switches shall not be permitted to be reconditioned.

Exception: Temporary connection of a portable generator without transfer equipment shall be permitted where conditions of maintenance and supervision ensure that only qualified persons service the installation and where the normal supply is physically isolated by a lockable disconnecting means or by disconnection of the normal supply conductors.

(B) Meter-Mounted Transfer Switches. Transfer switches installed between the utility meter and the meter enclosure shall be listed meter-mounted transfer switches and shall be approved. Meter-mounted transfer switches shall be of the manual type unless rated as determined by 702.4(B)(2).

Informational Note: For more information, see UL 1008M, *Transfer Switch Equipment, Meter Mounted.*

(C) Documentation. In other than dwelling units, the short-circuit current rating of the transfer equipment, based on the specific overcurrent protective device type and settings protecting the transfer equipment, shall be field marked on the exterior of the transfer equipment.

(D) Inadvertent Interconnection. Transfer equipment shall be suitable for the intended use and shall be listed, designed and installed so as to prevent the inadvertent interconnection of all sources of supply in any operation of the transfer equipment.

(E) Parallel Installation. Transfer equipment and electric power production systems installed to permit operation in parallel with the normal source shall also meet the requirements of Article 705.

702.6 Signals. Audible and visual signal devices shall be provided, where practicable, for the following purposes specified in 702.6(A) and (B).

(A) Malfunction. To indicate malfunction of the optional standby source.

(B) Carrying Load. To indicate that the optional standby source is carrying load.

Exception: Signals shall not be required for portable standby power sources.

702.7 Signs.

(A) Standby. A sign shall be placed at the service-entrance equipment for commercial and industrial installations that indicates the type and location of each on-site optional standby power source. For one- and two-family dwelling units, a sign shall be placed at the disconnecting means required in 230.85 that indicates the location of each permanently installed on-site optional standby power source disconnect or means to shut down the prime mover as required in 445.18(D).

(B) Grounding. Where removal of a grounding or bonding connection in normal power source equipment interrupts the grounding electrode conductor connection to the alternate power source(s) grounded conductor, a warning sign shall be installed at the normal power source equipment stating:

> WARNING: SHOCK HAZARD EXISTS IF GROUNDING
> ELECTRODE CONDUCTOR OR BONDING JUMPER
> CONNECTION IN THIS EQUIPMENT IS REMOVED
> WHILE ALTERNATE SOURCE(S) IS ENERGIZED.

The warning sign(s) or label(s) shall comply with 110.21(B).

(C) Power Inlet. Where a power inlet is used for a temporary connection to a portable generator, a warning sign shall be placed near the inlet to indicate the type of derived system that the system is capable of based on the wiring of the transfer equipment. The sign shall display one of the following warnings:

> WARNING: FOR CONNECTION OF A SEPARATELY
> DERIVED (BONDED NEUTRAL) SYSTEM ONLY
> or
> WARNING: FOR CONNECTION OF A NONSEPARATELY
> DERIVED (FLOATING NEUTRAL) SYSTEM ONLY

PART II. WIRING

702.10 Wiring Optional Standby Systems. The optional standby system wiring shall be permitted to occupy the same raceways, cables, boxes, and cabinets with other general wiring.

702.11 Portable Generator Grounding.

(A) Separately Derived System. Where a portable optional standby source is used as a separately derived system, it shall be grounded to a grounding electrode in accordance with 250.30.

(B) Nonseparately Derived System. Where a portable optional standby source is used as a nonseparately derived system, the equipment grounding conductor shall be bonded to the system grounding electrode.

702.12 Outdoor Generator Sets.

(A) Portable Generators Greater Than 15 kW and Permanently Installed Generators. Where an outdoor housed generator set is equipped with a readily accessible disconnecting means in accordance with 445.18, and the disconnecting means is located within sight of the building or structure supplied, an additional disconnecting means shall not be required where ungrounded conductors serve or pass through the building or structure. Where the generator supply conductors terminate at a disconnecting means in or on a building or structure, the disconnecting means shall meet the requirements of 225.36.

(B) Portable Generators 15 kW or Less. Where a portable generator, rated 15 kW or less, is installed using a flanged inlet or other cord- and plug-type connection, a disconnecting means shall not be required where ungrounded conductors serve or pass through a building or structure.

CHAPTER 28
CONDUIT AND TUBING FILL TABLES

INTRODUCTION

This chapter contains tables and notes from Chapter 9 — Tables, and Annex C — Conduit and Tubing Fill Tables for Conductors and Fixture Wires of the Same Size. These tables are used to determine the maximum number of conductors permitted in raceways (conduit and tubing). Tables 4 and 5 of Chapter 9 are used to determine the minimum size cylindrical raceway based on the same size conductors being installed or on combinations of different size conductors being installed. Annex C is permitted to be used when all the conductors are the same size. Table 8 — Conductor Properties (in Chapter 9) provides useful information such as size (AWG or kcmil), circular mils, overall area, and the resistance of copper and aluminum conductors.

CHAPTER 9 TABLES

Table 1 Percent of Cross Section of Conduit and Tubing for Conductors and Cables

Number of Conductors and/or Cables	Cross-Sectional Area (%)
1	53
2	31
Over 2	40

Informational Note No. 1: Table 1 is based on common conditions of proper cabling and alignment of conductors where the length of the pull and the number of bends are within reasonable limits. It should be recognized that, for certain conditions, a larger size conduit or a lesser conduit fill should be considered.

Informational Note No. 2: When pulling three conductors or cables into a raceway, if the ratio of the raceway (inside diameter) to the conductor or cable (outside diameter) is between 2.8 and 3.2, jamming can occur. While jamming can occur when pulling four or more conductors or cables into a raceway, the probability is very low.

Notes to Tables
(1) See Informative Annex C for the maximum number of conductors and fixture wires, all of the same size (total cross-sectional area including insulation) permitted in trade sizes of the applicable conduit or tubing.
(2) Table 1 applies only to complete conduit or tubing systems and is not intended to apply to sections of conduit or tubing used to protect exposed wiring and cable from physical damage.
(3) Equipment grounding or bonding conductors, where installed, shall be included when calculating conduit or tubing fill. The actual dimensions of the equipment grounding or bonding conductor (insulated or bare) shall be used in the calculation.
(4) Where conduit or tubing nipples having a maximum length not to exceed 600 mm (24 in.) are installed between boxes, cabinets, and similar enclosures, the nipples shall be permitted to be filled to 60 percent of their total cross-sectional area, and 310.15(C)(1) adjustment factors need not apply to this condition.

Conduit and Tubing Fill Tables

(5) For conductors not included in Chapter 9, such as multiconductor cables and optical fiber cables, the actual dimensions shall be used.

(6) For combinations of conductors of different sizes, use actual dimensions or Table 5 and Table 5A for dimensions of conductors and Table 4 for the applicable conduit or tubing dimensions.

(7) When calculating the maximum number of conductors or cables permitted in a conduit or tubing, all of the same size (total cross-sectional area including insulation), the next higher whole number shall be used to determine the maximum number of conductors permitted when the calculation results in a decimal greater than or equal to 0.8. When calculating the size for conduit or tubing permitted for a single conductor, one conductor shall be permitted when the calculation results in a decimal greater than or equal to 0.8.

(8) Where bare conductors are permitted by other sections of this *Code*, the dimensions for bare conductors in Table 8 shall be permitted.

(9) A multiconductor cable, optical fiber cable, or flexible cord of two or more conductors shall be treated as a single conductor for calculating percentage conduit or tubing fill area. For cables that have elliptical cross sections, the cross-sectional area calculation shall be based on using the major diameter of the ellipse as a circle diameter. Assemblies of single insulated conductors without an overall covering shall not be considered a cable when determining conduit or tubing fill area. The conduit or tubing fill for the assemblies shall be calculated based upon the individual conductors.

(10) The values for approximate conductor diameter and area shown in Table 5 are based on worst-case scenario and indicate round concentric-lay-stranded conductors. Solid and round concentric-lay-stranded conductor values are grouped together for the purpose of Table 5. Round compact-stranded conductor values are shown in Table 5A. If the actual values of the conductor diameter and area are known, they shall be permitted to be used.

Table 2 Radius of Conduit and Tubing Bends

Conduit or Tubing Size		One Shot and Full Shoe Benders		Other Bends	
Metric Designator	Trade Size	mm	in.	mm	in.
16	½	101.6	4	101.6	4
21	¾	114.3	4½	127	5
27	1	146.05	5¾	152.4	6
35	1¼	184.15	7¼	203.2	8
41	1½	209.55	8¼	254	10
53	2	241.3	9½	304.8	12
63	2½	266.7	10½	381	15
78	3	330.2	13	457.2	18
91	3½	381	15	533.4	21
103	4	406.4	16	609.6	24
129	5	609.6	24	762	30
155	6	762	30	914.4	36

Conduit and Tubing Fill Tables

Table 4 Dimensions and Percent Area of Conduit and Tubing (Areas of Conduit or Tubing for the Combinations of Wires Permitted in Table 1, Chapter 9)

Metric Designator	Trade Size	Over 2 Wires 40%		60%		1 Wire 53%		2 Wires 31%		Nominal Internal Diameter		Total Area 100%	
		mm²	in.²	mm²	in.²	mm²	in.²	mm²	in.²	mm	in.	mm²	in.²
				Article 358 — Electrical Metallic Tubing (EMT)									
16	½	78	0.122	118	0.182	104	0.161	61	0.094	15.8	0.622	196	0.304
21	¾	137	0.213	206	0.320	182	0.283	106	0.165	20.9	0.824	343	0.533
27	1	222	0.346	333	0.519	295	0.458	172	0.268	26.6	1.049	556	0.864
35	1¼	387	0.598	581	0.897	513	0.793	300	0.464	35.1	1.380	968	1.496
41	1½	526	0.814	788	1.221	696	1.079	407	0.631	40.9	1.610	1314	2.036
53	2	866	1.342	1299	2.013	1147	1.778	671	1.040	52.5	2.067	2165	3.356
63	2½	1513	2.343	2270	3.515	2005	3.105	1173	1.816	69.4	2.731	3783	5.858
78	3	2280	3.538	3421	5.307	3022	4.688	1767	2.742	85.2	3.356	5701	8.846
91	3½	2980	4.618	4471	6.927	3949	6.119	2310	3.579	97.4	3.834	7451	11.545
103	4	3808	5.901	5712	8.852	5046	7.819	2951	4.573	110.1	4.334	9521	14.753

(continued)

Table 4 Dimensions and Percent Area of Conduit and Tubing (Areas of Conduit or Tubing for the Combinations of Wires Permitted in Table 1, Chapter 9) — *Continued*

Metric Designator	Trade Size	Over 2 Wires 40%		60%		1 Wire 53%		2 Wires 31%		Nominal Internal Diameter		Total Area 100%	
		mm²	in.²	mm²	in.²	mm²	in.²	mm²	in.²	mm	in.	mm²	in.²
Article 362 — Electrical Nonmetallic Tubing (ENT)													
16	½	73	0.114	110	0.171	97	0.151	57	0.088	15.3	0.602	184	0.285
21	¾	131	0.203	197	0.305	174	0.269	102	0.157	20.4	0.804	328	0.508
27	1	215	0.333	322	0.499	284	0.441	166	0.258	26.1	1.029	537	0.832
35	1¼	375	0.581	562	0.872	497	0.770	291	0.450	34.5	1.36	937	1.453
41	1½	512	0.794	769	1.191	679	1.052	397	0.616	40.4	1.59	1281	1.986
53	2	849	1.316	1274	1.975	1125	1.744	658	1.020	52	2.047	2123	3.291
63	2½	—	—	—	—	—	—	—	—	—	—	—	—
78	3	—	—	—	—	—	—	—	—	—	—	—	—
91	3½	—	—	—	—	—	—	—	—	—	—	—	—
Article 348 — Flexible Metal Conduit (FMC)													
12	⅜	30	0.046	44	0.069	39	0.061	23	0.036	9.7	0.384	74	0.116
16	½	81	0.127	122	0.190	108	0.168	63	0.098	16.1	0.635	204	0.317
21	¾	137	0.213	206	0.320	182	0.283	106	0.165	20.9	0.824	343	0.533

Conduit and Tubing Fill Tables

Metric Designator	Trade Size	Over 2 Wires 40% mm²	Over 2 Wires 40% in.²	60% mm²	60% in.²	1 Wire 53% mm²	1 Wire 53% in.²	2 Wires 31% mm²	2 Wires 31% in.²	Nominal Internal Diameter mm	Nominal Internal Diameter in.	Total Area 100% mm²	Total Area 100% in.²
\multicolumn{13}{c}{Article 348 — Flexible Metal Conduit (FMC)}
27	1	211	0.327	316	0.490	279	0.433	163	0.253	25.9	1.020	527	0.817
35	1¼	330	0.511	495	0.766	437	0.677	256	0.396	32.4	1.275	824	1.277
41	1½	480	0.743	720	1.115	636	0.985	372	0.576	39.1	1.538	1201	1.858
53	2	843	1.307	1264	1.961	1117	1.732	653	1.013	51.8	2.040	2107	3.269
63	2½	1267	1.963	1900	2.945	1678	2.602	982	1.522	63.5	2.500	3167	4.909
78	3	1824	2.827	2736	4.241	2417	3.746	1414	2.191	76.2	3.000	4560	7.069
91	3½	2483	3.848	3724	5.773	3290	5.099	1924	2.983	88.9	3.500	6207	9.621
103	4	3243	5.027	4864	7.540	4297	6.660	2513	3.896	101.6	4.000	8107	12.566
\multicolumn{13}{c}{Article 342 — Intermediate Metal Conduit (IMC)}
12	⅜	—	—	—	—	—	—	—	—	—	—	—	—
16	½	89	0.137	133	0.205	117	0.181	69	0.106	16.8	0.660	222	0.342
21	¾	151	0.235	226	0.352	200	0.311	117	0.182	21.9	0.864	377	0.586
27	1	248	0.384	372	0.575	329	0.508	192	0.297	28.1	1.105	620	0.959
35	1¼	425	0.659	638	0.988	564	0.873	330	0.510	36.8	1.448	1064	1.647
41	1½	573	0.890	859	1.335	759	1.179	444	0.690	42.7	1.683	1432	2.225
53	2	937	1.452	1405	2.178	1241	1.924	726	1.125	54.6	2.150	2341	3.630
63	2½	1323	2.054	1985	3.081	1753	2.722	1026	1.592	64.9	2.557	3308	5.135
78	3	2046	3.169	3069	4.753	2711	4.199	1586	2.456	80.7	3.176	5115	7.922
91	3½	2729	4.234	4093	6.351	3616	5.610	2115	3.281	93.2	3.671	6822	10.584
103	4	3490	5.452	5235	8.179	4624	7.224	2705	4.226	105.4	4.166	8725	13.631

(continued)

Table 4 Dimensions and Percent Area of Conduit and Tubing (Areas of Conduit or Tubing for the Combinations of Wires Permitted in Table 1, Chapter 9) — *Continued*

Metric Designator	Trade Size	Over 2 Wires 40%		60%		1 Wire 53%		2 Wires 31%		Nominal Internal Diameter		Total Area 100%	
		mm²	in.²	mm²	in.²	mm²	in.²	mm²	in.²	mm	in.	mm²	in.²
		\multicolumn{13}{c}{Article 356 — Liquidtight Flexible Nonmetallic Conduit (LFNC-A*)}
12	⅜	50	0.077	75	0.115	66	0.102	39	0.060	12.6	0.495	125	0.192
16	½	80	0.125	121	0.187	107	0.165	62	0.097	16.0	0.630	201	0.312
21	¾	139	0.214	208	0.321	184	0.283	107	0.166	21.0	0.825	346	0.535
27	1	221	0.342	331	0.513	292	0.453	171	0.265	26.5	1.043	552	0.854
35	1¼	387	0.601	581	0.901	513	0.796	300	0.466	35.1	1.383	968	1.502
41	1½	520	0.807	781	1.211	690	1.070	403	0.626	40.7	1.603	1301	2.018
53	2	863	1.337	1294	2.006	1143	1.772	669	1.036	52.4	2.063	2157	3.343

*Corresponds to 356.2(1).

Metric Designator	Trade Size	Over 2 Wires 40%		60%		1 Wire 53%		2 Wires 31%		Nominal Internal Diameter		Total Area 100%	
		mm²	in.²	mm²	in.²	mm²	in.²	mm²	in.²	mm	in.	mm²	in.²
		\multicolumn{13}{c}{Article 356 — Liquidtight Flexible Nonmetallic Conduit (LFNC-B*)}
12	⅜	49	0.077	74	0.115	65	0.102	38	0.059	12.5	0.494	123	0.192
16	½	81	0.125	122	0.188	108	0.166	63	0.097	16.1	0.632	204	0.314
21	¾	140	0.216	210	0.325	185	0.287	108	0.168	21.1	0.830	350	0.541

Conduit and Tubing Fill Tables

Metric Designator	Trade Size	Over 2 Wires 40%		60%		1 Wire 53%		2 Wires 31%		Nominal Internal Diameter		Total Area 100%	
		mm²	in.²	mm²	in.²	mm²	in.²	mm²	in.²	mm	in.	mm²	in.²
		Article 356 — Liquidtight Flexible Nonmetallic Conduit (LFNC-B*)											
27	1	226	0.349	338	0.524	299	0.462	175	0.270	26.8	1.054	564	0.873
35	1¼	394	0.611	591	0.917	522	0.810	305	0.474	35.4	1.395	984	1.528
41	1½	510	0.792	765	1.188	676	1.050	395	0.614	40.3	1.588	1276	1.981
53	2	836	1.298	1255	1.948	1108	1.720	648	1.006	51.6	2.033	2091	3.246

*Corresponds to 356.2(2).

		Article 356 — Liquidtight Flexible Nonmetallic Conduit (LFNC-C*)											
12	⅜	47.7	0.074	71.5	0.111	63.2	0.098	36.9	0.057	12.3	0.485	119.19	0.185
16	½	77.9	0.121	116.9	0.181	103.2	0.160	60.4	0.094	15.7	0.620	194.778	0.302
21	¾	134.6	0.209	201.9	0.313	178.4	0.276	104.3	0.162	20.7	0.815	336.568	0.522
27	1	215.0	0.333	322.5	0.500	284.9	0.442	166.6	0.258	26.2	1.030	537.566	0.833
35	1¼	380.4	0.590	570.6	0.884	504.1	0.781	294.8	0.457	34.8	1.370	951.039	1.474
41	1½	509.2	0.789	763.8	1.184	674.7	1.046	394.6	0.612	40.3	1.585	1272.963	1.973
53	2	847.6	1.314	1271.4	1.971	1123.1	1.741	656.9	1.018	51.9	2.045	2119.063	3.285

*Corresponds to 356.2(3).

(continued)

Table 4 Dimensions and Percent Area of Conduit and Tubing (Areas of Conduit or Tubing for the Combinations of Wires Permitted in Table 1, Chapter 9) — *Continued*

Metric Designator	Trade Size	Over 2 Wires 40%		60%		1 Wire 53%		2 Wires 31%		Nominal Internal Diameter		Total Area 100%	
		mm²	in.²	mm²	in.²	mm²	in.²	mm²	in.²	mm	in.	mm²	in.²
Article 350 — Liquidtight Flexible Metal Conduit (LFMC)													
12	⅜	49	0.077	74	0.115	65	0.102	38	0.059	12.5	0.494	123	0.192
16	½	81	0.125	122	0.188	108	0.166	63	0.097	16.1	0.632	204	0.314
21	¾	140	0.216	210	0.325	185	0.287	108	0.168	21.1	0.830	350	0.541
27	1	226	0.349	338	0.524	299	0.462	175	0.270	26.8	1.054	564	0.873
35	1¼	394	0.611	591	0.917	522	0.810	305	0.474	35.4	1.395	984	1.528
41	1½	510	0.792	765	1.188	676	1.050	395	0.614	40.3	1.588	1276	1.981
53	2	836	1.298	1255	1.948	1108	1.720	648	1.006	51.6	2.033	2091	3.246
63	2½	1259	1.953	1888	2.929	1668	2.587	976	1.513	63.3	2.493	3147	4.881
78	3	1931	2.990	2896	4.485	2559	3.962	1497	2.317	78.4	3.085	4827	7.475
91	3½	2511	3.893	3766	5.839	3327	5.158	1946	3.017	89.4	3.520	6277	9.731
103	4	3275	5.077	4912	7.615	4339	6.727	2538	3.935	102.1	4.020	8187	12.692
129	5	—	—	—	—	—	—	—	—	—	—	—	—
155	6	—	—	—	—	—	—	—	—	—	—	—	—

Conduit and Tubing Fill Tables

Metric Designator	Trade Size	Over 2 Wires 40%		60%		1 Wire 53%		2 Wires 31%		Nominal Internal Diameter		Total Area 100%	
		mm²	in.²	mm²	in.²	mm²	in.²	mm²	in.²	mm	in.	mm²	in.²
		\multicolumn{12}{c}{Article 344 — Rigid Metal Conduit (RMC)}
12	⅜	—	—	—	—	—	—	—	—	—	—	—	—
16	½	81	0.125	122	0.188	108	0.166	63	0.097	16.1	0.632	204	0.314
21	¾	141	0.220	212	0.329	187	0.291	109	0.170	21.2	0.836	353	0.549
27	1	229	0.355	344	0.532	303	0.470	177	0.275	27.0	1.063	573	0.887
35	1¼	394	0.610	591	0.916	522	0.809	305	0.473	35.4	1.394	984	1.526
41	1½	533	0.829	800	1.243	707	1.098	413	0.642	41.2	1.624	1333	2.071
53	2	879	1.363	1319	2.045	1165	1.806	681	1.056	52.9	2.083	2198	3.408
63	2½	1255	1.946	1882	2.919	1663	2.579	972	1.508	63.2	2.489	3137	4.866
78	3	1936	3.000	2904	4.499	2565	3.974	1500	2.325	78.5	3.090	4840	7.499
91	3½	2584	4.004	3877	6.006	3424	5.305	2003	3.103	90.7	3.570	6461	10.010
103	4	3326	5.153	4990	7.729	4408	6.828	2578	3.994	102.9	4.050	8316	12.882
129	5	5220	8.085	7830	12.127	6916	10.713	4045	6.266	128.9	5.073	13050	20.212
155	6	7528	11.663	11292	17.495	9975	15.454	5834	9.039	154.8	6.093	18821	29.158
		\multicolumn{12}{c}{Article 352 — Rigid PVC Conduit (PVC), Schedule 80}
12	⅜	—	—	—	—	—	—	—	—	—	—	—	—
16	½	56	0.087	85	0.130	75	0.115	44	0.067	13.4	0.526	141	0.217
21	¾	105	0.164	158	0.246	139	0.217	82	0.127	18.3	0.722	263	0.409
27	1	178	0.275	267	0.413	236	0.365	138	0.213	23.8	0.936	445	0.688
35	1¼	320	0.495	480	0.742	424	0.656	248	0.383	31.9	1.255	799	1.237
41	1½	442	0.684	663	1.027	585	0.907	342	0.530	37.5	1.476	1104	1.711

(continued)

Chapter 28

Table 4 Dimensions and Percent Area of Conduit and Tubing (Areas of Conduit or Tubing for the Combinations of Wires Permitted in Table 1, Chapter 9) — Continued

Metric Designator	Trade Size	Over 2 Wires 40% mm²	Over 2 Wires 40% in.²	60% mm²	60% in.²	1 Wire 53% mm²	1 Wire 53% in.²	2 Wires 31% mm²	2 Wires 31% in.²	Nominal Internal Diameter mm	Nominal Internal Diameter in.	Total Area 100% mm²	Total Area 100% in.²
\multicolumn{13}{l}{Article 352 — Rigid PVC Conduit (PVC), Schedule 80 — Continued}
53	2	742	1.150	1113	1.725	983	1.523	575	0.891	48.6	1.913	1855	2.874
63	2½	1064	1.647	1596	2.471	1410	2.183	825	1.277	58.2	2.290	2660	4.119
78	3	1660	2.577	2491	3.865	2200	3.414	1287	1.997	72.7	2.864	4151	6.442
91	3½	2243	3.475	3365	5.213	2972	4.605	1738	2.693	84.5	3.326	5608	8.688
103	4	2907	4.503	4361	6.755	3852	5.967	2253	3.490	96.2	3.786	7268	11.258
129	5	4607	7.142	6911	10.713	6105	9.463	3571	5.535	121.1	4.768	11518	17.855
155	6	6605	10.239	9908	15.359	8752	13.567	5119	7.935	145.0	5.709	16513	25.598
\multicolumn{13}{l}{Articles 352 and 353 — Rigid PVC Conduit (PVC), Schedule 40, and HDPE Conduit (HDPE)}
12	⅜	—	—	—	—	—	—	—	—	—	—	—	0.285
16	½	74	0.114	110	0.171	97	0.151	57	0.088	15.3	0.602	184	0.508
21	¾	131	0.203	196	0.305	173	0.269	101	0.157	20.4	0.804	327	0.508
27	1	214	0.333	321	0.499	284	0.441	166	0.258	26.1	1.029	535	0.832
35	1¼	374	0.581	561	0.872	495	0.770	290	0.450	34.5	1.360	935	1.453

Conduit and Tubing Fill Tables

Metric Designator	Trade Size	Over 2 Wires 40%		60%		1 Wire 53%		2 Wires 31%		Nominal Internal Diameter		Total Area 100%	
		mm²	in.²	mm²	in.²	mm²	in.²	mm²	in.²	mm	in.	mm²	in.²
Articles 352 and 353 — Rigid PVC Conduit (PVC), Schedule 40, and HDPE Conduit (HDPE)													
41	1½	513	0.794	769	1.191	679	1.052	397	0.616	40.4	1.590	1282	1.986
53	2	849	1.316	1274	1.975	1126	1.744	658	1.020	52.0	2.047	2124	3.291
63	2½	1212	1.878	1817	2.817	1605	2.488	939	1.455	62.1	2.445	3029	4.695
78	3	1877	2.907	2816	4.361	2487	3.852	1455	2.253	77.3	3.042	4693	7.268
91	3½	2511	3.895	3766	5.842	3327	5.161	1946	3.018	89.4	3.521	6277	9.737
103	4	3237	5.022	4855	7.532	4288	6.654	2508	3.892	101.5	3.998	8091	12.554
129	5	5099	7.904	7649	11.856	6756	10.473	3952	6.126	127.4	5.016	12748	19.761
155	6	7373	11.427	11060	17.140	9770	15.141	5714	8.856	153.2	6.031	18433	28.567

(continued)

Table 4 Dimensions and Percent Area of Conduit and Tubing (Areas of Conduit or Tubing for the Combinations of Wires Permitted in Table 1, Chapter 9) — *Continued*

Metric Designator	Trade Size	Over 2 Wires 40%		60%		1 Wire 53%		2 Wires 31%		Nominal Internal Diameter		Total Area 100%	
		mm²	in.²	mm²	in.²	mm²	in.²	mm²	in.²	mm	in.	mm²	in.²
Article 352 — Type A, Rigid PVC Conduit (PVC)													
16	½	100	0.154	149	0.231	132	0.204	77	0.119	17.8	0.700	249	0.385
21	¾	168	0.260	251	0.390	222	0.345	130	0.202	23.1	0.910	419	0.650
27	1	279	0.434	418	0.651	370	0.575	216	0.336	29.8	1.175	697	1.084
35	1¼	456	0.707	684	1.060	604	0.937	353	0.548	38.1	1.500	1140	1.767
41	1½	600	0.929	900	1.394	795	1.231	465	0.720	43.7	1.720	1500	2.324
53	2	940	1.459	1410	2.188	1245	1.933	728	1.131	54.7	2.155	2350	3.647
63	2½	1406	2.181	2109	3.272	1863	2.890	1090	1.690	66.9	2.635	3515	5.453
78	3	2112	3.278	3169	4.916	2799	4.343	1637	2.540	82.0	3.230	5281	8.194
91	3½	2758	4.278	4137	6.416	3655	5.668	2138	3.315	93.7	3.690	6896	10.694
103	4	3543	5.489	5315	8.234	4695	7.273	2746	4.254	106.2	4.180	8858	13.723
129	5	—	—	—	—	—	—	—	—	—	—	—	—
155	6	—	—	—	—	—	—	—	—	—	—	—	—

Conduit and Tubing Fill Tables

Article 352 — Type EB, Rigid PVC Conduit (PVC)

Metric Designator	Trade Size	Over 2 Wires 40%		60%		1 Wire 53%		2 Wires 31%		Nominal Internal Diameter		Total Area 100%	
		mm²	in.²	mm²	in.²	mm²	in.²	mm²	in.²	mm	in.	mm²	in.²
16	½	—	—	—	—	—	—	—	—	—	—	—	—
21	¾	—	—	—	—	—	—	—	—	—	—	—	—
27	1	—	—	—	—	—	—	—	—	—	—	—	—
35	1¼	—	—	—	—	—	—	—	—	—	—	—	—
41	1½	—	—	—	—	—	—	—	—	—	—	—	—
53	2	999	1.550	1499	2.325	1324	2.053	774	1.201	56.4	2.221	2498	3.874
63	2½	—	—	—	—	—	—	—	—	—	—	—	—
78	3	2248	3.484	3373	5.226	2979	4.616	1743	2.700	84.6	3.330	5621	8.709
91	3½	2932	4.546	4397	6.819	3884	6.023	2272	3.523	96.6	3.804	7329	11.365
103	4	3726	5.779	5589	8.669	4937	7.657	2887	4.479	108.9	4.289	9314	14.448
129	5	5726	8.878	8588	13.317	7586	11.763	4437	6.881	135.0	5.316	14314	22.195
155	6	8133	12.612	12200	18.918	10776	16.711	6303	9.774	160.9	6.336	20333	31.530

Table 5 Dimensions of Insulated Conductors and Fixture Wires

Type: FFH-2, RFH-1, RFH-2, RFHH-2, RHH*, RHW*, RHW-2*, RHH, RHW, RHW-2, SF-1, SF-2, SFF-1, SFF-2, TF, TFF, THHW, THW, THW-2, TW, XF, XFF

Type	Size (AWG or kcmil)	Approximate Area mm²	Approximate Area in.²	Approximate Diameter mm	Approximate Diameter in.
RFH-2, FFH-2, RFHH-2	18	9.355	0.0145	3.454	0.136
	16	11.10	0.0172	3.759	0.148
RHH, RHW, RHW-2	14	18.90	0.0293	4.902	0.193
	12	22.77	0.0353	5.385	0.212
	10	28.19	0.0437	5.994	0.236
	8	53.87	0.0835	8.280	0.326
	6	67.16	0.1041	9.246	0.364
	4	86.00	0.1333	10.46	0.412
	3	98.13	0.1521	11.18	0.440
	2	112.9	0.1750	11.99	0.472
	1	171.6	0.2660	14.78	0.582
	1/0	196.1	0.3039	15.80	0.622
	2/0	226.1	0.3505	16.97	0.668
	3/0	262.7	0.4072	18.29	0.720
	4/0	306.7	0.4754	19.76	0.778

Conduit and Tubing Fill Tables

Type	Size (AWG or kcmil)	Approximate Area mm²	Approximate Area in.²	Approximate Diameter mm	Approximate Diameter in.
Type: FFH-2, RFH-1, RFH-2, RFHH-2, RHH*, RHW*, RHW-2*, RHH, RHW, RHW-2, SF-1, SF-2, SFF-1, SFF-2, TF, TFF, THHW, THW, THW-2, TW, XF, XFF					
	250	405.9	0.6291	22.73	0.895
	300	457.3	0.7088	24.13	0.950
	350	507.7	0.7870	25.43	1.001
	400	556.5	0.8626	26.62	1.048
	500	650.5	1.0082	28.78	1.133
	600	782.9	1.2135	31.57	1.243
	700	874.9	1.3561	33.38	1.314
	750	920.8	1.4272	34.24	1.348
	800	965.0	1.4957	35.05	1.380
	900	1057	1.6377	36.68	1.444
	1000	1143	1.7719	38.15	1.502
	1250	1515	2.3479	43.92	1.729
	1500	1738	2.6938	47.04	1.852
	1750	1959	3.0357	49.94	1.966
	2000	2175	3.3719	52.63	2.072
SF-2, SFF-2	18	7.419	0.0115	3.073	0.121
	16	8.968	0.0139	3.378	0.133
	14	11.10	0.0172	3.759	0.148

(continued)

Table 5 Dimensions of Insulated Conductors and Fixture Wires — Continued

Type: FFH-2, RFH-1, RFH-2, RHH*, RHW*, RHW-2*, RHH, RHW, RHW-2, SF-1, SF-2, SFF-1, SFF-2, TF, TFF, THHW, THW, THW-2, TW, XF, XFF — *Continued*

Type	Size (AWG or kcmil)	Approximate Area mm²	Approximate Area in.²	Approximate Diameter mm	Approximate Diameter in.
SF-1, SFF-1	18	4.194	0.0065	2.311	0.091
RFH-1, TF, TFF, XF, XFF	18	5.161	0.0088	2.692	0.106
TF, TFF, XF, XFF	16	7.032	0.0109	2.997	0.118
TW, XF, XFF, THHW, THW, THW-2	14	8.968	0.0139	3.378	0.133
TW, THHW, THW, THW-2	12	11.68	0.0181	3.861	0.152
	10	15.68	0.0243	4.470	0.176
	8	28.19	0.0437	5.994	0.236
RHH*, RHW*, RHW-2*	14	13.48	0.0209	4.140	0.163
RHH*, RHW*, RHW-2*, XF, XFF	12	16.77	0.0260	4.623	0.182

Conduit and Tubing Fill Tables

Type	Size (AWG or kcmil)	Approximate Area mm²	Approximate Area in.²	Approximate Diameter mm	Approximate Diameter in.
Type: RHH*, RHW*, RHW-2*, THHN, THHW, THW, THW-2, TFN, TFFN, THWN, THWN-2, XF, XFF					
RHH*, RHW*, RHW-2*, XF, XFF	10	21.48	0.0333	5.232	0.206
RHH*, RHW*, RHW-2*	8	35.87	0.0556	6.756	0.266
TW, THW, THHW, THW-2, RHH*, RHW*, RHW-2*	6	46.84	0.0726	7.722	0.304
	4	62.77	0.0973	8.941	0.352
	3	73.16	0.1134	9.652	0.380
	2	86.00	0.1333	10.46	0.412
	1	122.6	0.1901	12.50	0.492
	1/0	143.4	0.2223	13.51	0.532
	2/0	169.3	0.2624	14.68	0.578
	3/0	201.1	0.3117	16.00	0.630
	4/0	239.9	0.3718	17.48	0.688
	250	296.5	0.4596	19.43	0.765
	300	340.7	0.5281	20.83	0.820
	350	384.4	0.5958	22.12	0.871
	400	427.0	0.6619	23.32	0.918

(continued)

Table 5 Dimensions of Insulated Conductors and Fixture Wires — Continued

Type	Size (AWG or kcmil)	Approximate Area mm²	Approximate Area in.²	Approximate Diameter mm	Approximate Diameter in.
Type: RHH*, RHW*, RHW-2*, THHN, THHW, THW, THW-2, TFN, TFFN, THWN, THWN-2, XF, XFF — Continued						
TW, THW, THHW, THW-2, RHH*, RHW*, RHW-2*	500	509.7	0.7901	25.48	1.003
	600	627.7	0.9729	28.27	1.113
	700	710.3	1.1010	30.07	1.184
	750	751.7	1.1652	30.94	1.218
	800	791.7	1.2272	31.75	1.250
	900	874.9	1.3561	33.38	1.314
	1000	953.8	1.4784	34.85	1.372
	1250	1200	1.8602	39.09	1.539
	1500	1400	2.1695	42.21	1.662
	1750	1598	2.4773	45.11	1.776
	2000	1795	2.7818	47.80	1.882

Conduit and Tubing Fill Tables

Type	Size (AWG or kcmil)	Approximate Area mm²	Approximate Area in.²	Approximate Diameter mm	Approximate Diameter in.
Type: RHH*, RHW*, RHW-2*, THHN, THW, THW-2, TFN, TFFN, THWN, THWN-2, XF, XFF — Continued					
TFN, TFFN	18	3.548	0.0055	2.134	0.084
	16	4.645	0.0072	2.438	0.096
THHN, THWN, THWN-2	14	6.258	0.0097	2.819	0.111
	12	8.581	0.0133	3.302	0.130
	10	13.61	0.0211	4.166	0.164
	8	23.61	0.0366	5.486	0.216
	6	32.71	0.0507	6.452	0.254
	4	53.16	0.0824	8.230	0.324
	3	62.77	0.0973	8.941	0.352
	2	74.71	0.1158	9.754	0.384
	1	100.8	0.1562	11.33	0.446
	1/0	119.7	0.1855	12.34	0.486
	2/0	143.4	0.2223	13.51	0.532
	3/0	172.8	0.2679	14.83	0.584
	4/0	208.8	0.3237	16.31	0.642
	250	256.1	0.3970	18.06	0.711
	300	297.3	0.4608	19.46	0.766

(continued)

Table 5 Dimensions of Insulated Conductors and Fixture Wires — *Continued*

Type	Size (AWG or kcmil)	Approximate Area mm²	Approximate Area in.²	Approximate Diameter mm	Approximate Diameter in.
Type: FEP, FEPB, PAF, PAFF, PF, PFA, PFAH, PFF, PGF, PGFF, PTF, PTFF, TFE, THHN, THWN, TWHN-2, Z, ZF, ZFF, ZHF					
THHN, THWN, THWN-2	350	338.2	0.5242	20.75	0.817
	400	378.3	0.5863	21.95	0.864
	500	456.3	0.7073	24.10	0.949
	600	559.7	0.8676	26.70	1.051
	700	637.9	0.9887	28.50	1.122
	750	677.2	1.0496	29.36	1.156
	800	715.2	1.1085	30.18	1.188
	900	794.3	1.2311	31.80	1.252
	1000	869.5	1.3478	33.27	1.310
PF, PGFF, PGF, PFF, PTF, PAF, PTFF, PAFF	18	3.742	0.0058	2.184	0.086
	16	4.839	0.0075	2.489	0.098
PF, PGFF, PGF, PFF, PTF, PAF, PTFF, PAFF, TFE, FEP, PFA, FEPB, PFAH	14	6.452	0.0100	2.870	0.113

Conduit and Tubing Fill Tables

Type	Size (AWG or kcmil)	Approximate Area mm²	Approximate Area in.²	Approximate Diameter mm	Approximate Diameter in.
Type: FEP, FEPB, PAF, PAFF, PF, PFA, PFAH, PFF, PGF, PGFF, PTF, PTFF, TFE, THHN, THWN, THWN-2, Z, ZF, ZFF, ZHF					
TFE, FEP, PFA, FEPB, PFAH	12	8.839	0.0137	3.353	0.132
	10	12.32	0.0191	3.962	0.156
	8	21.48	0.0333	5.232	0.206
	6	30.19	0.0468	6.198	0.244
	4	43.23	0.0670	7.417	0.292
	3	51.87	0.0804	8.128	0.320
	2	62.77	0.0973	8.941	0.352
TFE, PFAH, PFA	1	90.26	0.1399	10.72	0.422
TFE, PFA, PFAH, Z	1/0	108.1	0.1676	11.73	0.462
	2/0	130.8	0.2027	12.90	0.508
	3/0	158.9	0.2463	14.22	0.560
	4/0	193.5	0.3000	15.70	0.618
ZF, ZFF, ZHF	18	2.903	0.0045	1.930	0.076
	16	3.935	0.0061	2.235	0.088
Z, ZF, ZFF, ZHF	14	5.355	0.0083	2.616	0.103
Z	12	7.548	0.0117	3.099	0.122
	10	12.32	0.0191	3.962	0.156

(continued)

Table 5 Dimensions of Insulated Conductors and Fixture Wires — Continued

Type	Size (AWG or kcmil)	Approximate Area mm²	Approximate Area in.²	Approximate Diameter mm	Approximate Diameter in.
Type: FEP, FEPB, PAF, PAFF, PF, PFA, PFAH, PFE, PGF, PGFF, PT, PTF, PTFE, THHN, TWHN-2, Z, ZF, ZFF — Continued					
Z	8	19.48	0.0302	4.978	0.196
	6	27.74	0.0430	5.944	0.234
	4	40.32	0.0625	7.163	0.282
	3	55.16	0.0855	8.382	0.330
	2	66.39	0.1029	9.195	0.362
	1	81.87	0.1269	10.21	0.402
Type: KF-1, KF-2, KFF-1, KFF-2, XHH, XHHW, XHHW-2, ZW					
XHHW, ZW, XHHW-2, XHH	14	8.968	0.0139	3.378	0.133
	12	11.68	0.0181	3.861	0.152
	10	15.68	0.0243	4.470	0.176
	8	28.19	0.0437	5.994	0.236
	6	38.06	0.0590	6.960	0.274
	4	52.52	0.0814	8.179	0.322
	3	62.06	0.0962	8.890	0.350
	2	73.94	0.1146	9.703	0.382

Conduit and Tubing Fill Tables

Type	Size (AWG or kcmil)	Approximate Area mm²	Approximate Area in.²	Approximate Diameter mm	Approximate Diameter in.
	Type: KF-1, KF-2, KFF-1, KFF-2, XHH, XHHW, XHHW-2, ZW				
XHHW, XHHW-2, XHH	1	98.97	0.1534	11.23	0.442
	1/0	117.7	0.1825	12.24	0.482
	2/0	141.3	0.2190	13.41	0.528
	3/0	170.5	0.2642	14.73	0.580
	4/0	206.3	0.3197	16.21	0.638
	250	251.9	0.3904	17.91	0.705
	300	292.6	0.4536	19.30	0.760
	350	333.3	0.5166	20.60	0.811
	400	373.0	0.5782	21.79	0.858
	500	450.6	0.6984	23.95	0.943
	600	561.9	0.8709	26.75	1.053
	700	640.2	0.9923	28.55	1.124
	750	679.5	1.0532	29.41	1.158
	800	717.5	1.1122	30.23	1.190
	900	796.8	1.2351	31.85	1.254
	1000	872.2	1.3519	33.32	1.312
	1250	1108	1.7180	37.57	1.479
	1500	1300	2.0156	40.69	1.602

(continued)

Table 5 Dimensions of Insulated Conductors and Fixture Wires — *Continued*

Type	Size (AWG or kcmil)	Approximate Area mm²	Approximate Area in.²	Approximate Diameter mm	Approximate Diameter in.
Type: KF-1, KF-2, KFF-1, KFF-2, XHH, XHHW, XHHW-2, ZW — *Continued*					
XHHW, XHHW-2, XHH	1750	1492	2.3127	43.59	1.716
	2000	1682	2.6073	46.28	1.822
KF-2, KFF-2	18	2.000	0.003	1.575	0.062
	16	2.839	0.0043	1.88	0.074
	14	4.129	0.0064	2.286	0.090
	12	6.000	0.0092	2.743	0.108
	10	8.968	0.0139	3.378	0.133
KF-1, KFF-1	18	1.677	0.0026	1.448	0.057
	16	2.387	0.0037	1.753	0.069
	14	3.548	0.0055	2.134	0.084
	12	5.355	0.0083	2.616	0.103
	10	8.194	0.0127	3.226	0.127

*Types RHH, RHW, and RHW-2 without outer covering.

Table 5A Compact Copper and Aluminum Building Wire Nominal Dimensions** and Areas

Size (AWG or kcmil)	Bare Conductor Diameter		Type RHH* Approximate Diameter		Type RHH* Approximate Area		Types THW and THHW Approximate Diameter		Types THW and THHW Approximate Area		Type THHN Approximate Diameter		Type THHN Approximate Area		Type XHHW Approximate Diameter		Type XHHW Approximate Area		Size (AWG or kcmil)
	mm	in.	mm	in.	mm²	in.²	mm	in.	mm²	in.²	mm	in.	mm²	in.²	mm	in.	mm²	in.²	
8	3.404	0.134	6.604	0.260	34.25	0.0531	6.477	0.255	32.90	0.0510	—	—	—	—	5.690	0.224	25.42	0.0394	8
6	4.293	0.169	7.493	0.295	44.10	0.0683	7.366	0.290	42.58	0.0660	6.096	0.240	29.16	0.0452	6.604	0.260	34.19	0.0530	6
4	5.410	0.213	8.509	0.335	56.84	0.0881	8.509	0.335	56.84	0.0881	7.747	0.305	47.10	0.0730	7.747	0.305	47.10	0.0730	4
2	6.807	0.268	9.906	0.390	77.03	0.1194	9.906	0.390	77.03	0.1194	9.144	0.360	65.61	0.1017	9.144	0.360	65.61	0.1017	2
1	7.595	0.299	11.81	0.465	109.5	0.1698	11.81	0.465	109.5	0.1698	10.54	0.415	87.23	0.1352	10.54	0.415	87.23	0.1352	1
1/0	8.534	0.336	12.70	0.500	126.6	0.1963	12.70	0.500	126.6	0.1963	11.43	0.450	102.6	0.1590	11.43	0.450	102.6	0.1590	1/0
2/0	9.550	0.376	13.72	0.540	147.8	0.2290	13.84	0.545	150.5	0.2332	12.57	0.495	124.1	0.1924	12.45	0.490	121.6	0.1885	2/0
3/0	10.74	0.423	14.99	0.590	176.3	0.2733	14.99	0.590	176.3	0.2733	13.72	0.540	147.7	0.2290	13.72	0.540	147.7	0.2290	3/0
4/0	12.07	0.475	16.26	0.640	207.6	0.3217	16.38	0.645	210.8	0.3267	15.11	0.595	179.4	0.2780	14.99	0.590	176.3	0.2733	4/0
250	13.21	0.520	18.16	0.715	259.0	0.4015	18.42	0.725	266.3	0.4128	17.02	0.670	227.4	0.3525	16.76	0.660	220.7	0.3421	250
300	14.48	0.570	19.43	0.765	296.5	0.4596	19.69	0.775	304.3	0.4717	18.29	0.720	262.6	0.4071	18.16	0.715	259.0	0.4015	300
350	15.65	0.616	20.57	0.810	332.3	0.5153	20.83	0.820	340.7	0.5281	19.56	0.770	300.4	0.4656	19.30	0.760	292.6	0.4536	350
400	16.74	0.659	21.72	0.855	370.5	0.5741	21.97	0.865	379.1	0.5876	20.70	0.815	336.5	0.5216	20.32	0.800	324.3	0.5026	400

(continued)

Table 5A Compact Copper and Aluminum Building Wire Nominal Dimensions** and Areas — Continued

Size (AWG or kcmil)	Bare Conductor Diameter		Type RHH*				Types THW and THHW				Type THHN				Type XHHW			
			Approximate Diameter		Approximate Area		Approximate Diameter		Approximate Area		Approximate Diameter		Approximate Area		Approximate Diameter		Approximate Area	
	mm	in.	mm	in.	mm²	in.²	mm	in.	mm²	in.²	mm	in.	mm²	in.²	mm	in.	mm²	in.²
500	18.69	0.736	23.62	0.930	438.2	0.6793	23.88	0.940	447.7	0.6939	22.48	0.885	396.8	0.6151	22.35	0.880	392.4	0.6082
600	20.65	0.813	26.29	1.035	542.8	0.8413	26.67	1.050	558.6	0.8659	25.02	0.985	491.6	0.7620	24.89	0.980	486.6	0.7542
700	22.28	0.877	27.94	1.100	613.1	0.9503	28.19	1.110	624.3	0.9676	26.67	1.050	558.6	0.8659	26.67	1.050	558.6	0.8659
750	23.06	0.908	28.83	1.135	652.8	1.0118	29.21	1.150	670.1	1.0386	27.31	1.075	585.5	0.9076	27.69	1.090	602.0	0.9331
900	25.37	0.999	31.50	1.240	779.3	1.2076	31.09	1.224	759.1	1.1766	30.33	1.194	722.5	1.1196	29.69	1.169	692.3	1.0733
1000	26.92	1.060	32.64	1.285	836.6	1.2968	32.64	1.285	836.6	1.2968	31.88	1.255	798.1	1.2370	31.24	1.230	766.6	1.1882

*Types RHH and RHW without outer coverings.
**Dimensions are from industry sources.

Conduit and Tubing Fill Tables

Table 8 Conductor Properties

Size (AWG or kcmil)	Area mm²	Area Circular mils	Stranding Qty	Stranding Diameter mm	Stranding Diameter in.	Overall Diameter mm	Overall Diameter in.	Overall Area mm²	Overall Area in.²	Copper Uncoated ohm/km	Copper Uncoated ohm/kFT	Copper Coated ohm/km	Copper Coated ohm/kFT	Aluminum ohm/km	Aluminum ohm/kFT
18	0.823	1620	1	—	—	1.02	0.040	0.823	0.001	25.5	7.77	26.5	8.08	42.0	12.8
18	0.823	1620	7	0.39	0.015	1.16	0.046	1.06	0.002	26.1	7.95	27.7	8.45	42.8	13.1
16	1.31	2580	1	—	—	1.29	0.051	1.31	0.002	16.0	4.89	16.7	5.08	26.4	8.05
16	1.31	2580	7	0.49	0.019	1.46	0.058	1.68	0.003	16.4	4.99	17.3	5.29	26.9	8.21
14	2.08	4110	1	—	—	1.63	0.064	2.08	0.003	10.1	3.07	10.4	3.19	16.6	5.06
14	2.08	4110	7	0.62	0.024	1.85	0.073	2.68	0.004	10.3	3.14	10.7	3.26	16.9	5.17
12	3.31	6530	1	—	—	2.05	0.081	3.31	0.005	6.34	1.93	6.57	2.01	10.45	3.18
12	3.31	6530	7	0.78	0.030	2.32	0.092	4.25	0.006	6.50	1.98	6.73	2.05	10.69	3.25
10	5.261	10380	1	—	—	2.588	0.102	5.26	0.008	3.984	1.21	4.148	1.26	6.561	2.00
10	5.261	10380	7	0.98	0.038	2.95	0.116	6.76	0.011	4.070	1.24	4.226	1.29	6.679	2.04

(continued)

Table 8 Conductor Properties — Continued

Size (AWG or kcmil)	Area		Conductors						Direct-Current Resistance at 75°C (167°F)					
			Stranding		Overall				Copper				Aluminum	
			Diameter		Diameter		Area		Uncoated		Coated			
	mm²	Circular mils	Qty	mm	in.	mm	in.	mm²	in.²	ohm/km	ohm/kFT	ohm/km	ohm/kFT	ohm/km	ohm/kFT
8	8.367	16510	1	—	—	3.264	0.128	8.37	0.013	2.506	0.764	2.579	0.786	4.125	1.26
8	8.367	16510	7	1.23	0.049	3.71	0.146	10.76	0.017	2.551	0.778	2.653	0.809	4.204	1.28
6	13.30	26240	7	1.56	0.061	4.67	0.184	17.09	0.027	1.608	0.491	1.671	0.510	2.652	0.808
4	21.15	41740	7	1.96	0.077	5.89	0.232	27.19	0.042	1.010	0.308	1.053	0.321	1.666	0.508
3	26.67	52620	7	2.20	0.087	6.60	0.260	34.28	0.053	0.802	0.245	0.833	0.254	1.320	0.403
2	33.62	66360	7	2.47	0.097	7.42	0.292	43.23	0.067	0.634	0.194	0.661	0.201	1.045	0.319
1	42.41	83690	19	1.69	0.066	8.43	0.332	55.80	0.087	0.505	0.154	0.524	0.160	0.829	0.253
1/0	53.49	105600	19	1.89	0.074	9.45	0.372	70.41	0.109	0.399	0.122	0.415	0.127	0.660	0.201
2/0	67.43	133100	19	2.13	0.084	10.62	0.418	88.74	0.137	0.3170	0.0967	0.329	0.101	0.523	0.159
3/0	85.01	167800	19	2.39	0.094	11.94	0.470	111.9	0.173	0.2512	0.0766	0.2610	0.0797	0.413	0.126
4/0	107.2	211600	19	2.68	0.106	13.41	0.528	141.1	0.219	0.1996	0.0608	0.2050	0.0626	0.328	0.100
250	127	—	37	2.09	0.082	14.61	0.575	168	0.260	0.1687	0.0515	0.1753	0.0535	0.2778	0.0847
300	152	—	37	2.29	0.090	16.00	0.630	201	0.312	0.1409	0.0429	0.1463	0.0446	0.2318	0.0707

Conduit and Tubing Fill Tables

Size (AWG or kcmil)	Area		Conductors						Direct-Current Resistance at 75°C (167°F)					
			Stranding		Overall				Copper				Aluminum	
				Diameter		Diameter		Area		Uncoated		Coated		
	mm²	Circular mils	Qty	mm	in.	mm	in.	mm²	in.²	ohm/km	ohm/kFT	ohm/km	ohm/kFT	ohm/km	ohm/kFT
350	177	—	37	2.47	0.097	17.30	0.681	235	0.364	0.1205	0.0367	0.1252	0.0382	0.1984	0.0605
400	203	—	37	2.64	0.104	18.49	0.728	268	0.416	0.1053	0.0321	0.1084	0.0331	0.1737	0.0529
500	253	—	37	2.95	0.116	20.65	0.813	336	0.519	0.0845	0.0258	0.0869	0.0265	0.1391	0.0424
600	304	—	61	2.52	0.099	22.68	0.893	404	0.626	0.0704	0.0214	0.0732	0.0223	0.1159	0.0353
700	355	—	61	2.72	0.107	24.49	0.964	471	0.730	0.0603	0.0184	0.0622	0.0189	0.0994	0.0303
750	380	—	61	2.82	0.111	25.35	0.998	505	0.782	0.0563	0.0171	0.0579	0.0176	0.0927	0.0282
800	405	—	61	2.91	0.114	26.16	1.030	538	0.834	0.0528	0.0161	0.0544	0.0166	0.0868	0.0265
900	456	—	61	3.09	0.122	27.79	1.094	606	0.940	0.0470	0.0143	0.0481	0.0147	0.0770	0.0235
1000	507	—	61	3.25	0.128	29.26	1.152	673	1.042	0.0423	0.0129	0.0434	0.0132	0.0695	0.0212
1250	633	—	91	2.98	0.117	32.74	1.289	842	1.305	0.0338	0.0103	0.0347	0.0106	0.0554	0.0169
1500	760	—	91	3.26	0.128	35.86	1.412	1011	1.566	0.02814	0.00858	0.02814	0.00883	0.0464	0.0141
1750	887	—	127	2.98	0.117	38.76	1.526	1180	1.829	0.02410	0.00735	0.02410	0.00756	0.0397	0.0121
2000	1013	—	127	3.19	0.126	41.45	1.632	1349	2.092	0.02109	0.00643	0.02109	0.00662	0.0348	0.0106

Notes:

1. These resistance values are valid **only** for the parameters as given. Using conductors having coated strands, different stranding type, and, especially, other temperatures changes the resistance.

2. Equation for temperature change: $R_2 = R_1 [1 + a(T_2 - 75)]$, where $\alpha_{cu} = 0.00323$, $\alpha_{AL} = 0.00330$ at 75°C.

3. Conductors with compact and compressed stranding have about 9 percent and 3 percent, respectively, smaller bare conductor diameters than those shown. See Table 5A for actual compact cable dimensions.

4. The IACS conductivities used: bare copper = 100%, aluminum = 61%.

5. Class B stranding is listed as well as solid for some sizes. Its overall diameter and area are those of its circumscribing circle.

Informational Note: The construction information is in accordance with NEMA WC70-2009 or ANSI/UL 1581-2017. The resistance is calculated in accordance with *National Bureau of Standards Handbook 100*, dated 1966, and Handbook 109, dated 1972.

Table 9 Alternating-Current Resistance and Reactance for 600-Volt Cables, 3-Phase, 60 Hz, 75°C (167°F) — Three Single Conductors in Conduit

Ohms to Neutral per Kilometer
Ohms to Neutral per 1000 Feet

Size (AWG or kcmil)	X_L (Reactance) for All Wires		Alternating-Current Resistance for Uncoated Copper Wires			Alternating-Current Resistance for Aluminum Wires			Effective Z at 0.85 PF for Uncoated Copper Wires			Effective Z at 0.85 PF for Aluminum Wires			Size (AWG or kcmil)
	PVC, Aluminum Conduits	Steel Conduit	PVC Conduit	Aluminum Conduit	Steel Conduit	PVC Conduit	Aluminum Conduit	Steel Conduit	PVC Conduit	Aluminum Conduit	Steel Conduit	PVC Conduit	Aluminum Conduit	Steel Conduit	
14	0.190	0.240	10.2	10.2	10.2	—	—	—	8.9	8.9	8.9	—	—	—	14
	0.058	0.073	3.1	3.1	3.1	—	—	—	2.7	2.7	2.7	—	—	—	
12	0.177	0.223	6.6	6.6	6.6	10.5	10.5	10.5	5.6	5.6	5.6	9.2	9.2	9.2	12
	0.054	0.068	2.0	2.0	2.0	3.2	3.2	3.2	1.7	1.7	1.7	2.8	2.8	2.8	
10	0.164	0.207	3.9	3.9	3.9	6.6	6.6	6.6	3.6	3.6	3.6	5.9	5.9	5.9	10
	0.050	0.063	1.2	1.2	1.2	2.0	2.0	2.0	1.1	1.1	1.1	1.8	1.8	1.8	
8	0.171	0.213	2.56	2.56	2.56	4.3	4.3	4.3	2.26	2.26	2.30	3.6	3.6	3.6	8
	0.052	0.065	0.78	0.78	0.78	1.3	1.3	1.3	0.69	0.69	0.70	1.1	1.1	1.1	
6	0.167	0.210	1.61	1.61	1.61	2.66	2.66	2.66	1.44	1.48	1.48	2.33	2.36	2.36	6
	0.051	0.064	0.49	0.49	0.49	0.81	0.81	0.81	0.44	0.45	0.45	0.71	0.72	0.72	
4	0.157	0.197	1.02	1.02	1.02	1.67	1.67	1.67	0.95	0.95	0.98	1.51	1.51	1.51	4
	0.048	0.060	0.31	0.31	0.31	0.51	0.51	0.51	0.29	0.29	0.30	0.46	0.46	0.46	

3	0.154 0.047	0.194 0.059	0.82 0.25	0.82 0.25	0.82 0.25	1.31 0.40	1.35 0.41	1.31 0.40	0.75 0.23	0.79 0.24	0.79 0.24	1.21 0.37	1.21 0.37	1.21 0.37	3
2	0.148 0.045	0.187 0.057	0.62 0.19	0.66 0.20	0.66 0.20	1.05 0.32	1.05 0.32	1.05 0.32	0.62 0.19	0.62 0.19	0.66 0.20	0.98 0.30	0.98 0.30	0.98 0.30	2
1	0.151 0.046	0.187 0.057	0.49 0.15	0.52 0.16	0.52 0.16	0.82 0.25	0.85 0.26	0.82 0.25	0.52 0.16	0.52 0.16	0.52 0.16	0.79 0.24	0.79 0.24	0.82 0.25	1
1/0	0.144 0.044	0.180 0.055	0.39 0.12	0.43 0.13	0.39 0.12	0.66 0.20	0.69 0.21	0.66 0.20	0.43 0.13	0.43 0.13	0.43 0.13	0.62 0.19	0.66 0.20	0.66 0.20	1/0
2/0	0.141 0.043	0.177 0.054	0.33 0.10	0.33 0.10	0.33 0.10	0.52 0.16	0.52 0.16	0.52 0.16	0.36 0.11	0.36 0.11	0.36 0.11	0.52 0.16	0.52 0.16	0.52 0.16	2/0
3/0	0.138 0.042	0.171 0.052	0.253 0.077	0.269 0.082	0.259 0.079	0.279 0.085	0.295 0.090	0.282 0.086	0.289 0.088	0.302 0.092	0.308 0.094	0.43 0.13	0.43 0.13	0.46 0.14	3/0
4/0	0.135 0.041	0.167 0.051	0.203 0.062	0.220 0.067	0.207 0.063	0.233 0.071	0.249 0.076	0.236 0.072	0.243 0.074	0.256 0.078	0.262 0.080	0.36 0.11	0.36 0.11	0.36 0.11	4/0
250	0.135 0.041	0.171 0.052	0.171 0.052	0.187 0.057	0.177 0.054	0.200 0.061	0.217 0.066	0.207 0.063	0.217 0.066	0.230 0.070	0.240 0.073	0.308 0.094	0.322 0.098	0.33 0.10	250
300	0.135 0.041	0.167 0.051	0.144 0.044	0.161 0.049	0.148 0.045	0.177 0.054	0.190 0.058	0.180 0.055	0.194 0.059	0.207 0.063	0.213 0.065	0.269 0.082	0.282 0.086	0.289 0.088	300
350	0.131 0.040	0.164 0.050	0.125 0.038	0.141 0.043	0.128 0.039	0.148 0.045	0.161 0.049	0.155 0.047	0.174 0.053	0.190 0.058	0.197 0.060	0.240 0.073	0.253 0.077	0.262 0.080	350

(continued)

Table 9 Alternating-Current Resistance and Reactance for 600-Volt Cables, 3-Phase, 60 Hz, 75°C (167°F) — Three Single Conductors in Conduit — *Continued*

Ohms to Neutral per Kilometer
Ohms to Neutral per 1000 Feet

Size (AWG or kcmil)	X_L (Reactance) for All Wires		Alternating-Current Resistance for Uncoated Copper Wires			Alternating-Current Resistance for Aluminum Wires			Effective Z at 0.85 PF for Uncoated Copper Wires			Effective Z at 0.85 PF for Aluminum Wires			Size (AWG or kcmil)
	PVC, Aluminum Conduits	Steel Conduit	PVC Conduit	Aluminum Conduit	Steel Conduit	PVC Conduit	Aluminum Conduit	Steel Conduit	PVC Conduit	Aluminum Conduit	Steel Conduit	PVC Conduit	Aluminum Conduit	Steel Conduit	
400	0.131	0.161	0.108	0.125	0.115	0.177	0.194	0.180	0.161	0.174	0.184	0.217	0.233	0.240	400
	0.040	0.049	0.033	0.038	0.035	0.054	0.059	0.055	0.049	0.053	0.056	0.066	0.071	0.073	
500	0.128	0.157	0.089	0.105	0.095	0.141	0.157	0.148	0.141	0.157	0.164	0.187	0.200	0.210	500
	0.039	0.048	0.027	0.032	0.029	0.043	0.048	0.045	0.043	0.048	0.050	0.057	0.061	0.064	
600	0.128	0.157	0.075	0.092	0.082	0.118	0.135	0.125	0.131	0.144	0.154	0.167	0.180	0.190	600
	0.039	0.048	0.023	0.028	0.025	0.036	0.041	0.038	0.040	0.044	0.047	0.051	0.055	0.058	
750	0.125	0.157	0.062	0.079	0.069	0.095	0.112	0.102	0.118	0.131	0.141	0.148	0.161	0.171	750
	0.038	0.048	0.019	0.024	0.021	0.029	0.034	0.031	0.036	0.040	0.043	0.045	0.049	0.052	

			Ohms to Neutral per Kilometer / Ohms to Neutral per 1000 Feet									
	X_L (Reactance) for All Wires		Alternating-Current Resistance for Uncoated Copper Wires			Alternating-Current Resistance for Aluminum Wires			Effective Z at 0.85 PF for Uncoated Copper Wires			Effective Z at 0.85 PF for Aluminum Wires		
Size (AWG or kcmil)	PVC, Aluminum Conduits	Steel Conduit	PVC Conduit	Aluminum Conduit	Steel Conduit	PVC Conduit	Aluminum Conduit	Steel Conduit	PVC Conduit	Aluminum Conduit	Steel Conduit	PVC Conduit	Aluminum Conduit	Steel Conduit	Size (AWG or kcmil)
1000	0.121	0.151	0.049	0.062	0.059	0.075	0.089	0.082	0.105	0.118	0.131	0.128	0.138	0.151	1000
	0.037	0.046	0.015	0.019	0.018	0.023	0.027	0.025	0.032	0.036	0.040	0.039	0.042	0.046	

Notes:

1. These values are based on the following constants: UL-Type RHH wires with Class B stranding, in cradled configuration. Wire conductivities are 100 percent IACS copper and 61 percent IACS aluminum, and aluminum conductivity is 45 percent IACS. Capacitive reactance is ignored, since it is negligible at these voltages. These resistance values are valid only at 75°C (167°F) and for the parameters as given, but are representative for 600-volt wire types operating at 60 Hz.

2. *Effective Z* is defined as $R \cos(\theta) + X \sin(\theta)$, where θ is the power factor angle of the circuit. Multiplying current by effective impedance gives a good approximation for line-to-neutral voltage drop. Effective impedance values shown in this table are valid only at 0.85 power factor. For another circuit power factor (PF), effective impedance (Ze) can be calculated from R and X_L values given in this table as follows: $Ze = R \times PF + X_L \sin(\arccos(PF))$.

Table 10 Conductor Stranding

Conductor Size		Number of Strands		
		Copper		Aluminum
AWG or kcmil	mm²	Class B[a]	Class C	Class B[a]
24–30	0.20–0.05	[b]	—	—
22	0.32	7	—	—
20	0.52	10	—	—
18	0.82	16	—	—
16	1.3	26	—	—
14–2	2.1–33.6	7	19	7[c]
1–4/0	42.4–107	19	37	19
250–500	127–253	37	61	37
600–1000	304–508	61	91	61
1250–1500	635–759	91	127	91
1750–2000	886–1016	127	271	127

[a]Conductors with a lesser number of strands shall be permitted based on an evaluation for connectability and bending.

[b]Number of strands vary.

[c]Aluminum 14 AWG (2.1 mm²) is not available.

With the permission of Underwriters Laboratories, Inc., material is reproduced from UL Standard 486A-486B, *Wire Connectors*, which is copyrighted by Underwriters Laboratories, Inc., Northbrook, Illinois. While use of this material has been authorized, UL shall not be responsible for the manner in which the information is presented, nor for any interpretations thereof. For more information on UL or to purchase standards, visit our Standards website at www.comm-2000.com or call 1-888-853-3503.

Table 11(A) and Table 11(B)

For listing purposes, Table 11(A) and Table 11(B) provide the required power source limitations for Class 2 and Class 3 power sources. Table 11(A) applies for alternating-current sources, and Table 11(B) applies for direct-current sources.

The power for Class 2 and Class 3 circuits shall be either (1) inherently limited, requiring no overcurrent protection, or (2) not inherently limited, requiring a combination of power source and overcurrent protection. Power sources designed for interconnection shall be listed for the purpose.

Conduit and Tubing Fill Tables — Table 11(A) and Table 11(B)

Table 11(A) Class 2 and Class 3 Alternating-Current Power Source Limitations

Power Source	Inherently Limited Power Source (Overcurrent Protection Not Required)					Not Inherently Limited Power Source (Overcurrent Protection Required)				
	Class 2			Class 3		Class 2			Class 3	
	0 through 20*	Over 20 and through 30*	Over 30 and through 150	Over 30 and through 100	Over 100 and through 150	0 through 20*	Over 20 and through 30*	Over 30 and through 100	Over 30 and through 100	Over 100 and through 150
Source voltage V_{max} (volts) (see Note 1)	0 through 20*	Over 20 and through 30*	Over 30 and through 150	Over 30 and through 100	Over 100 and through 150	0 through 20*	Over 20 and through 30*	Over 30 and through 100	Over 30 and through 100	Over 100 and through 150
Power limitations VA_{max} (volt-amperes) (see Note 1)	—	—	—	—	—	250 (see Note 3)	250	250	250	N.A.
Current limitations I_{max} (amperes) (see Note 1)	8.0	8.0	0.005	$150/V_{max}$	—	$1000/V_{max}$	$1000/V_{max}$	$1000/V_{max}$	$1000/V_{max}$	1.0
Maximum overcurrent protection (amperes)	—	—	—	—	—	5.0	100	100	100	1.0
Power source maximum nameplate rating	VA (volt-amperes)	$5.0 \times V_{max}$	100	$0.005 \times V_{max}$	100	$5.0 \times V_{max}$	100	100	100	100
	Current (amperes)	5.0	$100/V_{max}$	0.005	$100/V_{max}$	5.0	$100/V_{max}$	$100/V_{max}$	$100/V_{max}$	$100/V_{max}$

Note: Notes for this table can be found following Table 11(B).

*Voltage ranges shown are for sinusoidal ac in indoor locations or where wet contact is not likely to occur. For nonsinusoidal or wet contact conditions, see Note 2.

Table 11(A) and Table 11(B)

Table 11(B) Class 2 and Class 3 Direct-Current Power Source Limitations

Power Source		Inherently Limited Power Source (Overcurrent Protection Not Required)				Not Inherently Limited Power Source (Overcurrent Protection Required)			
		Class 2			Class 3	Class 2			Class 3
		0 through 20*	Over 20 and through 30*	Over 30 and through 60*	Over 60 and through 150	Over 60 and through 100	0 through 20*	Over 20 and through 60*	Over 60 and through 100	Over 100 and through 150
Source voltage V_{max} (volts) (see Note 1)		—	—	—	—	—	250 (see Note 3)	250	250	N.A.
Power limitations VA_{max} (volt-amperes) (see Note 1)		8.0	8.0	$150/V_{max}$	0.005	$150/V_{max}$	$1000/V_{max}$	$1000/V_{max}$	$1000/V_{max}$	1.0
Current limitations I_{max} (amperes) (see Note 1)		—	—	—	$0.005 \times V_{max}$	—	5.0	5.0	$100/V_{max}$	1.0
Maximum overcurrent protection (amperes)		—	—	100	0.005	100	$5.0 \times V_{max}$	100	100	100
Power source maximum nameplate rating	VA (volt-amperes)	$5.0 \times V_{max}$	—	100	$0.005 \times V_{max}$	100	$5.0 \times V_{max}$	$100/V_{max}$	$100/V_{max}$	$100/V_{max}$
	Current (amperes)	5.0	$100/V_{max}$	$100/V_{max}$	0.005	$100/V_{max}$	5.0	$100/V_{max}$	$100/V_{max}$	$100/V_{max}$

*Voltage ranges shown are for continuous dc in indoor locations or where wet contact is not likely to occur. For interrupted dc or wet contact conditions, see Note 4.

Conduit and Tubing Fill Tables — Table 11(A) and Table 11(B)

Notes for Table 11(A) and Table 11(B)

1. V_{max}, I_{max}, and VA_{max} are determined with the current-limiting impedance in the circuit (not bypassed) as follows:

   V_{max}: Maximum output voltage regardless of load with rated input applied.

   I_{max}: Maximum output current under any noncapacitive load, including short circuit, and with overcurrent protection bypassed if used. Where a transformer limits the output current, I_{max} limits apply after 1 minute of operation. Where a current-limiting impedance, listed for the purpose, or as part of a listed product, is used in combination with a non-power-limited transformer or a stored energy source, e.g., storage battery, to limit the output current, I_{max} limits apply after 5 seconds.

   VA_{max}: Maximum volt-ampere output after 1 minute of operation regardless of load and overcurrent protection bypassed if used.

2. For nonsinusoidal ac, V_{max} shall not be greater than 42.4 volts peak. Where wet contact (immersion not included) is likely to occur, Class 3 wiring methods shall be used or V_{max} shall not be greater than 15 volts for sinusoidal ac and 21.2 volts peak for nonsinusoidal ac.

3. If the power source is a transformer, VA_{max} is 350 or less when V_{max} is 15 or less.

4. For dc interrupted at a rate of 10 to 200 Hz, V_{max} shall not be greater than 24.8 volts peak. Where wet contact (immersion not included) is likely to occur, Class 3 wiring methods shall be used, or V_{max} shall not be greater than 30 volts for continuous dc; 12.4 volts peak for dc that is interrupted at a rate of 10 to 200 Hz.

As part of the listing, the Class 2 or Class 3 power source shall be durably marked where plainly visible to indicate the class of supply and its electrical rating. A Class 2 power source not suitable for wet location use shall be so marked.

Exception: Limited power circuits used by listed information technology equipment.

Overcurrent devices, where required, shall be located at the point where the conductor to be protected receives its supply and shall not be interchangeable with devices of higher ratings. The overcurrent device shall be permitted as an integral part of the power source.

Table 12(A) and Table 12(B)

For listing purposes, Table 12(A) and Table 12(B) provide the required power source limitations for power-limited fire alarm sources. Table 12(A) applies for alternating-current sources, and Table 12(B) applies for direct-current sources. The power for power-limited fire alarm circuits shall be either (1) inherently limited, requiring no overcurrent protection, or (2) not inherently limited, requiring the power to be limited by a combination of power source and overcurrent protection.

As part of the listing, the PLFA power source shall be durably marked where plainly visible to indicate that it is a power-limited fire alarm power source. The overcurrent device, where required, shall be located at the point where the conductor to be protected receives its supply and shall not be interchangeable with devices of higher ratings. The overcurrent device shall be permitted as an integral part of the power source.

Conduit and Tubing Fill Tables — Table 12(A) and Table 12(B)

Table 12(A) PLFA Alternating-Current Power Source Limitations

Power Source		Inherently Limited Power Source (Overcurrent Protection Not Required)			Not Inherently Limited Power Source (Overcurrent Protection Required)		
		0 through 20	Over 20 and through 30	Over 30 and through 100	0 through 20	Over 20 and through 100	Over 100 and through 150
Circuit voltage V_{max} (volts) (see Note 1)		—	—	—	250 (see Note 2)	250	N.A.
Power limitations VA_{max} (volt-amperes) (see Note 1)		8.0	8.0	150V_{max}	1000V_{max}	1000V_{max}	1.0
Current limitations I_{max} (amperes) (see Note 1)		—	—	—	5.0	100/V_{max}	1.0
Maximum overcurrent protection (amperes)		—	100	100	5.0 × V_{max}	100	100
Power source maximum nameplate ratings	VA (volt-amperes)	5.0 × V_{max}	100	100	5.0 × V_{max}	100	100
	Current (amperes)	5.0	100/V_{max}	100/V_{max}	5.0	100/V_{max}	100/V_{max}

Note: Notes for this table can be found following Table 12(B).

Table 12(B) PLFA Direct-Current Power Source Limitations

Power Source		Inherently Limited Power Source (Overcurrent Protection Not Required)			Not Inherently Limited Power Source (Overcurrent Protection Required)		
		0 through 20	Over 20 and through 30	Over 30 and through 100	0 through 20	Over 20 and through 100	Over 100 and through 150
Circuit voltage V_{max} (volts) (see Note 1)		—	—	—	250 (see Note 2)	250	N.A.
Power limitations VA_{max} (volt-amperes) (see Note 1)		8.0	—	$150/V_{max}$	$1000/V_{max}$	$1000/V_{max}$	1.0
Current limitations I_{max} (amperes) (see Note 1)		8.0	8.0	—	5.0	$100/V_{max}$	1.0
Maximum overcurrent protection (amperes)		—	—	100	$5.0 \times V_{max}$	100	100
Power source maximum nameplate ratings	VA (volt-amperes)	$5.0 \times V_{max}$	100	$100/V_{max}$	5.0	$100/V_{max}$	$100/V_{max}$
	Current (amperes)	5.0					

Notes for Table 12(A) and Table 12(B)

1. V_{max}, I_{max}, and VA_{max} are determined as follows:

   V_{max}: Maximum output voltage regardless of load with rated input applied.

   I_{max}: Maximum output current under any noncapacitive load, including short circuit, and with overcurrent protection bypassed if used. Where a transformer limits the output current, Imax limits apply after 1 minute of operation. Where a current-limiting impedance, listed for the purpose, is used in combination with a nonpower-limited transformer or a stored energy source, e.g., storage battery, to limit the output current, I_{max} limits apply after 5 seconds.

   VA_{max}: Maximum volt-ampere output after 1 minute of operation regardless of load and overcurrent protection bypassed if used. Current limiting impedance shall not be bypassed when determining I_{max} and VA_{max}.

2. If the power source is a transformer, VA_{max} is 350 or less when V_{max} is 15 or less.

INDEX

Accessible
- Boxes, conduit bodies, handhole enclosures 158
- Communications circuits 300
- Definitions 6
- Disconnecting means 69–70
- Grounding electrode conductors/bonding jumpers 119
- Overcurrent protection/devices 91–93
- Switches 86

Air-conditioning equipment 289–297
- Ampacity and rating 291
- Branch circuit conductors 293–294
- Branch circuit selection current 290
- Cord-connected 292
- Disconnecting means 292
- Load calculation 49–50
- Overload protection 295
- Receptacle outlet 242
- Room (*see* Room air conditioners)
- Short-circuit/ground-fault protection 293

Ampacity
- Air-conditioning and refrigerating equipment 291
- Branch circuits 229–230
- Conductors 215–221, 229–230
- Definition 6
- Feeders 78
- Metal-clad cable (Type MC) 171
- Service-drop conductors 59, 62
- Underground feeder and branch-circuit cable (Type UF) 172–173

Appliances 269–277
See also specific appliances
- Definition 6
- Disconnecting means 275–277
- Flexible cords 272–274
- Installation 270–275
- Load calculation for small 43
- Overcurrent protection 271–272

Index

Approved/approval
    Conductors and equipment 22
    Definition 6
Arc-fault circuit interrupter (AFCI)
    Branch circuits 227–229
    Definition 6
    Receptacle protection 245
Attachment plug (plug cap, plug), definition 7
Authority having jurisdiction (AHJ) 7
Automatic, definition 7

Basements, receptacles 241
Bathrooms
    Definition 7
    Disconnecting means 69
    Luminaires 259
    Receptacles 240, 247–248
    Switches 250
Bonding
    Definition 7
    Electrically conductive materials 104
    Enclosures 124
    Equipment 104
    Equipotential 346–349
    Intersystem termination 123–124
    Jumpers 109, 112
    Photovoltaic systems 388–392
    Service equipment 122
    Structural metal 128–129
    Swimming pools, fountains, and similar installations 356, 361, 364
    Water piping systems 127–128
Bonding conductors/jumpers 124
    Communications equipment 301–303
    Connection to electrodes 121
    Definition 8
Bonding jumper, equipment
    Connecting receptacle 140–141
    Definition 7

Index

Bonding jumper, main	107
Definition	8
Boxes	
Accessible	158
Ceiling	156
Ceiling-suspended fan outlets	156
Continuity/attachment of grounding conductors to	141–142
Covers	155
Cutout, definition	10
Damp/wet locations	144
Depth	154–155
Enclosing flush devices	151–152
Fill calculations	145–149
Floor	156
Flush-mounted installations	151
Installation	144–161
Metal	149–150
Nonmetallic	144
Number of conductors in	145
Outlets	155–156
Pendant	153
Pull/junction	157–158
Repairing noncombustible surfaces	151
Round	144
Supports	151–154
Surface extensions	151
Swimming pools, fountains, and similar installations	343–345, 360–361
Wall	155
Wiring methods and	209–210
Branch circuits	223–234
AFCI protection	227–229
Appliances	270–271
Appliances, definition	8
Buildings with more than one occupancy	233
Definition	8
General-purpose, definition	8
GFCI protection	225–226

Index

Branch circuits *(continued)*
   Identification 224–225
   Individual, definition 8
   Load calculations 40–41
   Multiple 225
   Multiwire 224
   Multiwire, definition 8
   Overcurrent protective devices 16, 92, 231
   Permissible loads 232–233
   Ratings 229–233
   Receptacles 231–232
   Required outlets 236–242
   Space-heating equipment 280
Building(s)
   Definition 8
   Receptacles in accessory 241
   Separate 325–329

Cabinet, definition 8
Cable(s)
   Communications equipment 319–321
   Mechanical continuity 209–210
   Metal-clad [*see* Metal-clad cable (Type MC)]
   Non-metallic-sheathed [*see* Non-metallic-sheathed
     (Types NM, NMC, and NMS)]
   Protection against physical damage 200–203
   Service-entrance [*see* Service-entrance cable
     (Types SE and USE)]
   Space-heating equipment 283–287
   Underground feeder and branch-circuit [*see* Underground
     feeder and branch-circuit cable (Type UF)]
Ceiling-suspended fans 156, 274–275
Central vacuum outlet assemblies 272
Circuit breakers
   Ampere ratings 90, 91
   Definition 8–9
   Indicating 95
   Multiwire branch circuits 91
   Ungrounded conductors 90, 91

Index

Closet storage space	257–260
Clothes closets, definition	9

Communications equipment

Circuits	299–323
Definition	9
Dwellings	303–306
Grounding methods	301–303
Installation	303–317
Listing requirements	317–319
Raceways	304, 317
Concealed, definition	9

Conductors

Ampacities	215–221, 230
Bare, definition	10
Branch circuits	92, 230, 293–294
Copper	23
Copper-clad aluminum, definition	10
Covered, definition	10
Entering boxes, conduit bodies, fittings	149–150
Grounded	12
Insulated	218
Insulated, definition	10
Neutral, definition	15
Number in boxes and conduit bodies	145
Overcurrent protection/devices	88
Parallel	213
Protection against physical damage	200–203
Service	18, 92, 206
Sizes	23
Small and overcurrent protection	88–90
Solid/stranded	218
Ungrounded and overcurrent protection	90–91
Uses permitted	212–213
Wiring methods and	200, 212–221

Conduit and tubing (Tables)

Conductor properties	439–441
Cross section for conductors	412
Dimensions and percent area	415–425
Dimensions of insulated conductors, fixture wires	426–436

Index

Conduit and tubing (Tables) *(continued)*
   Radius of bends 414
Conduit bodies
   Accessible 158
   Damp/wet locations 144
   Definition 10
   Fill 148, 149
   Installation 144–161
   Metal 149–150
   Number of conductors in 145
   Supports 151–154
   Used as pull/junction boxes 157–158
   Wiring methods and 209
Connectors, pressure (solderless), definition 10
Continuous load (definition) 10
Cooking appliances, load calculations 40–41, 44
Cooking ranges
   Grounding to frame 139
   Receptacles 229–230
Cooking units, counter-mounted
   Definition 10
   Flexible cord connection 273–274
   Receptacles 238–239
Cord-and-plug-connected equipment, grounding 130, 138

Damp locations
   Boxes, conduit bodies, fittings 144
   Conductors 212
   Definition 15
   Luminaires 259
   Overcurrent protection/devices 93
   Panelboards 84
   Receptacles 247
   Switches 85, 251
Dead front, definition 10
Dedicated equipment space 31–32
Definitions 6–20
Demand factor, definition 10

Index

Device, definition 11
Disconnecting means
   Accessible location 69
   Air-conditioning refrigerating equipment 292
   Appliances 275–278
   Definition 11
   Grounding electrode conductors and multiple 115–117
   Grouping 70
   Identification 68–69, 71
   Maximum number 69–70
   Multiwire branch circuits 224–225
   Photovoltaic systems 378–381, 399
   Rating 71
   Separate buildings/structures 326, 328–329
   Service equipment 68–73, 115–117
   Space-heating equipment 281–283
   Swimming pools, fountains, and similar installations 335
   Temporary installations 35
Dishwashers 273–274
Dry location, definition. *See also* Laundry
Dryers, electric clothes. *See also* Laundry
   Grounding to frame 139
   Load calculations 40–41, 43–44
Dwelling units
   Branch circuits 225–226, 227–229, 233
   Closet storage space 258, 259
   Definition 11
   Existing 50–51
   Feeders 78–79
   Installation of communications equipment 303–306
   Lighting outlets 256–257
   Load calculations 41, 48–53
   Multifamily 51–53
   Receptacle outlets 236–242, 247–248
   Space-heating equipment 282
Dwellings
   Definitions 11
   Disconnecting means 72–73

Index

Edison-base fuses. *See* Fuses, Edison-base type
Electrical metallic tubing (Type EMT)
    Bends 179
    Couplings and connectors 180
    Definition 178
    Grounding 180
    Permitted/non-permitted uses 178–179
    Securing and supporting 179
Electrical nonmetallic tubing (Type ENT)
    Bends 189
    Bushings 190
    Definition 186
    Grounding 190
    Joints 190
    Permitted/non-permitted uses 188–189
    Securing and supporting 190
    Trimming 189
Enclosed, definition 11
Enclosures
    Bonding 124
    Definition 11
    Electrical continuity 207
    Grounding electrode conductors 117
    Grounding of service 122
    Overcurrent protection/devices 93
    Raceway supported 152–153
    Supported by concrete/masonry 153
    Swimming pools, fountains, and similar
        installations 343–345, 360–361
    Switches 85
Energized
    Definition 12
Equipment
    Bonding 104–105, 138–139
    Definition 12
    Deteriorating agents and 23–24
    Electrical connections 24
    Examination of 22–23
    Grounding 104, 129–142

Index

Installation, general provisions	21–32
Integrity	24
Limited access to	30
Mounting	25
Spaces about	28–32

Equipment grounding conductors
- Connections 137
- Cord-and-plug-connected 138
- Fixed equipment 138
- Installation 133
- Size 134

Exposed, definitions 12

Feeders 77–79, 215
- Definition 12
- Diagrams of 79
- Ground-fault circuit interrupter (GFCI) 79
- Grounding conductors 79
- Load calculations 42–53
- Neutral load 48
- Overcurrent protection 79, 92
- Pool equipment 346
- Rating/size 78

Festoon lighting, definition 12

Fittings
- Definition 12
- Installation 144–161
- Insulated 202–203
- Rigid polyvinyl chloride conduit 186
- Wiring methods and 209–210

Flexible metal conduit (Type FMC)
- Bends 191
- Couplings and connectors 193
- Definition 191
- Grounding and bonding 193–194
- Number of conductors 191
- Permitted/non-permitted uses 191
- Securing and supporting 193
- Trimming 191

461

Index

Fuses
- Ampere ratings 90, 91
- Edison-base type 94
- Plug type 95
- Type S 95

Garage(s)
- Receptacles 241

General use snap switch, definition 19
See also Snap switches

Ground-fault circuit interrupter (GFCI)
- Branch circuit protection 225–226
- Definition 12
- Feeders 79
- Swimming pools, fountains, and
  similar installations 337, 338, 351, 359

Ground-fault current path, effective 105, 121
Ground rings 111, 113

Grounded conductors 97–101
- Connection to supply system 98
- Definition 12
- Identification 98–101, 132–133
- Insulation 98, 100–101
- Sizing for raceways 108
- Terminals 101
- Use for grounding equipment 139–140

Grounded/grounding
- AC systems 106–109, 118–119
- Communications equipment 301–303
- Definition 12
- Electrical nonmetallic tubing 190
- Electrical system 104
- Feeders 79
- Flexible metal conduit 193–194
- Liquidtight flexible metal conduit 195
- Liquidtight flexible nonmetallic conduit 197
- Objectionable current 105
- Panelboards 84–85
- Photovoltaic systems 388–392

Index

    Receptacles 244
    Rigid polyvinyl chloride conduit 186
    Separate buildings/structures 327–329
    Solidly, definition 12
    Swimming pools, fountains, and similar installations 332, 356, 362
Grounding conductors. *See also* Bonding conductors/jumpers;
    Grounding electrode conductor(s)
    Connections 105–106
    Continuity/attachment to boxes 141–142
    Disconnection 70
    Equipment (*see* Equipment grounding conductors)
    Equipment (EGC), definition 12
    Feeder taps and 135
    Identification at terminals 136–137
    Rigid metal conduit as 183
    Types 130–132
Grounding electrode conductor(s) 108
    Aluminum/copper-clad aluminum not usable 114
    Communications equipment 301–303
    Connection to electrodes 121
    Continuous 115
    Definition 13
    Enclosures 117
    Installation 114–115
    Installation of taps 116
    Material 114–115
    Protection against physical damage 115
    Separate buildings/structures 329
    Size 118–119
Grounding electrode(s)
    Common 113–114
    Concrete encased 110–111
    Definition 13
    Ground rings 111, 113
    Interior water pipe 120
    Metal in-ground support structure 110
    Metal frame of building 120
    Plate electrodes 112
    Rod/pipe electrodes 111

Index

Grounding electrode(s) *(continued)*
   Separate buildings/structures..327
   System 109–110
   System installation 112–113
   Systems/materials not permitted 111
   Underground water pipe..110, 113
Guarded
   Definition...13
   Equipment..32, 68
Guest room, definition 13

Handhole enclosures..158–159
   Accessible 158
   Definition...13
   Installation 144–161
Heating. *See also* Space-heating equipment
   Central, branch circuit 272
   Load calculation 49–50
   Receptacle outlet 242
Hermetic refrigerant motor-compressor.......................290, 291, 292
   Definition...14

Identification/identified
   Branch circuits 224–225
   Circuit breakers 95–96
   Definition...14
   Disconnecting means...68–69
   Grounded/grounding conductors...............98–101, 132–133, 136–137
   Photovoltaic systems...381–383, 392–394
   Plug fuses...94–95
   Service equipment..67–68
   Space-heating cables 283
   Switchboards and panelboards..82
   Underground service conductors 206
Illumination
   Lamp protection 35
   Working spaces 30–32
Installation requirements...21–32
Intersystem bonding termination, definition 14

Index

Jumpers, bonding 107, 140
   Main 8, 109

Kitchen
   Definition 14
   Receptacles outlets 238–240
Kitchen waste disposers 273

Labeled, definition 14
Laundry
   Load calculation 40–41, 43–44
   Receptacles 241
Leakage-current detector-interrupter (LCDI) 290
Lighting. *See also* Luminaires
   Cove 260
   Load calculations 42
   Loads and switches 250–251
   Locations 259–260
   Outlets 14, 256–257, 261
   Outline, definition 16
   Track 267–268
Liquidtight flexible metal conduit (Type LFMC)
   Bends 194
   Couplings and connectors 195
   Definition 194
   Grounding and bonding 195
   Permitted/non-permitted uses 194
   Securing and supporting 194–195
Liquidtight flexible nonmetallic conduit (Type LFNC)
   Bends 197
   Couplings and connectors 197
   Definition 196
   Grounding and bonding 197
   Permitted/non-permitted uses 196–197
   Securing and supporting 197
   Trimming 197
Listed, definition 15
Loads, continuous, definition 10

Index

Luminaires
- Clearance 266
- Closets 260
- Connected together 265
- Definition 15
- Electric-discharge and LED 261
- Flush/Recessed 265–266
- Grounding 130, 263–264
- Load calculations 41
- Locations 259–260
- Near combustible material 260
- Supports 153, 261–263
- Swimming pools, fountains, and similar installations 336–343, 352, 355, 359–360
- Used as raceways 210
- Wet/damp locations 259
- Wiring 264–265, 266

Metal-clad cable (Type MC)
- Ampacity 171
- Bending radius 170
- Definition 168
- Permitted/non-permitted uses 168–169
- Protection 169
- Securing/supporting 170–171

Metal well casings, grounding 130
Motor-operated water pumps, grounding 130
Motors, load calculations 42
Multioutlet assembly, definition 15

Neutral conductors 98–99
See also Grounded conductors
- Definition 16

Neutral point, definition 16

Non-metallic-sheathed cable (Types NM, NMC, and NMS)
- Ampacity 167–168
- Bending radius 166
- Boxes and fittings 167
- Definitions 164

Index

Exposed work	165–166
Permitted/non-permitted uses	164–165
Protection	165–166
Securing/supporting	166–167

Outlets
Boxes	155–156
Branch circuits	236–242
Ceiling	156
Definition	16
Floor	237
Lighting	14, 261
Outdoor	240
Receptacle, definition	18
Wall	155
Outline lighting, definition	16
Overcurrent, definition	16
Overcurrent protection/devices	87–96
Ampere ratings	90
Appliances	271–272
Branch-circuits	16, 92
Circuit breakers (*see* Circuit breakers)	
Conductors and	88
Damp/wet locations	93
Disconnecting/guarding	93
Enclosures	93
Feeders	78
Fuses (*see* Fuses)	
Location	91–93
Locked	74
Near easily ignitible material	93
Over steps	93
Panelboards	83
Photovoltaic systems	373–375
Protection from physical damage	93
Small conductors	88–89
Space-heating equipment	283
Supplementary	90
Supplementary, definition	16

Index

Overcurrent protection/devices *(continued)*
   Use of higher rating 88
   Vertical position 93
Overload protection 16, 295

Panelboards 82–86
   Damp/wet locations 84
   Definition 17
   Grounding 84–85
   Overcurrent protection 83–84
   Support for busbars and conductors 82
Photovoltaic systems 367–403
   Circuit requirements 370–378
   Definitions 369
   Disconnecting means 378–381
   Grounding/bonding 388–392
   Marking/labelling 392–394
   Microgrid systems 403
   Power sources 394–403
   Storage systems 394
   Wiring methods 381–388
Plug fuses (*see* Fuses, Plug type)
Pool lifts 364–365
Power sources 394–403
Premises wiring (system), definition 17

Raceways 177–197
   Communications equipment 304, 317–318
   Definition 17
   Electrical continuity 207
   Electrical metallic tubing [*see* Electrical metallic tubing (Type EMT)]
   Electrical nonmetallic tubing [*see* Electrical nonmetallic tubing (Type ENT)]
   Enclosures not used as 85
   Feeders in 79
   Flexible metal conduit [*see* Flexible metal conduit (Type FMC)]
   Liquidtight flexible metal conduit [*see* Liquidtight flexible metal conduit (Type LFMC)]

Index

Liquidtight flexible nonmetallic conduit [see Liquidtight flexible nonmetallic conduit (Type LFNC)]
Protection against physical damage 200–203
Rigid metal conduit [see Rigid metal conduit (Type RMC)]
Rigid polyvinyl chloride conduit [see Rigid polyvinyl chloride conduit (Type PVC)]
Seals 206
Wet locations 207
Wiring methods and 211
Rainproof/raintight, definitions 17
Range hoods 274
Receptacles 231–248
Branch circuits 231–232
Damp/wet locations 247–248
Definition 17
Dwelling units 236–242
Faceplates (coverplates) 246
Floor 237
Grounding type 244
Installation requirements 243–244
Mounting 244
Rating/type 243–244
Swimming pools, fountains, and
    similar installations 336–337, 352, 354–355
Tamper-resistant 248
Temporary installations 35
Refrigeration equipment 289–297
Ampacity and rating 291
Branch circuit conductors 293–294
Branch circuit selection current 290
Cord-connected 292
Disconnecting means 292
Overload protection 295
Receptacle outlet 242
Short-circuit/ground-fault protection 293
Remote-control circuit, definition 18
Rigid metal conduit (Type RMC)
Bends 181

Index

Rigid metal conduit (Type RMC) *(continued)*
   Bushings 183
   Couplings and connectors 183
   Definition 180
   Grounding 183
   Permitted uses 180–181
   Reaming and threading 181
   Securing and supporting 181–182

Rigid polyvinyl chloride conduit (Type PVC)
   Bends 185
   Bushings 186
   Definition 183
   Expansion fittings 185
   Grounding 186
   Joints 186
   Permitted/non-permitted uses 183–185
   Securing and supporting 185
   Trimming 185

Room air conditioners 293, 296–297
Rules, mandatory/permissive 2–3

Separately derived system, definition 18
Service cable, definition 18
Service conductors 219–221
   Definition 18

Service conductors, overhead
   Clearances 59–60
   Definition 18
   Drip loops 67
   Insulation/size 58
   Means of attachment 60
   Service heads/goosenecks 67
   Support 58, 60

Service conductors, underground
   Definition 18
   Identification 203
   Size/rating 62

Service drops, definition 18

Index

Service-entrance cable (Types SE and USE)
    Definitions 173
    Permitted/non-permitted uses 173–175
Service-entrance cables, protection against damage 66
Service-entrance conductors
    Considered outside building 57
    Insulation...63
    Number 62–63
    Overhead system, definition 18
    Protection against damage 66
    Size/rating 64
    Splices 65
    Support 66
    Underground system, definition 19
    Wiring methods 64–65
Service equipment
    Definition 19
    Disconnecting means 68–73
    Enclosed/guarded 67–68
    Identification 68
    Overcurrent protection 68–73, 92
Service heads, bushed openings 67
Service lateral, definition 19
Service point, definition 19
Service raceways
    Drainage 66
    Grounded 121–122
    Other conductors in 57
    Seals 57
Services 55–75
    Definition 18
    Load calculations 42–53
    Neutral load 48
    Number 56–57
    Supply to one building not through another 57
Snap switches
    Defined 19
    Faceplates 251

Index

Snap switches *(continued)*
    Grounding 251–252
    Mounting 252–253
    Rating and use 253–254
Space-heating equipment 279–287
    Branch circuits 280
    Cables 283–287
    Disconnecting means 281–282
    Installation 280–281
    Load calculations 43
    Thermostatically controlled switching devices 282–283
Standby systems 405–409
    Definition 406
Structure, definition 19
Surge arrester, definition 19
Swimming pools, fountains, and similar installations 331–365
    Clearances 332–335
    Corrosive environment 335
    Disconnecting means 335
    Equipotential bonding 346–349
    Fountains 359–363
    GFCI protection 337, 338, 351, 359
    Grounding 332, 356, 362
    Hydromassage bathtubs 363–364
    Junction boxes and enclosures 344–345
    Luminaires 337–343, 352, 355, 359–360
    Motors 335–336
    Permanently installed pools 335–351
    Pool lifts 364–365
    Receptacles 336, 352, 354–355
    Spas/hot tubs 353–356
    Specialized equipment 349–351
    Storable pools 351–353
Switchboards 82–86
    Identification 82
Switches 249–254
    Accessibility/grouping 86
    Damp/wet locations 251
    Definitions 19

Index

Enclosures	85
Grounded conductors and	250
Indicating	86
Lighting loads	250–251
Position/connection	86
Snap (*see* Snap switches)	
Three-way and four-way	250
Unit, as disconnecting means	276–277

Temporary installations
- Branch circuits 34–35
- Disconnecting means 35
- Feeders 34
- Ground-fault protection 37–38
- Protection from damage 36
- Receptacles 35
- Splices 36
- Support systems 36–37
- Time constraints 34

Transfer switch, definition 19
Trash compactors 273–274
Type S fuses (*see* Fuses, Type S)

Underground feeder and branch-circuit cable (Type UF)
- Ampacity 176
- Bending radius 176
- Definition 175
- Permitted/non-permitted uses 176

Ungrounded
- Definition 20
- Service conductors 75

Unused openings 24
Utilization equipment, definition 20

Voltage
- Circuit, definition 20
- Nominal, definition 20

Voltage to ground, definition 20

Index

Wall-mounted ovens	274
Water heaters	272
Water pipe, underground as grounding electrode	110, 113
Watertight, definition	19
Weatherproof, definition	19
Wet locations	203
Boxes, conduit bodies, fittings	144
Conductors	212
Definition	15
Electrical metallic tubing	178
Luminaires	259
Overcurrent protection/devices	93
Panelboards	84
Raceways	207
Receptacles	35, 247–248
Switches	85, 251
Wiring methods	200–211
Boxes	209–210
Conductors	200, 212–221
Conduit bodies	209–210
Electrical continuity of raceways/enclosures	207
Fittings	209–210
Induced currents in enclosures/raceways	211
Mechanical and electrical continuity of conductors	209
Mechanical continuity of raceways/cables	208–209
Photovoltaic systems	381–388
Protection against physical damage	200–203
Raceway/cable to open or concealed wiring	210
Raceway installations	211
Securing and supporting	207–208
Underground installations	203–207
Working space, about electrical equipment	28–32
Workmanlike installation	24